Global Power Shift

Comparative Analysis and Perspectives

Ample empirical evidence points to recent power shifts in multiple areas of international relations taking place between industrialized countries and emerging powers, as well as between states and non-state actors. However, there is a dearth of theoretical interpretation and synthesis of these findings, and a growing need for coherent approaches to understand and measure the transformation. The central issues to be addressed include theoretical questions and empirical puzzles: How can studies of global power shift and the rise of 'emerging powers' benefit from existing theories, and which alternative aspects and theoretical approaches might be suitable? How can the meanings, perceptions, dynamics, and consequences of global power shift be determined and assessed? This edited series will include highly innovative research on these topics. It aims to bring together scholars from all major world regions as well as different disciplines, including political science, economics and human geography. The overall aim is to discuss and possibly blend their different approaches and provide new frameworks for understanding global affairs and the governance of global power shifts.

More information about this series at
http://www.springer.com/series/10201

Sarah Kirchberger

Assessing China's Naval Power

Technological Innovation, Economic Constraints, and Strategic Implications

Springer

Sarah Kirchberger
Department of Chinese Language and Culture
Asia-Africa-Institute
University of Hamburg
Hamburg, Germany

ISSN 2198-7343 ISSN 2198-7351 (electronic)
Global Power Shift
ISBN 978-3-662-47126-5 ISBN 978-3-662-47127-2 (eBook)
DOI 10.1007/978-3-662-47127-2

Library of Congress Control Number: 2015942898

Springer Heidelberg New York Dordrecht London

Printed on acid-free paper

Springer-Verlag GmbH Berlin Heidelberg is part of Springer Science+Business Media (www.springer.com)

Dedicated to the memory of my grandfather
L.I. Suurla

Acknowledgements

Much of the research on the global naval arms production processes contained in this study is based on practical and theoretical insight gained during three years working as a naval analyst with shipbuilder Blohm + Voss, where I was charged with studying naval developments around the world. In that role, I was in a position to observe several ongoing naval projects in Western countries in various stages of completion; conduct interviews with shipbuilding and naval electronics executives from various countries and industry backgrounds, in some cases with work experience reaching back several decades, in a transnational setting; interact with navy delegations from a number of countries including China; work with an interdisciplinary team of experts on a wide range of analyses of producers' strategies, naval customer countries' political and security environments, and whole regions' and subregions' naval development tendencies; and conduct postmortems on a number of failed projects. As someone who had to learn the art of judging warships from scratch in a relatively short time by conducting analyses for people trying to solve practical problems, I sympathize with everyone who cannot tell a corvette from a frigate. Therefore, in the present study I tried to avoid obscure jargon as much as possible and have taken pains to explain some basic facts and premises that are often left unsaid in the more specialized literature. While experts may want to skip these parts of the study, I still recommend consulting them in order to better understand the evaluations later made.

Some of the ideas put forward in this volume have been previously published in a short book chapter, which in turn was based on a longer conference paper.[1] Nevertheless, most of the content is entirely original, and many of the earlier

[1] For the conference paper, see Kirchberger, Sarah. 2011. China's Rising Naval Power and Its Impact on Global Power Shift. Paper presented at the ISA Annual Convention 2011 at Montréal, Quebec, Canada, March 16–19, 2011. 31 pp. For the book chapter, see Kirchberger, Sarah. 2012. Evaluating Maritime Power: The Example of China. In *Power in the Twenty-first Century. International Security and International Political Economy in a Changing World,* ed. Enrico Fels, Jan-Frederik Kremer and Katharina Kronenberg, 151–175. Berlin et al.: Springer.

conclusions and interpretations had to be refined, if not entirely revised, after reviewing the rapidly expanding knowledge available on the topic during the past four years.

Hiroki Takeuchi was the first to propose turning the draft conference paper into a full-fledged book project during ISA 2011 at Montréal. His encouragement and constructive criticism were of crucial importance at that point, and this book would probably not exist without it. Further and equally important support was extended to me by Enrico Fels and Maximilian Mayer of Bonn University, and I have to thank Enrico for the invitation to publish a short version of the 2011 paper in an edited volume. I am grateful to Gu Xuewu of Bonn University for the invitation to join his panel at the DVPW conference in October 2011 in Munich, and to him and Enrico for the invitation to publish this study in their Global Power Shift series with Springer. The staff at Springer, especially Barbara Fess and Johannes Glaeser, have done a wonderful job of bringing this book into existence. Many thanks are also due to Christopher Reid for his superb copyediting.

Gudrun Wacker of Berlin's Stiftung Wissenschaft und Politik (SWP) and Nicola Spakowski of Freiburg University have, on different occasions in 2011 and 2012, commented on draft paper presentations and offered constructive criticism. So has Shi Ming, who was available to me several times for inspiring exchanges of ideas on the subject matter of China's defense development, and he in particular has broadened my understanding of domestic Chinese discourses by pointing out interesting Chinese language materials.

With regard to understanding the technical aspects of the topic at hand, I owe an enormous debt of gratitude to naval architect Jörg Möller of ThyssenKrupp Marine Systems' Blohm + Voss naval yard, who during 2007 through 2010 patiently shared his expert knowledge with a young and fairly clueless colleague. Together with Henning Waschk and Norbert Schlichthorst, he effectively tutored me in all kinds of related knowledge areas. After I left the company to join the faculty of the University of Hamburg in 2010, Jörg generously agreed to support this research project by commenting on drafts and giving his opinion on technical data during several interviews. He has also read through the entire manuscript. The support of a technical expert of his caliber and experience has been invaluable; I do not know how to properly thank him for all the time and thought he invested. For comments on Chap. 5, I am also grateful to Uwe Sievert, Andreas Grudda, and Bernd Kulmus. It goes without saying that all technical errors remaining in the text are purely my own responsibility.

Several other former colleagues at ThyssenKrupp Marine, among them retired officers of the German, Turkish, Greek, Swedish, and South African naval forces, have during my time there explained obscure technical facts and naval strategy matters, or given their insight on the multifaceted phenomenon of global naval industry development, or worked with me on puzzling aspects of defense industrial development from a national perspective. These include Thomas Scheiter, Oliver Andresen, Reinhard Mehl, Thomas Ruckert, Wolfgang Bohlayer, Patrick Kaeding, Karl-Otto Sadler, Jonathan Kamerman, John Nilsson, Ahmed Tüfekcioglu, Ioannis Manolemis, Ektor Kalathas, and Sonja Langner, to name just a few in no particular order. I would also like to collectively thank the numerous representatives of other international shipbuilding and systems companies who I met and worked with during those years, and also the active service members of various naval forces who I had the chance to meet, exchange ideas, and discuss ships with. It was a pleasure working with and learning from all of these people of various professional backgrounds and nationalities, and their viewpoints on a multitude of naval issues are in some form or other reflected in this study.

In 2012, I was extremely fortunate to meet Chang Ching when this project was already well underway. His support enabled me to present some preliminary results on Taiwan and engage in discussions with a number of Taiwanese experts, among them Alex Huang, Arthur Ding, and Eric Shih, during several presentations and workshops in Taipei and the surrounding area. Their questions and comments were invaluable for the development of this study. Lately, Joachim Gutow of the Military Academy of the German Armed Forces (FüAK) has been available to me several times for interesting discussions of Asia's naval development, and so has fellow sinologist Oliver Corff. Both have given me comments on parts of this study for which I'm immensely grateful.

Many thanks are due to my colleagues at the University of Hamburg. It is by no means natural for scholars in the humanities to accept research topics as martial as this one, and I am only too keenly aware of how much this project must have strained some colleagues' capacity for intellectual tolerance. Completing a book project next to normal term work is especially challenging, and I fondly remember all the large and small instances of practical support extended to me. At the Asia-Africa-Institute (AAI) I am especially grateful to Ni Shaofeng, who read through the entire manuscript, pointed me to relevant materials, and offered support and encouragement throughout; to Hans Stumpfeldt and Michael Friedrich, who many times gave me interesting materials, and sometimes carried them to me from afar; to Ruth Cremerius, whose benevolent presence and serene judgment helped immensely when trying to balance research work, teaching load, and other academic responsibilities; and to Monika, Kai, Karin, Charlotte, and Liu Dongdong, for being exceptionally supportive colleagues. At the Department of Economics, Michael Funke never forgot to point out articles on a great variety of related topics to me and has offered valuable comments on Chaps. 2 and 3 of this study. Unfortunately, it is with sorrow that I have to acknowledge the intellectual debt to my esteemed late teacher during undergraduate years at the University of Hamburg, archaeologist Helmut Ziegert, who in the 1990s taught me a method of structuring research and organizing data. His emphasis on the need to think interdisciplinary and "laterally" has been a major source of inspiration for the approach chosen in this volume. I will continue to miss the lively discussions with him.

Last but not least, I am indebted to my friends and family for their practical and spiritual support throughout these past four years. My father has read through the entire manuscript. It must have been his lifelong enthusiasm for all things maritime that got me interested in naval matters in the first place, not least because I had the opportunity to spend large portions of childhood aboard yachts and ferries. The injured party has been my 11-year-old, who on various occasions expressed indignation at being "neglected" due to her mother's "excessive workloads." She may have a point there. I have to thank her for graciously allowing me to continue writing this "boring and pointless" book at least during the wee hours, and to her father and grandparents for helping out with childcare whenever necessary. Jonas, Lydia, Albrecht, Christina, and Niklas have been patient with me for years whenever I went off on some tangent of naval shipbuilding and have been supportive when needed. Finally, ever since Jan came into the picture, he has contributed in numerous ways to this study's stock of ideas, found interesting materials for me to look at, and helped with many thoughtful suggestions. Most of all, I am grateful for his steady encouragement, because his unwavering faith in this project, and in my ability to carry it through, has helped me over many a hurdle.

Acronyms and Abbreviations

AAD	Advanced air defense
AAW	Anti-air warfare
ADIZ	Air defense identification zone
AESA	Active electronically scanned array
AEW	Airborne early warning
AIP	Air-independent propulsion
ANCS	Advanced naval combat system
APAR	Active phased array radar
ASBM	Anti-ship ballistic missile
ASCM	Anti-ship cruise missile
ASEAN	Association of Southeast Asian Nations
ASM	Anti-ship missile
ASuW	Anti-surface warfare
ASW	Anti-submarine warfare
AVIC	Aviation Industry Corporation of China
B2G	Business-to-government
BICC	Bonn International Center for Conversion
bn	billion
BRIC(S)	Brazil, Russia, India, China (and South Africa)
BTI	Bertelsmann Transformation Index
C	Cruiser
C2	Command and control
C4ISR	Command, control, communications, computers, intelligence, surveillance and reconnaissance
CAJ	China Academic Journals
CATOBAR	Catapult assisted take-off but arrested recovery
CCP	Chinese Communist Party
CDS	Combat direction system
CEC/CEDA	China Enterprise Confederation/China Enterprise Directors Association

CEO	Chief Executive Officer
CETC	China Electronics Technology Group Corp.
CG	Coast guard *or* Guided-missile cruiser
CIA	Central Intelligence Agency
CIWS	Close-in weapon system
CM	Cruise missile
CMC	Central Military Commission
CMS	Combat management system
CNP	Comprehensive National Power
CNPEC	China Nuclear Power Engineering Co.
CODAD	Combined diesel and diesel
CODAG	Combined diesel and gas
CODLAG	Combined diesel-electric and gas
CODOG	Combined diesel or gas
COGAG	Combined gas and gas
COSCO	China Ocean Shipping (Group) Company
COSTIND	Commission for Science, Technology and Industry for National Defense
CPMIEC	China Precision Machinery Import-Export Corporation
CS	Combat system
CSI	Combat system integration
CSIC	China Shipbuilding Industry Corporation
CSOC	China Shipbuilding & offshore International Co.
CSSC	China State Shipbuilding Corporation
CSTC	China Shipbuilding Trading Co.
CTOL	Conventional take-off and landing
CV	Aircraft carrier
CVA	Attack aircraft carrier
CVL	Conventionally powered aircraft carrier
CVN	Nuclear-powered aircraft carrier
CVW	Carrier air wing
dB	decibel
DCNS	Direction des Constructions Navales
DD	Destroyer
DDG	Guided-missile destroyer
EADS	European Aeronautic Defence and Space Company (since 2013 Airbus Group)
EDA	European Defence Agency
EEZ	Exclusive economic zone
EMALS	Electromagnetic aircraft launch system
EMPAR	European multifunction phased array radar
EU	European Union
EW	Electronic warfare
FAC	Fast Attack Craft

FC	Fire control
FDI	Foreign direct investment
FF	Frigate
FFG	Guided-missile frigate
FFL	Light frigate
fl	full load
FMF	Foreign Military Financing
FMS	Foreign Military Sales
FREMM	Frégate européenne multi-mission/Fregata europea multi-missione
FS	Corvette
G.E.	General Electric Co.
G2G	Government-to-government
G7	Group of seven
GDP	Gross domestic product
GMI	Global Militarization Index
GNP	Gross national product
GRT	Gross registered tons
GT	Gas turbine
HR	Human resources
Hz	Hertz
IC	Industrialized country
IPO	Initial public offering
IWS	Integrated warfare system
JMSDF	Japan Maritime Self-defense Force
km	Kilometer (1 km = 0.53996 nm)
LAN	Local area network
LHD	Landing helicopter dock
LNG	Liquefied natural gas
l.o.a.	Length over all
LOS	Line-of-sight
LPD	Landing platform dock
LST	Tank landing ship
m	meter
MCMV	Mine countermeasures vessel
MESMA	Module d'Energie Sous-Marine Autonome
MoD	Ministry of Defense
NATO	North Atlantic Treaty Organisation
NBC	Nuclear, biological and chemical (warfare)
NCO	Non-commissioned officer
NFU	No first use (of nuclear weapons)
NIC	Newly-industrialized (or industrializing) country
nm	Nautical miles (1 nm = 1.825 km)
NPT	Non-Proliferation Treaty
ONI	Office of Naval Intelligence

OPEC	Organization of the Petroleum Exporting Countries
OPV	Offshore Patrol Vessel
OTH	Over-the-horizon
OTHR	Over-the-horizon radar
PAAMS	Principal anti air missile system
PACOM	United States Pacific Command
PB	Patrol boat
PC	Patrol craft
PLA	People's Liberation Army
PLAN	People's Liberation Army Navy
PLANAF	People's Liberation Army Naval Air Force
PPP	Purchasing power parity
PRC	People's Republic of China
PV	Patrol vessel
R&D	Research and development
RAM	Rolling airframe missile
RIMPAC	Rim of the Pacific Exercise
RMA	Revolution in military affairs
RMB	Renminbi
ROC	Republic of China
RSC(T)	Regional security complex (theory)
S&T	Science and technology
SAC	Second artillery corps
SAM	Surface-to-air missile
SAR	Search-and-rescue
SCO	Shanghai Cooperation Organisation
SIPRI	Stockholm International Peace Research Institute
SLOC	Sea lines of communication
SM	Standard missile
sq	Square
SS	Submarine
SSB	Ballistic missile submarine
SSBN	Ballistic missile submarine, nuclear
SSDS	Ship self-defense system
SSGN	Cruise missile submarine, nuclear-powered
SSK	Attack submarine, conventional
SSM	Surface-to-surface missile
SSN	Attack submarine, nuclear
STOBAR	Short take-off but arrested recovery
STOVL	Short take-off and vertical landing
t	ton
TKMS	ThyssenKrupp Marine Systems
TRA	Taiwan Relations Act
UNCTAD	United Nations Conference on Trade and Development

U.S.	United States
US$	U.S. Dollar
USM	Underwater-to-surface missile
U.S.N.	U.S. Navy
UHF	Ultra-high frequency
UK	United Kingdom
UN	United Nations
UNCLOS	United Nations Convention on the Law of the Sea
USSR	Union of Soviet Socialist Republics (or Soviet Union)
VDS	Variable-depth sonar
VHF	Very high frequency
VLS	Vertical launch system
WTO	World Trade Organisation

Contents

Chapter 1
Introduction

> 兵者，國之大事也。死生之地，存亡之道，不可不察也。
> *"Warfare is a principal matter of the state. It is the territory of life and death, the road to survival or disaster. Its study cannot be neglected."*
> —*Sunzi bingfa*, sect. 1, my translation. Chinese text cited from Lau (1992: 1).

Naval power is one of the component parts of national power. Though not limited to countries with maritime borders, it is an essential concern of such states. In an increasingly globalized trade environment that links countries around the globe through a vast network of maritime trade routes, the study of naval power should form a necessary part not only of military and strategic studies, but of power shift studies in general.

Naval power is also a unique, multi-faceted phenomenon that poses a number of methodological problems to the researcher. It has technological, economic, political, as well as geostrategic core aspects that require specialized knowledge in several of these fields. In the case of China, one of the world's most dynamic rising powers, the analyst's task is made even more challenging by a relative dearth of transparency in military affairs. Conducting a fair evaluation of China's naval development and future strategic aims is therefore rendered more difficult than usual, leading to persistent disagreements within the research community regarding the actual level of China's capability, its future developmental trajectory, and the resulting strategic implications for various interested parties.

Naval build-ups are often discussed in popular discourse as well as in specialized publications in terms of newly procured weapon systems, i.e. vessels, aircraft, or missiles. However, questions concerning what kinds of additional capabilities these new systems yield, what kinds of further missions they enable, or how they fit in with the overall fleet profile are asked much less frequently. Yet it is precisely this kind of analysis that is needed to judge the nature of a power shift resulting from the procurement of new weapons systems and the build-up of related industrial infrastructures. To take up a real-world example, the acquisition of a small aircraft carrier in 1997 did not elevate the Royal Thai Navy to the status of a blue-water force on par with India, Brazil, or Britain. It would seem that it takes more than just

S. Kirchberger, *Assessing China's Naval Power*, Global Power Shift,
DOI 10.1007/978-3-662-47127-2_1

a single ship to achieve that level of capability. Yet, while Thailand's carrier—dubbed by some observers "the most expensive Royal yacht" because of its limited and largely ceremonial mission history—received very little news coverage, China's commissioning of the ageing ex-Soviet carrier *Varyag* (rechristened the *Liaoning* in 2012) was unanimously met with concern both in Asia and abroad. This was the case, even though as of this writing its air wing still awaits completion and the capability of its weapon and sensor outfit remains largely unknown. Is this ship really a game changer in the Taiwan conflict and various island conflict scenarios, as some observers presume, or is it rather a giant, costly, but strategically worthless floating target? While both positions can be found in press reports and scholarly analyses, they cannot both be correct.

Analyzing China's naval build-up appears not only to be a question of studying military hardware, training, logistics, and strategy, or of delineating alliances, economic interdependencies, and balances of power. It is also a question of taking seriously the issue of threat perceptions both within China and abroad. Apart from technological and geostrategic considerations, naval power has a symbolic aspect which is currently most evident in the domestic Chinese discourse on aircraft carrier programs. This deserves careful analysis because it touches on the way reality is constructed by different observers. Naval vessels are often perceived as instruments of diplomacy, but they are also employed by governments for internal propaganda purposes. Reports of their domestic development or foreign deployment can play a role not only in regime legitimacy discussions, but also influence mutual threat perceptions between rival powers. China and Japan, China and India, and China and the U.S. continually find themselves within mutually intensifying threat perception dynamics that can easily spiral out of control. In such a case, material facts regarding new technologies and faulty or exaggerated evaluations of such facts spread on the Internet or in the mass media can quickly go their disparate ways. Meaningfully evaluating the technical basis behind threat perceptions therefore requires not only some degree of technical understanding. It means taking both sides—China's and its respective counterpart's perspective—equally into account.

In the case of China, several key geostrategic issues are intimately connected with the question of naval power and therefore deserve special consideration. Among these are the as-yet unresolved Taiwan question, the Spratly and Paracel Islands issue, and the Senkaku/Diaoyutai Islands issue. Another important issue is the so-called "String of Pearls" strategy that China supposedly employs by building its own network of future naval bases to gain a greater foothold within the wider Asia Pacific and broaden its geographic reach in the face of a perceived "encirclement" by the U.S. and its allies. All these issues are in some way related to, or have an influence upon, China's heavy dependence on maritime trade routes for the bulk of its foreign trade. They also all touch upon questions of national identity and China's search for a more dominant role in the world.

China scholars cannot help noting that China's projected rise to great power status is also a matter of regime survival. The uneasy balance struck between the ruling Chinese Communist Party (CCP) and China's population after violent protests in the spring of 1989 is being maintained in no short measure by constant appeals to nationalist ideals that generate expectations of success within the Chinese populace. Enhancing China's symbolic, economic, and military power profile and gaining respect abroad have therefore become the Chinese leadership's prime tools for maintaining legitimacy at home. It is not hard to see why the Chinese government has increasingly viewed naval power issues to be critical. Judging from China's actions and pronouncements over the past months and years, the takeover of a new leadership generation in 2012 appears to have exacerbated the need to demonstrate a commitment towards building a powerful navy, perhaps not least to help ensure the continued loyalty of the People's Liberation Army (PLA) towards the CCP leadership. As will be shown in Sect. 2.3, China has far less leeway for decision-making in this particular sphere of development than other rising powers due to a rather complex, even somewhat unfortunate maritime geography.

China's ongoing naval modernization has raised concerns across the whole Asia Pacific region. As one journalist observed in 2013, while the Chinese leaders publicly stress their desire for China to "rise peacefully" and work towards a "harmonious world," they sometimes have a rather odd way of showing it. When a secretly built Chinese submarine base was spotted at Hainan island on satellite images in 2008, countries as diverse as Australia, Vietnam, and India made moves to bolster their submarine procurement programs. Even the Philippine Navy, a small force still operating many obsolete vessels dating from the 1940s, has announced its intention to add a submarine component by 2020. Other regional navies have similar plans.[1] However valid these threat perceptions may have been, it is necessary to realize that China began its present build-up from a very low starting point. The current modernization effort is therefore largely catch-up development. World naval history clearly shows that building a powerful navy is usually an extremely costly long-term endeavor that is not merely accomplished by introducing a handful of modern vessels into the fleet. The real challenge lies in integrating and sustaining these new systems in the existing fleet structure by providing the necessary logistics, maintenance, and training. Another critical step is to achieve interoperability between the new systems and the rest of the fleet by actually testing them under real-life conditions during large-scale fleet exercises. Despite its economic superpower status, it is easy to forget that China continues to face tough and costly domestic development challenges partially linked to factors beyond its immediate control (see Sect. 2.2). China's continued ability to invest in its naval modernization as heavily as it has these past few years is far from secure, at least in the long term, given other pressing state spending priorities.[2]

Notwithstanding—or indeed perhaps because of—a myriad of rising domestic challenges, China has generally become more assertive in its dealings with the

[1] For the discovery itself see 'Secret Sanya—China's new nuclear naval base revealed' (2008). The reaction of regional powers and the U.S. was widely covered in various press reports during May and June 2008, i.e. *Daily Yomiuri,* 5 June 2008; *The Indian Express* online, May 06, 2008; *Times of India,* 3 May 2008. For the Philippine procurement plan see Grevatt (2011).

[2] This is a point granted even by Holmes and Yoshihara (2008: 106) who otherwise tend to discount skeptical views of China's naval progress.

outside world, especially since the 2008 financial crisis. This trend can also be seen in naval matters and the country's maritime behavior in particular. In November 2007, a Chinese *Type 039* diesel-electric submarine that managed to surface without prior detection in the middle of a U.S. Navy Pacific exercise area between Taiwan and Southern Japan, and within torpedo range of the aircraft carrier *USS Kitty Hawk*, sent shockwaves through the U.S. naval community. No matter whether this incident was accidental or intended to send a message,[3] its implication was clear: No surface ship will ever be fully immune to the threat posed by small and relatively cheap diesel-electric submarines, which the Chinese happen to possess in a very large fleet. This reality was made painfully clear when the South Korean corvette *Cheonan* sank in 2010, presumably after being hit by a North Korean torpedo fired from a submerged submarine. Forty-six sailors were killed. While most observers called on China to officially condemn this act of aggression, it shielded North Korea from international censure, even in the face of damning material evidence.[4] The above-mentioned incidents illustrate that naval power cannot be assessed simply by counting vessels, or, for that matter, by comparing the technological sophistication levels of individual pieces of naval hardware. Navies are complex systems, not mere assemblages of individual parts, and there is an asymmetric threat element to consider in naval strategy as well.

Small conventionally powered submarines—dubbed by some the "weapon of choice for the maritime underdog"—are one of the key components of China's fleet. These vessels pose a special problem to even the world's most powerful navies. They are far more accessible to smaller navies than their extremely expensive, giant nuclear-powered counterparts, yet they can also easily evade detection and inflict great damage on even the most powerful surface ships.[5] Thus, a small number of such relatively inexpensive vessels can serve to blockade large maritime areas, making military intervention from outside powers (e.g. in the event of a Taiwan crisis) exceedingly difficult (Howarth 2006: 70). In recent years, China has greatly expanded its naval presence in the South China and East China Seas and thereby challenged American supremacy in small, but unmistakable ways. The rationale underlying these developments seems to be an area denial strategy enacted by increasingly powerful missiles, naval vessels, and aircraft.

---

[3] Cf. Hickley 2007. According to Cole (2010: 205, fn. 15) U.S.N. personnel indicated afterwards that the surfacing may have been accidental, and that the carrier did not detect it because it was operating without its usual escort ships. It remains unclear whether detection would have been possible under better circumstances.

[4] Cf. a report in *The Economist*, May 22, 2010, entitled 'A guilty verdict for North Korea: Their number is up'. Details of the multinational investigation are discussed in Falletti (2010). Other recent examples of sophisticated naval vessels worsted by asymmetrical enemies included the U.S. Navy's Arleigh Burke class destroyer *USS Cole* that fell victim to an Al-Qaeda attack in 2000 and the Israeli corvette *Hanit* that suffered heavy damage from a Chinese-built missile fired from Lebanon in 2006.

[5] The vulnerability against modern diesel subs was one of the factors, aside from cost overruns, for the U.S. decision to truncate the new *Zumwalt* class destroyer program to just three ships when originally 32 were planned.

As Till has observed, with regard to navies, there is no such thing as absolute power:

> The power of one navy is a relative thing, which can only be assessed in comparison with that of another navy, given the commitments they both face. A large navy faced with huge commitments may in fact prove surprisingly vulnerable to a much smaller navy with very limited commitments (Till 2009: 118).

Clearly overall naval capability is determined by several interrelated factors, among which technology is only one. Other factors include logistics, training, maritime strategy, regular exercise experience, and, last but not least, the existence of an indigenous industrial base with the capability of providing maintenance and repairs, refits, and the new building of vessels for sustaining the operational status of the fleet.

Just the same, naval strategy is of little consequence, no matter how sophisticated it may be, if the necessary conditions and tools for implementing it are not in place. Naval weapon systems—a large category of highly diverse technologies comprising vessels, sensors, effectors, countermeasures, and naval aircraft as well as data links and computerized combat direction systems—are essentially the tools of action in this strategic field. Over the past few decades there have been massive changes worldwide in the nature and quality of these technologies, with a general tendency in Western countries towards the development of so-called "network-centric warfare." This involves capable platforms that can not only automatically defend themselves and other ships against incoming threats, but can also generate, display, and share large amounts of tactical data between disparate units, resulting in a shared situational awareness between vessels, aircraft, and ground stations and thus a powerful alliance.[6] Their operation, however, requires much more intensively trained soldiers and regular practice under real-life conditions than the older technologies. These developments—which have transformed the strategic function of warships from the individually operating gun-carrying armored troop transport vessels of the past into the lean-manned, highly automated, integrated and interlinked combatants of today—have been possible mainly due to striking advances in sensor and data-processing technologies. This in turn has been driven to a large extent by the evolution of ever more lethal missile systems.

China's "critical" naval technologies[7] for some reason have not yet received the necessary systematic attention within general power shift studies, apart perhaps from the more narrow field of nuclear submarines, which contributes directly to the study of nuclear deterrence. Analyzing the various technological changes underlying political strategy shifts, however, would yield a better understanding of the limits and consequences of power shifts. While a state's ideology, core interests, political values, and resulting grand strategy may inform a leadership's outlook and

---

[6] An excellent book-length treatment of network-centric warfare, also called "picture-centric warfare", is provided by veteran naval analyst Friedman (2009).

[7] See Bimber and Popper (1994) for a thorough discussion of the issue of criticality in the field of technology.

orientation towards the world, its possible range of action is to a large extent defined by the level of its technology, which in turn is partly determined by the development of its economy. Taken together, these factors delineate the limits of possible future trajectories of development.

Therefore, rather than simply enumerating ships, missiles, or aircraft, or concentrating on isolated problems such as individual weapon systems development, more systematic questions need to be asked. Just what kinds of technologies deserve to be regarded as "critical" for understanding shifting naval power relations, and what is their relation to other technologies contributing to naval power? More interdisciplinary research bridging the social sciences and engineering communities is clearly needed. However, as Crosston (2014) notes in a recent book chapter on cyber warfare, intellectual barriers between scholars of different disciplines seem to persist. This has accordingly led to a "technical-political gap" in understanding the real-world consequences of technological innovations. As this situation limits the understanding of a subject much too grave to treat lightly, it is especially deplorable, even dangerous. The present study hopes to make a small contribution to bridging this gap with regard to naval technology innovation in China.

1.1 Scope of Research

One characteristic that stands out in the Western literature on China's naval rise is the fact that the same data can be interpreted very differently by various observers, resulting in a confusing array of contradictory analyses.[8] While a certain range of interpretation is to be expected, the divergence of expert opinion in some instances is difficult to comprehend. One common reason seems to be that many studies lack a historical-comparative perspective, as well as clear-cut evaluative criteria. A question rarely asked, but useful to tackle, would have to be: What yardsticks should be used for measuring China's naval progress? And, along similar lines: How can regular replacement procurement and irregular capacity build-ups be distinguished from each other using objective criteria? This gives rise to yet another related question: What level of naval development can be considered "normal" under any given circumstances, and at which point does a build-up such as China's truly begin to become "excessive"?

Evidence of technological progress in China's naval modernization effort is in itself often deemed worrisome. The thin line separating alarmism from demonization is easily crossed, especially in the news articles of various Western media outlets (cf. the analysis by Yang and Liu 2012). Be that as it may, an overly simplistic perspective will not suffice. As naval equipment is of finite durability, naval forces around the globe are constantly undergoing technological change. It is

[8] Cf. the discussion in Holslag (2010: 11–12).

neither rational nor practical to expect a nation to cling to yesterday's obsolete technology when fleet renewals must take place in fixed intervals to maintain current capabilities and newly introduced vessels are typically kept in service for decades. Zero technological change during procurement would therefore mean a de facto decrease in military capability through automatic depreciation, much in the same way that money automatically diminishes in value through inflation if it is not invested with adequate returns. That said, qualitative and quantitative improvements in other regional navies' capabilities—e.g. India's, Pakistan's, South Korea's, or Japan's—tend to receive far fewer critical comments. They are rarely seen as controversial, except by their respective strategic rivals, while technological progress made by China is usually greeted with concern both within the region and abroad. As analysts, we need to ask ourselves where this fundamental difference in the perception of facts—this bias of automatically seeing China as a military threat—stems from and what its real-world consequences might be. By heightening mutual threat perceptions, are we creating the very monster we are afraid of?

In order to distinguish more clearly fact from perception, this study will *not* start by discussing threat perceptions, or by pitting U.S. naval capabilities against China's. Instead, it will first try to establish some basic principles for evaluating naval power in a comparative perspective. Taking ascertainable data as the material "basis" and treating threat perceptions as the resulting and dependent "superstructure," it will then try to show where the main differences in evaluation can originate and what analysts can do to maintain an independent perspective in the midst of wildly conflicting evaluations of the available evidence. Toward this end, it will be necessary to give an overview of the basic types of naval hardware in use by China and other comparable navies, and to discuss their life cycles, strategic functions, interrelations, and industrial production processes (see Chap. 3).

Just how powerful have China's naval forces become until now? This issue is hotly contested by Western and Chinese naval analysts alike who offer a wide range of opinions. Recent emphasis seems to lie more on the significant technological advances of Chinese naval shipbuilding during the last few years, despite the Western arms embargo.[9] Taking the above-mentioned considerations into account, the major research questions to be addressed in this study include the following:

- How can "sea power" be measured comparatively?
- How does China's navy compare globally?
- What outside factors influence China's naval build-up?
- What effect does the U.S./EU arms embargo have on China's naval development?
- What consequences can be expected for the global power shift?

---

[9] Holmes and Yoshihara (2008: 7–8) distinguish between proponents of a "blue-water hypothesis", which expects China to challenge U.S. hegemony in the Pacific area, and a "coastal defense hypothesis", which expects China's navy to stay confined within its own Exclusive Economic Zone (EEZ). Holmes and Yoshihara themselves argue in favour of third hypothesis occupying the middle ground between the two extremes.

The aim here is to provide an empirically based assessment of China's naval build-up as it currently stands in relation to comparable cases and according to transparent evaluation criteria. It hopes to make a contribution towards a more multi-faceted understanding of China's naval build-up and its consequences for the Asia Pacific region. This can only be achieved if a number of non-military factors, such as domestic policy imperatives and economic interdependencies, are equally taken into account.

1.2 The State of the Field

There has been a noticeable upsurge in recent years in Western scholarship on China's naval build-up due to increased public interest and intensified contact between Western navy personnel and China's navy, but also due to changed perceptions about China's rise in the aftermath of the 2008 financial crisis. When the national economies of industrialized Western nations stumbled one by one, a seemingly invulnerable China was increasingly difficult to ignore on the world stage. This resulted in greater assertiveness on the part of Chinese leaders in negotiations with foreigners, a decline in respect for Western superiority among China's intellectuals (as reflected in the ongoing "China model" debate on the supposed superiority of China's chosen development path), and continuous military modernization unfettered by financial bottlenecks (McGregor 2010: ix–xii).

Apart from Cole's classic monograph *The Great Wall at Sea*, which already saw a second edition in 2010, there have been several book-length English language studies in the past few years that have analyzed different aspects of China's naval modernization. The monograph by Bussert and Elleman (2011) on China's combat system development represents a milestone in our still tentative understanding of China's naval technology. Other recent monographs and edited volumes discuss specialized topics such as aircraft carrier development (Erickson and Goldstein 2011), nuclear submarine development (Howarth 2006; Erickson et al. 2007), and maritime strategy (Holmes and Yoshihara 2008; Yoshihara and Holmes 2010), among others. Chinese language monographs exclusively focused on aspects of China's naval development published in Hong Kong, Taiwan and Mainland China include Lin (2008), Yi (2011), Wu (2010), Ping (2010) and Hu (2012).

Notwithstanding this progress, however, there remain glaring holes in the current research. The majority of English language studies concentrate on aspects of naval strategy, mutual threat perceptions, and possible conflict scenarios, while the systemic problems involved in warship production and the associated material, financial, technological, and political requirements currently remain overlooked aspects in the case of China.

In some of these areas, there is a notable absence of hard data. So far, very little reliable information on China's defense electronics is publicly available, whereas actual performance data is often impossible to verify.[10] This knowledge gap is a significant problem, especially when it comes to evaluating the military-strategic implications. Analyses must therefore proceed by piecing together disparate fragments of available information and trying to make up for any missing pieces with cautious estimations, cross-country comparisons, and plausibility checks—a difficult task that is fraught with possible errors. Some analysts simply skip over this daunting challenge and thus imply a security of knowledge that does not really exist.

Another troubling tendency involves the confusion of fact with rumor. As Cole states, analytic caution should be a priority:

> A conservative attitude is appropriate when evaluating Chinese policy and military developments; it is not helpful to 'confuse aspiration with accomplishment,' a caution echoed in another analyst's view that a 'Chinese penchant for metaphorical writing. . .makes literal interpretation of interesting rumors problematic' (Cole 2010: xvi).

As Bussert and Elleman (2011: 19) note, when it comes to military matters, Chinese leaders do not publicize accidents. Friedman, while commenting on publicized performance data of Chinese missile technology under development, remarked that as a result of comparing material evidence with such pronouncements, it seems that "the Chinese media habitually claim nonexistent military capabilities" (Friedman 2006: 513).

Apart from the Chinese government's own interest in publicizing its accomplishments, there is also a fast-growing Internet community, both in China and abroad, engaged in spreading news and rumors about military developments. As Bussert and Elleman warned in 2011, extraordinary claims regarding Chinese military developments are quite often untrue:

> Stories of dramatic technical advances in the military capabilities of the People's Republic of China (PRC) seem to crop up regularly in the Western press, including an endless list of startling 'revelations' about China's military developments that are usually exaggerated or perhaps disinformation. Past experience suggests that most such reports should be discounted; the more dramatic the story, the less likely it is to prove true (Bussert and Elleman 2011: 2).

Therefore it is necessary to keep in mind that in a party-controlled Leninist polity such as China's, news releases on military breakthroughs have a definite propagandistic purpose. They are geared toward creating a feeling of pride and unity among the populace and therefore aimed at a domestic audience just as much

[10] The ubiquitous use of terms such as "probably", "presumably", "perhaps", "almost certainly", and "one interpretation would be. . ." in the China entry of Friedman's normally pretty matter-of-fact treatment of the World's Naval Weapon Systems is telling. The content of the article illustrates even more clearly the painstaking process that lies behind his China-related evaluations of interpreting protruding parts such as antennae, of drawing conclusions from analogy, and of gathering visual information piecemeal from photographs, models, and published information (cf. Friedman 2006: 61–63).

as at foreign powers.[11] Moreover, despite the CCP's centralist leadership structure, it is far from monolithic. Numerous constituencies vie for power that push their agendas even within the exclusive Central Committee, composed of about 200 full members who represent diverse subsystems of the party-state. It is therefore important to recognize the simultaneous existence of several strategic arenas for China's leaders—intra-party, domestic, and international—that follow different rules.

The points just highlighted indicate only some of the difficulties research is currently facing. Another problem concerns the quantitative preponderance of U.S. scholarship, which has resulted in an unavoidable analytic bias. So far, the vast majority of studies on China's naval rise have been conducted by U.S.-based analysts, and, not surprisingly, many of these concentrate on bilateral U.S.-China comparisons of military capabilities and maritime strategies, often with a view to possible conflict scenarios over Taiwan. This is only to be expected given the fact that the U.S. Navy would almost certainly become involved in a military conflict should cross-strait tensions escalate.[12]

From a theory viewpoint, it seems that many, if not most analyses of China's naval rise conducted in the U.S., China, and other Asia Pacific countries are primarily informed by "defensive realism" (cf. Blazevic 2009). The traditional categories of realist analysis—"deterrence, containment, power alliances and balance of power politics" are thus prevalent in these studies, but the major assumption seems to be that "nations gain power for self-preservation", not necessarily for projecting their power offensively (Blazevic 2009: 60). Recently, an increasing number of studies have focused on mutual threat perceptions, and used constructivist approaches including discourse and content analysis of textual evidence, often including numerous Chinese language publications, for gaining insight into perception shifts.[13] Compared with the amount of literature published on these topics, only very few studies have undertaken in-depth evaluations of the material aspects of China's naval rise—technological and economic factors, incentives and constraints.

---

[11] Cf. Young (2013: 40–41) and Cheung (2011: 132–133). In IR theory, the "two-level games" approach delineated by Putnam and Robert (1988) could serve as a useful analytic lens for tackling this problem of an authoritarian leadership catering to various audiences at once.

[12] After shifting diplomatic recognition from the Republic of China on Taiwan to the People's Republic of China in 1979 and cancelling the Mutual Defense Treaty with Taipei, the U.S. congress passed the Taiwan Relations Act (TRA) which compels the U.S. government to "consider any effort to determine the future of Taiwan by other than peaceful means, including by boycotts or embargoes, a threat to the peace and security of the Western Pacific area and of grave concern to the United States" (Taiwan Relations Act 1979, Sect. 2, 2.4). The U.S. have therefore become Taiwan's de facto guardian power against Chinese aggression as well as part of a strategic triangle Beijing–Taipei–Washington despite the intentional vagueness of the formulation quoted.

[13] Examples include Swaine and Michael (2013), Zhu and Lu (2013), Yang and Liu (2012) and Hughes (2011), among others.

Another problem faced by research concerns regional bias and skewed yardsticks. A number of U.S.-centric analyses openly state a political preference towards "containment" of China's emerging capabilities (cf. e.g. Yoshihara and Holmes 2010; O'Rourke 2008). In a few extreme cases of state-sponsored scholarship, discrepancies between the presented empirical data and the conclusions are staggering enough to suggest that political pressure might have played a role regarding the desired outcome.[14] Moreover, most studies tend to either treat China as a singular case or to pit it against its main strategic rival, the U.S—even though, in qualitative as in quantitative terms, the U.S. Navy is the most historically anomalous example of a naval power and can therefore hardly serve as a useful yardstick for evaluating any rising power's (let alone China's) naval development status. While many of the currently available studies offer valuable analyses of specific issues, the above-mentioned drawbacks can lead to skewed interpretations of the strategic consequences of China's rise for the Asia-Pacific region and the rest of the world, especially if they are not supplemented by studies with a different bias.

In sum, all the indicated problems impede our current ability to map China's progress against universal and historical standards, something the social sciences should always insist on. They also undermine the chances of conducting meaningful cross-country comparisons with other rising powers. Moreover, the quantitative dominance of Anglo-American scholarship and its (understandable) fixation on bilateral U.S.-China relations results in the dependency of European and other non-U.S. decision-makers, scholars, and publicists on sometimes less-than-transparent evaluations. Given the fact that decision-makers in Europe, India, Southeast Asia, and other regions have a need to form their own opinions of China's strategic development, it seems desirable to supplement the existing research with further studies from other parts of the world, especially from countries that have no vested strategic interests in the various East Asian conflict scenarios. In light of the European Union's current arms embargo policy, an issue that is the source of ongoing discussions among EU members, Europeans would do well to conduct their own independent research on this topic to supplement the existing repository of knowledge.

[14] Cf. the U.S. Air Force sponsored analysis by Medeiros et al. (2005) that gives excellent data but often draws seemingly odd conclusions from said facts. Another example of perhaps politically motivated misrepresentations of facts was discussed by nuclear weapons specialist Hans Kristensen who reports that the "U.S.-China Commission established by Congress [. . .] stated in 2006 that China is pursuing measures to try to 'control' the seas in the Western Pacific, although 'controlling' the seas is a daunting technological and operational task, and that China continues to 'expand' its submarine force. [. . .] Although China is modernizing its submarine force, it is not 'expanding' it. Since the mid-1980s, the force has been in steady decline from nearly 120 boats to roughly 55 operational submarines today" (Kristensen 2007).

1.3 Sources

The technical and visual data used for this study was taken from openly accessible sources such as the regularly updated warship and naval weapons systems catalogues *Jane's Fighting Ships*, the U.S. Naval Institute's *Combat Fleets of the World* (Wertheim 2007, 2013) and *World Naval Weapon Systems* (Friedman 2006) and other relevant works. Even between these reputable sources, however, technical details for the exact same systems can often be at odds. In such cases, and when data was otherwise unavailable, these references were cross-checked for plausibility and, when possible, cautiously supplemented with data from relatively reliable online portals such as sinodefense.com, globalsecurity.com, etc.[15] Still, there remains a gap at present, especially with regard to performance data. The lack of access to detailed official Chinese primary data on warship production is due to the severe stance taken by the Chinese authorities towards "violations of state secrets," which makes true fieldwork in China on this kind of topic all but unthinkable. To be fair, however, technical performance data of Western warships is also to some degree classified, and non-transparency is more or less the normal state of being in the field of weapons production worldwide. In the words of Neuman (2009: 74), "[m]ost governments provide little information about the manufacture of their military equipment, and those that do use different, often idiosyncratic, definitions of arms production." The same goes for arms manufacturers, which, as a rule, closely safeguard their technical and business knowledge from outside observers. Nevertheless, the Chinese government draws the line earlier than is customary in the West, which means that Chinese naval vessels, especially the new designs, containing a mixture of imported, reverse-engineered, newly developed, and experimental weapons and electronics with largely unknown specifications and capabilities, are to a large extent "black boxes." There remains therefore a huge gap in public knowledge concerning the main data of these weapon platforms that cannot be easily bridged, at least for the foreseeable future and in the absence of "empirical" observations of their true performance during conflict. As a result, only indirect conclusions based on observable phenomena are possible, leaving actual strategic capabilities of these systems largely a subject of conjecture.

The contribution this study intends to make to the research therefore *cannot* for the reasons mentioned consist in sensational revelations regarding technical specifics of new, secret, or experimental Chinese weapon systems. Instead, its main focus is methodological in nature: By offering a variety of perspectives on how to systematically interpret the currently available data—however incomplete and inaccurate it may occasionally be—and drawing on earlier conceptual work by naval analysts such as Friedman, Todd and Lindgren, Pugh, Till, and others, this study will delineate an analytical framework using China as a case study that can be

[15] Shambaugh (2004: xxv–xxviii) provides a very concise and useful overview over the various available types of source materials on Chinese military matters as of 2004. Many of his observations remain valid.

further refined, adapted and still used if and when more accurate datasets become available in the future. Therefore, the existing open-source data on China's naval development, despite the limitations outlined above, is sufficient for the purpose at hand.

For the parts of the study dealing with geopolitics and strategy, Chinese-language as well as Western materials in English, German and French were used. Chinese scholarship on naval development has become much more accessible over the past few years and I searched the *Chinese Academic Journals (CAJ)* database for articles shedding light on the issues currently being studied by Chinese naval engineers. Scholarship from Taiwan and Hong Kong also proved to be a useful source of information regarding the perception of China's naval and military development among non-Mainland Chinese.

As a former analyst in the naval shipbuilding industry, I not only had the chance to personally experience the inner-workings of the industry as a participant observer over the span of several years in various transnational cross-company settings, but I also made extensive use of former colleagues' technical expertise. Interviews with executives from the field of military shipbuilding, including several with first-hand project experience in Asia, were an extremely valuable source of information for cross-checking technical data and assessing the plausibility of some conclusions. Other interviewed sources include active and retired officers of Western navies who gave their perspectives on tactical matters, maritime geography, and practical naval strategy issues. In order to protect the privacy of these sources, I have, in accordance with their wishes, refrained from attributing quotes directly to them.

1.4 Methodological Tenets

This study aims to develop methods for comparatively evaluating China's naval power—its sources, strategic uses, and comparative limitations—and its complex interdependence with other aspects of state power. Above all, this study focuses on the issues of strategic industry alliances and technological dependence on industrial infrastructure development. It will also examine the limiting influence of maritime geography which is extremely relevant for explaining maritime behavior. Though subject to virtually no change, this factor is of prime importance especially for submarine operations and yet surprisingly little analyzed. Moreover, as is appropriate to the social sciences, this study will not start from an assumption of Chinese exceptionalism, but rather try to draw cross-country comparisons whenever possible in order to put China's naval build-up into perspective. This is one of the main analytical differences between this study and most published works on the subject to date.

The analytical frameworks most often selected for studying arms industries and related phenomena can be broadly classified into structural/dependency, economic/

development, dominance/world system, and historical/systemic approaches.[16] This study uses a combination of structural/dependency and historical/systemic approaches for analyzing various aspects of China's military modernization. Structural/dependency frameworks include the notion that technology is best conceptualized as a process rather than a product, and tend to emphasize structural inequality patterns between different classes of arms producing countries. Historical/systemic approaches, on the other hand, attempt to identify patterns of development across different time periods and world regions by using a comparative perspective. The goal is to identify:

> [. . .] a series of different 'tiers of production,' based on the diffusion of dominant military-related technologies and the skills, techniques, and infrastructure necessary for their production (Hoyt 2007: 12).

An approach combining these two perspectives seems useful, especially in the case of rising powers such as China which attempt to move up the "ladder of production" while still being embedded in historical export relations with technologically superior countries and thus finding themselves in a somewhat dependent position. These de-facto dependencies can give rise to areas of strategic vulnerability, and these have repercussions on the way a country's leadership perceives "national interest."

A central assumption of this study is that countries that are not at war will typically be reluctant to invest more than the necessary minimum of economic resources into their naval forces. Governments are usually under pressure to consider which part of their state expenditure can be dedicated to maritime security without compromising other state priorities. There are thus definite limits to military spending for any state. As Pugh has argued in his enlightening analysis of the cost of naval modernization:

> You can't spend the same money twice[.] What is spent on defence is not spent upon something else. More tanks, warships and aircraft mean fewer hospitals, poorer schools, worse roads, less social welfare and so on. Any state must balance many conflicting calls upon its resources (Pugh 1986: 6).

This point is highly relevant for a vast country such as China, which faces stark and costly challenges due to extremely skewed demographics and highly uneven development. These factors contribute to social welfare problems, heavy ecological degradation, food production limitations, and severe internal security issues which will certainly put the state budget under heavy pressure in the future.

Due to this "scarcity-of-resources" aspect of military build-ups, it is useful to compare state spending priorities to assess the leadership's intentions. A very similar line of reasoning is used by the authors of the *Global Militarization Index* (GMI) developed by the Bonn International Center for Conversion (BICC). This index measures the "relative importance of the military apparatus of a state in relation to the overall society" by evaluating "the behaviour of a state in distributing

---

[16] Cf. the very insightful overview provided by Hoyt (2007: 8–14).

its resources to the military sector in comparison to other sectors in society" (Grebe 2011: 7). The variables selected and weighted for this index include a country's defense expenditures as a percentage of the GDP and in relation to health spending; the number of its military and para-military personnel including reserves in relation to total population size and to the number of physicians; and the number of "heavy weapons"[17] in relation to total population (cf. Grebe 2011: 16–17). This highly interesting approach yields somewhat surprising results when applied to the Asia Pacific region, as will be discussed in more detail in Sect. 3.6.

Another methodological point this study hopes to make concerns the necessity of using appropriate yardsticks for drawing comparisons. This aspect is all too often overlooked, e.g. when analysts compare fleet sizes in terms of absolute vessel numbers regardless of the size of the maritime territory to be surveyed, or when they pit historically anomalous cases against more typical ones, or when they compare defense spending only nominally, not as a share of the country's GNP, or calculated per capita or per military manpower. Faulty yardsticks are often to blame for the more outrageously skewed evaluations that are especially found in the mass media. By looking at countries facing similar geostrategic situations and those that have developed economies that are roughly similar to China's, it should be possible to gain at least an approximate understanding of what constitutes a "typical" case of naval militarization level under such circumstances. This approach will then provide a means of identifying "atypical" cases in comparison with more typical ones.

The other BRIC states—Brazil, Russia, and India—seem appropriate for a comparative analysis with China for the following reasons: They are countries with large populations, vast territories and huge coastlines that aspire to play a dominant role in their respective world region, and have been developing rapidly during the past decade. Russia is already one of the most highly militarized countries of the world according the 2012 GMI, but appears to be on a quest to achieve an even greater level of naval competence after two decades of neglect. All these countries have similar strategic aims of becoming or remaining regional great powers that are increasingly less dependent on outside assistance and influence:

> While Russia tried to support its world political position by modernizing its army, Brazil continues to aim for a position of regional leadership, which also counts on military strength. India sees itself confronted with threats on various fronts: in its Kashmir conflict with Pakistan, the country keeps experiencing more intensive phases which also include armed violence. China, too, puts great political and military pressure on India with its competition for regional dominance. In addition to this, by massively expanding its military, the People's Republic intends to consolidate and expand its rising role in world politics. At the same time, Peking increases its number of arms to respond to threats from within, such as uprisings, to be capable of responding to the Tibet question and foreign

[17] These are defined as "any piece of military equipment which fits into either one of four categories: armored vehicles (armored personnel carriers, light tanks, main battle tanks), artillery (multiple rocket launchers, self-propelled artillery, towed artillery) above 100 mm caliber, combat aircraft (attack helicopters, fixed-wing fighter aircraft), and major fighting ships (submarines, major surface combatants above corvette size)" (Grebe 2011: 17).

> policy challenges, such as the conflict about the status of Taiwan and the dominance of the United States in the region; if necessary by military means (Grebe 2011: 13).

Including South Africa in this group of countries (then called "BRICS"), as has been sometimes done in recent years, seems less conclusive with regard to naval modernization because although South Africa plays a similarly dominant role vis-a-vis its neighbouring countries as the classic BRICs, its nominal defense spending level and total naval capability level, as well as total population and GDP size lag far behind the BRIC states, even though its military modernization is also on the rise. For completeness' sake I have included South Africa in some comparisons but do not find it to be a generally appropriate yardstick to evaluate China's naval status and progress against.

Another necessary dimension for analysis is the regional yardstick, inasmuch as China is a member of a regional East Asian—or even, a larger Asia Pacific—"security complex" in the sense put forward by Regional Security Complex Theory or "RSCT" (cf. Buzan and Wæver 2008: 43–51). This conceptual perspective is justified because, as Hoyt observes:

> Distinct regional 'security complexes' are formed in areas of cultural similarity, and are identifiable because the *primary* security relationships of states in the system are concerned with other states in the system. Relative indifference, as between Pakistan and Iran, signifies the effective boundary of a regional system. [...] powerful states may have interests in more than one security complex (Hoyt 2007: 3, emphasis in the original).

Member states of such a regional subsystem tend to be oriented towards each other in their threat perceptions as well as their strategic outlook, and they frequently base their arms procurement decisions on the prior decisions made by their respective strategic counterparts. They are thus to a large extent reacting to the "pressures of local geographical proximity" (Buzan and Wæver 2008: 45). It therefore makes sense to analyze their security and arms procurement strategies in relation to each other in order to assess the regional military balance. The group of countries most affected by, and in turn most able to affect, China's military rise are accordingly treated as parts of a regional "security complex" comprising, among others, Japan, South Korea, Taiwan, Malaysia, Indonesia, Singapore, Vietnam, and also the U.S. as the region's de facto "resident power." According to Buzan and Wæver, the East Asian Security Complex is unusual in the sense that it is a great power regional security complex which contains not just one, but two great powers (China and Japan) which "form the core of a bipolar great power R[egional] S[ecurity]C[omplex]," and this means that "their dynamics directly affect balancing calculations at the global level in ways that one would not expect from a standard RSC." Moreover, spillover effects of these security dynamics into the wider region are to be expected (Buzan and Wæver 2008: 59). Therefore, East Asia can be analyzed as a "Great Power Security Complex" according to RSCT, while East and South Asia (and the entire Asia Pacific region including Australia) can be conceptualized as a "Supercomplex" comprised of several regional security complexes (Buzan and Wæver 2008: 60–62). This conceptualization is also useful because of the existence of several pan-regional territorial conflicts causing constant friction within the region, such as the South China Sea issue, the Senkaku/Diaoyutai Islands issue, and the Taiwan question, and furthermore because of the existence of military alliances between a number of the affected nations.

What criteria can be used for assessing the relative strength of naval power across a wide variety of countries? This study adopts a multi-dimensional approach as suggested e.g. by Geoffrey Till (2009), in which several separate criteria are taken as equally relevant variables, because it enables the analyst to gain a more detailed and more accurate picture of the situation than would be possible by merely counting vessel numbers. The criteria deemed most relevant in this line of study for assessing a navy's actual level of sea power are:

- Force structure—type, age, and number of vessels
- Sustainability—the ability to keep naval forces on station
- Task and Reach—the Navy's primary function and maximum mission capability
- Flexibility—the number and variety of possible missions
- Access to state-of-the-art technology.

The practical difficulty here lies in finding ways of operationalizing these criteria in a meaningful manner. In the following chapters, various methods for operationalizing and comparatively evaluating these criteria will be demonstrated using China as the main case study, though comparisons will also be drawn throughout with other regional neighbors, the BRIC states, and the U.S.

In sum, the aim of this book is to analyze China's naval rise and its possible strategic consequences from a wide variety of analytical angles—technological, geostrategic, as well as political—while employing a strictly comparative perspective, and using transparent and explicit evaluation criteria. While the BRIC states will be used to establish a common baseline of what can currently be considered "normal" naval development under the conditions of fast-growing rising powers, a selection of China's neighboring states from the "Asian security complex" will also be evaluated in comparison with China to provide a regional yardstick. Since China is after all the main focus of this study, Chap. 2 will discuss China's strategic development imperatives, domestic politics constraints, and peculiar geostrategic situation, in addition to Chinese naval strategy discourses, in order to make more transparent the political framework of conditions upon which China's naval build-up will hinge over the long term.

Chapter 3 will then try to delineate a set of heuristic tools for evaluating total fleet capability and militarization levels in an international comparison, and explain the economic impact of naval modernization. It draws upon a ranking system developed by Todd and Lindberg, as well as approaches published by Till (2009), Pugh (1986), and Grebe (2011). For evaluating a navy's maximum level of capability, a system of naval mission types according to complexity published by Tellis et al. (2000) will be drawn upon and explained in Sect. 3.2. The approach chosen in this study for analyzing force structure will be delineated in that chapter as well. Section 3.2 will also explain the function of various vessel types in a fleet along with their cost impact and technical complexity characteristics. Fleet age structure can be analyzed visually by using a simple graphic device that shows vessels as separate arrows on a timeline and thus indicates a fleet's need for cyclical rejuvenation along with the distribution of age groups and ship classes within the fleet (cf. Sect. 3.3). Some criteria for evaluating different technologies' strategic values, and for comparing different vessels types' approximate capabilities, will be discussed in Sect. 3.4. The remainder of Chap. 3 discusses the cost impact of naval modernization in general and comparative terms.

Since naval development is a huge task that requires vast and steady financial investments and, in most cases, also durable international partnerships, I adopt an industry viewpoint in Chap. 4. So far, the systemic problems of warship production and the associated repercussions on China's naval build-up—material, financial, technological, and political—have been remarkably understudied, with very few pioneering efforts to speak of. The criterion of access to state-of-the-art technology (or, in China's case, lack of access to such technologies due to the U.S./EU arms embargo) will be analyzed by assessing the economic, political, and strategic consequences of the hierarchical structure of the global naval electronics industry for China (Sects. 4.2 and 4.3), and by studying embargo implications and coping strategies through strategic industry alliances and indigenous development (Sects. 4.4 through 4.6). Chapter 5 puts all these approaches to use by conducting a thorough analysis of the PLAN's fleet structure, naval weapon systems development tendencies, and operational capabilities. Chapter 6 then discusses the consequences of China's naval build-up from the point of view of other members of the Asia Pacific security complex and draws various cross-country comparisons for putting China's naval development into perspective in relation to various yardsticks.

References

'A guilty verdict for North Korea: Their number is up'. 2010. *The Economist*, May 22, 2010: 57.

Bimber, Bruce A. and Steven W. Popper. 1994. *What is a Critical Technology?* Santa Monica, CA: RAND.

Blazevic, Jason J. 2009. Defensive Realism in the Indian Ocean: Oil, Sea Lanes and the Security Dilemma. *China Security* Vol 5, No. 3: 59-71.

Bussert, James C. and Bruce A. Elleman. 2011. *People's Liberation Army Navy Combat Systems Technology, 1949-2010*. Annapolis: Naval Institute Press.

Buzan, Barry and Ole Wæver. 2008. *Regions and Powers: The Structure of International Security*. 6th ed. Cambridge: Cambridge University Press.

Cheung, Tai Ming. 2011. Engineering Human Souls: The Development of Chinese Military Journalism and the Emerging Defense Media Market. In *Changing Media, Changing China*, ed. Susan Shirk, 128-149. Oxford: Oxford University Press.

Cole, Bernard D. 2010. *The Great Wall at Sea. China's Navy in the Twenty-First Century*. 2nd ed. Annapolis: Naval Institute Press.

Crosston, Matthew. 2014. Phreak the Speak: The Flawed Communications within Cyber Intelligentsia. In *Cyberspace and International Relations: Theory, Prospects and Challenges*, eds. Jan-Frederik Kremer and Benedikt Müller, 253-267. Berlin et al: Springer.

Erickson, Andrew S. et al. (eds.). 2007. *China's Future Nuclear Submarine Force*. Annapolis: Naval Institute Press.

Erickson, Andrew S. and Lyle J. Goldstein (eds.). 2011. *Chinese Aerospace Power. Evolving Maritime Roles*. Annapolis: Naval Institute Press.

Falletti, Sebastien. 2010. Investigation Attributes Chon An Sinking to North Korean Torpedo. *Jane's Navy International* (el. ed.), May 21, 2010. http://www.janes.com.

Friedman, Norman. 2006. *The Naval Institute Guide to World Naval Weapon Systems*. 5th ed. Annapolis: Naval Institute Press.

Friedman, Norman. 2009. *Network-Centric Warfare: How Navies Learned to Fight Smarter Through Three World Wars*. Annapolis: Naval Institute Press.

Grebe, Jan. 2011. The Global Militarization Index (GMI) – A tool for evaluating development orientation of states as well as regional developments. BICC Occasional Paper, February 2011. Bonn: BICC, http://www.bicc.de/old-site/uploads/pdf/publications/papers/occ_paper_07/occasional_paper_VII_02_11_eng.pdf. Accessed September 7, 2013.

Grevatt, Jon. 2011. Philippines Eyes Submarine Acquisition by 2020. *Jane's Defence Weekly* (el. ed.), May 17, 2011. http://www.janes.com.

Hickley, Matthew. 2007. The uninvited guest: Chinese sub pops up in middle of U.S. Navy exercise, leaving military chiefs red-faced. *Daily Mail* online. November 10, 2007. http://www.dailymail.co.uk/news/article-492804/The-uninvited-guest-Chinese-sub-pops-middle-U-S-Navy-exercise-leaving-military-chiefs-red-faced.html. Accessed February 11, 2015.

Holmes, James R. and Toshi Yoshihara. 2008. *Chinese Naval Strategy in the 21st Century. The Turn to Mahan*, London and New York: Routledge.

Holslag, Jonathan. 2010. *Trapped Giant: China's Military Rise*. The International Institute of Strategic Studies (IISS), London: Routledge.

Howarth, Peter. 2006. *China's Rising Submarine Power. The PLA Navy's Submarine Challenge*. New York et al.: Routledge.

Hoyt, Timothy D. 2007. *Military Industry and Regional Defense Policy: India, Iraq, and Israel*. London and New York: Routledge.

Hu, Bo 胡波. 2012. *Zhongguo haiquan ce: waijiao, haiyang jingji ji hai shang liliang* (中国海权策: 外交, 海洋经济及海上力量 | China's sea power strategy: Foreign relations, maritime economy and maritime power). Beijing: Xinhua chubanshe.

Hughes, Christopher. 2011. Reclassifying Chinese Nationalism: The *Geopolitik* Turn. *Journal of Contemporary China* (2011), 20(71), September: 601-620.

Jane's Fighting Ships (el. ed.). Various years. http://www.janes.com.

Kristensen, Hans M. 2007. China's Submarine Fleet Continues Low Patrol Rate. FAS Strategic Security Blog from the Federation of American Scientists, February 6, 2007. http://blogs.fas.org/security/2007/02/post_2/. Accessed September 24, 2013.

Lau, D.C. 劉殿爵 (ed.). 1992. *Bingfa si zhong* (兵法四種 | A Concordance to the Militarists). The ICS Ancient Chinese Text Concordance Series - Chinese University of Hong Kong. Hong Kong: Commercial Press.

Lin, Yingyou 林穎佑. 2008. *Haijiang wan li: jiefangjun haijun zhanlüe* (海疆萬里: 解放軍海軍戰略 | A Maritime Border of Ten Thousand Miles: The PLA Navy's Strategy). Taipei: Shiying.

McGregor, Richard. 2010. *The Party. The Secret World of China's Communist Rulers*. New York: Harper Collins.

Medeiros, Evan S. et al. 2005. A New Direction for China's Defense Industry. RAND Project Air Force. http://www.rand.org/pubs/monographs/MG334.html. Accessed March 2, 2011.

Neuman, Stephanie G. 2009. Power, Influence, and Hierarchy: Defense Industries in a Unipolar World. In *The Modern Defense Industry. Political, Economic, and Technological Issues*, ed. Richard A. Bitzinger, 60-94. Santa Barbara, Cal. et al.: Praeger Security International.

O'Rourke, Ronald. 2008. *China Naval Modernization: Implications for U.S. Navy Capabilities*. New York: Nova Science Publishers.

Ping, Kefu 平可夫 (a.k.a. Pinkov, Andrei, a.k.a. Chang, Andrei). 2010. *Zhongguo zhizao hangkong mujian* (China constructs aircraft carriers | 中國製造航空母艦). Hong Kong: Kanwa Press.

Pugh, Philip. 1986. *The Cost of Seapower: The Influence of Money on Naval Affairs from 1815 to the Present Day*. London: Conway Maritime Press.

Putnam, Robert D. 1988. Diplomacy and Domestic Politics: The Logic of Two-Level Games. *International Organization* Vol. 42, No. 3 (Summer 1988): 427-460.

'Secret Sanya - China's new nuclear naval base revealed.' 2008. *Jane's Intelligence Review* (el. ed.), April 21, 2008. http://www.janes.com.

Shambaugh, David. 2004. Modernizing China's Military. Progress, Problems, and Prospects. Paperback ed. Berkeley et al.: University of California Press.

Swaine, Michael D. 2013. Chinese Views Regarding the Senkaku/Diaoyu Islands Dispute. *China Leadership Monitor* No. 41 (6 June 2013): 1-27. http://media.hoover.org/sites/default/files/documents/CLM41MS.pdf. Accessed January 20, 2014.

Taiwan Relations Act. 1979. Public Law 96-8, 96th congress. Full text available at http://photos.state.gov/libraries/ait-taiwan/171414/ait-pages/tra_e.pdf. Accessed February 16, 2015.

Tellis, Ashley J. et al. 2000. *Measuring National Power in the Postindustrial Age*. Santa Monica: RAND.

Till, Geoffrey. 2009. *Seapower: A Guide for the Twenty-First Century*. 2nd ed. London and New York: Routledge.

Wertheim, Eric. 2007. *The Naval Institute Guide to Combat Fleets of the World*. 15th ed. Annapolis: Naval Institute Press.

Wertheim, Eric. 2013. *The Naval Institute Guide to Combat Fleets of the World*. 16th ed. Annapolis: Naval Institute Press.

Wu, Donglin 吳東林. 2010. *Zhongguo haiquan yu hangkong mujian* 中國海權與航空母艦 (China's Sea Power and Aircraft Carrier). Taipei: Shiying.

Yang, Yi Edward and Xinsheng Liu. 2012. The 'China Threat' through the Lens of the US Print Media: 1992-2006. *Journal of Contemporary China* (2012), 21(76): 695-711.

Yi, Yang 弈洋. 2011. *Zhu jian: Zhongguo diyisou hangkong mujian danshengji* (鑄劍: 中國第一艘航空母艦誕生 | Casting swords: The birth of China's first aircraft carrier). Hong Kong: Tianxingjian chubanshe.

Yoshihara, Toshi and James R. Holmes. 2010. *Red Star Over the Pacific. China's Rise and the Challenge to U.S. Maritime Strategy*. Annapolis: Naval Institute Press.

Young, Doug. 2013. *The Party Line: How the Media Dictates Public Opinion in Modern China*. Singapore: John Wiley & Sons.

Zhu, Jiangnan and Jie Lu. 2013. One rising China, multiple interpretations: China's 60th anniversary celebration through the lens of the world's printed media. Journal of Contemporary China Vol. 22, No. 84: 1067-1088. http://dx.doi.org/10.1080/10670564.2013.795316.

Chapter 2
Naval Power and Its Role in China's Rise

> *"China is a big country, and other countries are small countries and that is just a fact."*
> —Yang Jiechi, China's then foreign minister, at the 17th ASEAN Regional Forum in 2010. Cited from 'China's Military Rise' (2012).

The present chapter describes the perceived need for naval modernization from Beijing's point of view. It also discusses the nature and functions of naval power—or "sea power"—in general and through Chinese eyes. Since Mahan's classic treatment of sea power is being widely discussed in China, this chapter will start from this notion and attempt to show the connections between sea power, the Chinese notion of "Comprehensive National Power" and China's self-perception as a "big" country, as well as the various frustrations and limitations imposed on China's naval ambitions by its difficult maritime geography, other development priorities, and international alliances situation.

2.1 Naval Power, Sea Power, Maritime Power, and National Power

While "sea power" seems to stress the naval component of the term and is almost interchangeable in general usage with "naval power," "maritime power" also includes non-military aspects such as the "ability of a state to use the sea to its optimum" and is thus broader in its meaning (cf. Sakhuja 2011: 4). This study is mostly concerned with the narrow field of naval power. Alfred Thayer Mahan's term "sea power" which he coined in his 1890 classic *The Influence of Sea Power Upon History, 1660–1783* was extremely influential, but not very clearly defined, and has consequently been interpreted by contemporary authors in a wide variety of senses.[1] Its popularity today is unbroken. In China, for instance, Mahan's concept of sea power has yielded extensive discussions, which have been described and

[1] Cf. Mahan (1987 [1890]: 26–69) and Till (2009: 20).

S. Kirchberger, *Assessing China's Naval Power*, Global Power Shift,
DOI 10.1007/978-3-662-47127-2_2

analyzed in detail by Holmes and Yoshihara (2008: 27–47) and need not be recounted here.[2]

What is the meaning of sea power within the context of this study? If power should be defined in a Weberian sense, as Till implies in his contemporary treatment of the subject, then sea power can be understood as "the capacity to influence the behaviour of other people by what you do at or from sea" (Till 2009: 83). It seems necessary here to discuss the relation between sea power thus understood and the military, specifically naval capability, and their combined influence on total national power.

Military power is viewed by some as "the ultimate yardstick of national power," because "countries subsist in an environment where internal and external threats to security are both common and ever-present." As a result, Tellis et al. write, "the effectiveness of their coercive arms becomes the ultimate measure of power" (Tellis et al. 2000: 133). Nonetheless, the actual translation of military capability into political power is far less immediate in practice than the statement above seems to suggest, especially in a world where armed conflict between great powers—the ultimate test of military power—has thankfully become extremely rare:

> The link between military forces [...] on the one hand, and power [...] on the other hand, is neither a direct nor a simple one. The immediate effect of using military force is purely physical: killing people and destroying things. Developing and successfully implementing a strategy that translates this capacity for harm into the desired political effect is difficult, and an uncertain proposition at the best of times (Frühling and O'Neil 2012: 83).

In China, the government's attempts to measure and define power through the use of various mathematical formulas is so wide-spread that several analysts have commented on the emergence of a concept called "Comprehensive National Power" (综合国力) or "CNP."[3] In his study of major Chinese think tanks, Leonard describes a seminar at the CCP's Central Party School he witnessed in 2005, where the well-known social scientist Hu Angang explained all the different power ranking systems then in use by China's various think tanks, each of which has apparently developed its very own power index. As Leonard concludes, "China must be the most self-aware rising power in history. [...] Measuring 'CNP' [...] has become a national obsession" (Leonard 2008: 84). This could very well be true.

---

[2] The first translation of Mahan's (chin. 漢 or 马汉) monumental work into Chinese seems to have been the 1954 edition prepared by Yang Zhenjia (楊鎮甲) published under the title *Haijun zhanlüe lun* ("A theory of naval strategy") by Zhonghua wenhua chuban in Taipei/Taiwan. It appears to have sold quickly, for it saw a third printing already in 1955. Much later, in 1998, a mainland Chinese version translated by An Changrong (安常容) and Cheng Zhongqin (成忠勤) was published in Beijing by the PLA Press under the more accurately translated title *Haiquan dui lishi de yingxiang: 1660–1783*.

[3] In a 2004 paper on the subject, Hu and Men define the term as follows: "CNP may be simply defined as the comprehensive capabilities of a country to pursue its strategic objectives by taking actions internationally and the core factors to the concept are strategic resources, strategic capabilities and strategic outcomes, with the strategic resources as the material base" (Hu and Men 2004: 3). A thorough discussion of the concept is provided by Pillsbury (2000: 203–258).

In 2006, a major government-sponsored academic study entitled *The Rise of Great Powers* (大国崛起) was completed by a group of China's top scholars after several years of work comprising hundreds of interviews with politicians and academics (Tang 2006). The study analyzed historical case studies of nine former rising powers—Portugal, Spain, the Netherlands, the UK, France, Germany, Japan, Russia, and the U.S.—with a view to what China could learn from their experiences. The study also gave rise to a very popular book series and a TV documentary series and thus created a sort of "hype" surrounding the issue of naval affairs. In sum, it "suggests that national power stems from economic development fueled by foreign trade, which can in turn be furthered by a strong navy" (Erickson and Goldstein 2009: 402).[4]

For reasons that will be explained in more detail below, many Chinese analysts feel that China is currently at a decided strategic disadvantage compared to other members of the "Asian security complex" that have close political and military ties with the U.S. This is especially true when it comes to securing vital sea lines and enforcing sovereignty over claimed offshore territories. As a consequence, the enhancement of China's sea power both symbolically and materially often claims center stage within the context of an envisaged rise in China's "Comprehensive National Power" status.

After concluding in their analysis that China lags behind other countries with regard to all the various component parts of "CNP," especially in military modernization, Hu and Men offer the following piece of advice to their leaders:

> National security and unification must be safeguarded. At the same time, defense and combat capability in high-tech conditions should be increased, the mechanization and informationization of the military be completed so as to achieve a leapfrog development of military modernization (Hu and Men 2004: 31).

From the point of view of China's military leaders, innovation management becomes a key area in achieving such a "leapfrog development" with the aim of passing over the normal stages of military-technological development which other nations have to slowly traverse.

2.1.1 The Importance of Critical Technologies

Mastery of certain advanced technologies appears to lie at the root of actual or potential military power. Some observers even go so far as to accord it central importance:

[4] The "Great Powers hype" has also translated to the realm of popular culture, as Gotelind Müller describes in her analysis of the above mentioned TV documentary series *Daguo jueqi*. Both the series and the study were extensively discussed among Chinese intellectuals, cadres, and the general public (cf. Müller 2013: 22–92).

> technology—understood as the material instantiation of knowledge, methods, resources, and innovation—[is] identified as the first and most important building block for the production of national power (Tellis et al. 2000: 53).

The strategic function of technological progress in the field of warfare is self-evident. Technological advances can act as "force multipliers" by neutralizing disadvantages such as numerical inferiority and by enhancing combat effectiveness (Hoyt 2007: 5). The connection between advanced technologies and military power is easily made since advances in technology are universally scrutinized for possible military applications by all the great and rising powers, and military research, for its part, traditionally drives much of the world's technological progress. Moreover, in some cases, innovative weapon systems can in effect "change the way war is fought" and thus effectively transform the global security environment (Mayer 2012: 290). Therefore, the study of "[. . .] technology as a building block of national power must include [. . .] the scrutiny of a country's ability to produce all the militarily critical technologies deemed to be vital today" (Tellis et al. 2000: 55). This implies that not all technology is of prime importance for assessing military and national power. Rather, it is primarily a country's ability to produce the so-called "critical" technologies that indicates the power fault lines within the current world order. Exactly what technologies should be labeled "critical" in any given context is subject to debate, but usually there exists at least "[. . .] a loose consensus in government, industry, and among technologists on which technologies today are deemed to be critical," with extensive lists drawn up regularly in the U.S. (Tellis et al. 2000: 53–54).[5] However, it is important to note that the quality of criticality is, as Bimber and Popper (1994: 20) note, "not inherent in the technology itself." Rather, it stems from "[. . .] the importance of the outputs of the system of which the technology is a constituent part, as well as from the significance the technology has for enabling that system." The stance cited here is a practical and functionality-based view of the issue. In order to determine a technology's degree of criticality in any given circumstance, questions such as "Critical for what, to whom, and by what standards?" need to be answered.

In an era characterized by rapid development, especially in the field of information technology, there exists an ever widening "knowledge gap" between "innovators" that drive technological advance and "laggard" countries that currently do not possess the ability to join in the race for progress. According to the above-cited study, the ability to innovate:

> [. . .] can be generated only as a result of attaining mastery over advanced scientific concepts and having both the resources and the ability to translate these concepts, first, into new components and, thereafter, into a larger socio-technical system built around the introduction of these new components (Tellis et al. 2000: 53).

[5] As Tellis et al. note: "The authoritative U.S. study [. . .] has identified about 2,060 militarily significant technologies, of which fully 656 were deemed to be critical for the purposes of developing advanced weaponry, all of which fell within eighteen broad technology areas which, in turn, are further divided into eighty-four subsections" (Tellis et al. 2000: 55).

As this description indicates, the financial resources needed to join the race can be formidable and this effectively limits the number of countries that can hope to join the ranks of the "first-tier producers" of sophisticated military technologies to the group of dynamically growing rising powers. By virtue of abundant resources gained from its successful external trade development over the past few decades, China finds itself among those rising powers that can direct large amounts of money to enhancing their technology profile. This effort not only can raise the quality of China's defense technology, but also reduce outside dependence on foreign suppliers of subsystems. Complete technological self-reliance would be an ideal situation from a security viewpoint, but in reality this ideal is hardly ever achievable for any one country.

The influence of technology upon international relations and military power still remains a somewhat understudied subject, notwithstanding the very real consequences of knowledge fault lines between countries, or the wide-spread implications and "momentous causal impact" that changing levels of technological innovation capability can have in the international system (Crosston 2014: 253–254). In Chap. 3, I will try to outline those areas in naval technological development that appear to be especially "critical" with regard to assessing naval power.

2.1.2 Sea Power and National Power

Regarding the function of sea power within the wider scope of military capability and national power, two aspects seem to be particularly important: First, that sea power is a relative rather than an absolute concept, or "something that some countries have more than others," but that most countries still possess to some degree (either through naval strength, shipbuilding capability, provision of services to seafarers, or through a combination of the above); and second, that sea power is "the product of an amalgam of interconnected constituents that are difficult to tease apart" (Till 2009: 22, 83). Such an understanding of sea power implies that it not only makes sense to directly compare different countries' levels of sea power to each other using certain explicit criteria, but also that it will be difficult, if not impossible, to assign an absolute numerical value to any one country's level of sea power.[6] It also follows that sea power relations between states may be subject to change over time. Just the same, it is nevertheless possible to determine different countries' maximum levels of naval capability by defining demarcated categories. While it may not be possible to further differentiate between navies that are deemed to fall within the same category, a crude relative hierarchy of tiers or ranks can still be established as a heuristic tool (see Chap. 3). It goes without saying that a direct

[6] Sakhuja's (2011: 20–23) pioneering attempt to create a maritime power formula notwithstanding, it remains to be seen whether naval power relations can actually be quantified like that in a meaningful way when put into practice, which he has opted not to demonstrate.

comparison between countries that are part of the same regional sub-system is especially useful because of the direct influence that their mutual threat perceptions and strategic rivalries have on each other.

Sea power usually does not figure as a variable within the various formulas that Western as well as Chinese scholars have developed for measuring total national power.[7] Usually more generalized indicators of military strength (e.g. total military manpower, defense budgets, nuclear weapons) are used in such calculations. This may be due to the difficulty of quantifying a vague and relative concept, as well as to its somewhat limited applicability, as naval power is less relevant for landlocked countries, even if some of these maintain small inland waterway navies. What's more, navies typically constitute only one branch of any country's total armed forces.

Nevertheless, for coastal countries with access to the high seas the existence or absence of sea power does play a role for determining overall national power. In the case of the handful of nuclear powers, nuclear ballistic missile submarines (SSBN) form an essential ingredient of a credible deterrence strategy. Their existence or absence is of crucial importance for assessing a country's realistic nuclear second-strike capability. This is only relevant, however, for the very few nations currently operating or building SSBN—the U.S., Russia, China, India, France, and Britain.

Apart from their purely military function, which is mainly useful in times of conflict, navies also regularly contribute to maintaining their state's territorial integrity by safeguarding maritime borders. In recent years, the protection of maritime resources and safeguarding of coastal installations against asymmetric threats such as terrorism, smuggling, illegal fishing, and waste dumping has become more relevant.[8] Another important responsibility of naval forces and coast guards is maritime search-and-rescue, as well as humanitarian aid during natural disasters or refugee crises. This was well demonstrated in Asia during the Tsunami catastrophes of 2004 and 2011, and during the civil war in Libya in 2011, when a Chinese warship first entered the Mediterranean Sea in order to help evacuate more than 35,000 Chinese citizens from the war zone (Lampton 2014: 162). On a more symbolic level, naval diplomacy and international maritime cooperation can also offer chances of furthering intraregional integration, or can serve to enhance a state's international power perception.[9] Last but not least, navies also have symbolic import that can be used to promote national pride within the domestic

---

[7] Pfetsch (1994: 146–157) gives a concise overview of various methodological approaches to measuring national power, starting with Morgenthau's *Politics among nations: The struggle for power and peace* (1948). Hu and Men (2004: 17–22) list many different mathematical formulas in use by Western and Chinese scholars. Likewise, Chap. 7 of Tellis et al. (2000), "Measuring military capability", give a very good overview and discussion of the relevant methodological questions (pp. 133–176).

[8] Potgieter (2009) gives a good overview of these new maritime challenges.

[9] Oddly illustrative of the last two points were Northrop Grumman's full-page ads in various naval technology magazines during 2008. The ads featured Northrops's latest nuclear carrier design with a byline reading "Meet the world's most persuasive diplomat."

population. In China, there has been a marked rise in recent years in television programs describing China's successes in naval development and other military technology areas, and related Web pages have proliferated in the Chinese Internet (cf. Cheung 2011: 140).

Even though it is hard to quantify the exact contribution of naval power to total national power, it is easy to recognize the consequences of deficient or absent naval power. Providing security from external threats has traditionally been seen as a basic core function of statehood as defined e.g. by Rice and Patrick (2008: 3) in their *Index of State Weakness*. Pugh, moreover, writes in *The Cost of Sea Power*: "A secure defence is one of the greatest benefits that a state can confer upon its populace," while "a state which neglects its defences invites attack" (Pugh 1986: 6). This truism is as old as warfare itself. There are dangers associated with very low levels of militarization, not only externally, but also internally, as noted by the editors of the BICC's *Global Militarization Index* (GMI):

> Adequate militarization can have positive effects if it prevents dangers and preserves security interests—both internally and internationally—and creates a good investment climate. If military resources of a country are too low, however, a government will possibly not be able to sufficiently enforce its state monopoly on the use of force (Grebe 2011: 8).

The protection of maritime borders against exterior threats and the maintenance of territorial integrity must therefore be placed among the central duties of the state.[10] It follows from this reasoning that coastal states with grave deficiencies in this area cannot be considered strong, no matter how well they might perform in other areas.[11]

In the age of globalization, trading nations have become increasingly dependent upon maritime transport, and China, as the world's new largest export nation, is no exception. The narrow Malacca Strait alone is transited by more than 60,000 vessels each year and provides passage to about half of the world's energy transports (Cole 2010: xxi). Apart from this, it is useful to reflect on the fact that seas and oceans cover nearly three quarters of the world's entire territory. Moreover, the sea is a fundamentally hostile environment that can never be truly controlled by humankind. It remains to this day, as Till (2009: 23) writes, "a dark, mysterious and dangerous place in which people cannot commonly live." Nevertheless, since antiquity, the sea has simultaneously served as a resource and a medium of transport, exchange, information, and dominion. Since the advent of globalization, maritime trade has become the primary means of transport for manufactured goods, raw materials, and oil (Till 2009: 25–33).

---

[10] According to Rice and Patrick (2008: 3), the core functions of statehood include "*fostering an environment conducive to sustainable and equitable economic growth; establishing and maintaining legitimate, transparent, and accountable political institutions; securing their populations from violent conflict and controlling their territory; and meeting the basic human needs of their population*" [italics in the original].

[11] China's experience in the nineteenth century is a good example of just such a scenario. Oriented towards the Northwestern continental threat but weak in sea power, China easily fell prey to comparatively small forces of British and French gun boats with superior capability.

2.1.3 China's Strategic Interest in Acquiring Sea Power

As one of the major trading nations, China feels increasingly compelled to protect its overseas investments as well as Chinese personnel deployed in African, Middle East, and Maghreb countries. China's participation in the multinational effort against piracy off the Horn of Africa, as well as its 2011 decision to send a frigate to Libya for the protection of Chinese laborers, are reflections of these concerns and have also signaled a new era of Chinese international maritime activity. In 2006, renowned Chinese strategic thinker Zhang Wenmu (张文木, *1957), a professor at the Center of Strategic Studies at the University of Aeronautics and Astronautics in Beijing, outlined the arguments in an ongoing debate within China on whether or not sea power should be more strongly developed, and towards what end:

> Opinions in China are greatly divided on whether or not, and if so, how China should strive for sea power. Given both the nature of global interdependence and the disastrous naval defeats of certain countries in history, some have put forward that it is unnecessary for China to emphasize sea power in the process of economic development. Others have stressed the importance of vastly strengthening China's navy in order to vie with other naval powers for hegemony. However, both views are inaccurate. China's sea power is uniquely defined. A traditional Western notion of sea power is the ability to control the sea, while China's concept of sea power is a marriage of the notion of equal sea rights and sea power. In the latter, the application of power on the seas cannot exceed the former but rather should serve the aim and scope of a nation's sea rights (Zhang 2006: 23).

In writings such as this, emphasis is not placed on the expansion of a sphere of military influence, but more on the defensive safeguarding of legitimate maritime interests. Chinese analysts, however, tend to include in their definition of China's legitimate "core concerns" (核心利益) the control of conflicted territories such as the Spratly and Paracel Islands, the Senkaku/Diaoyutai Islands, and the island of Taiwan. The boundaries between a so-called "hegemonial" Western concept of sea power and China's supposedly more "peaceful" approach, as described above by Zhang Wenmu, are therefore somewhat blurred—at least in the eyes of non-Chinese observers.

2.2 The Impact of China's Domestic Politics on Naval Modernization

In order to put naval modernization into perspective in relation to other strategic developmental goals, it is useful to consider the trajectory of China's "grand strategy" articulated by Chinese government officials as well as prominent academics.[12] An example of the latter is found in the following excerpt from an article on China's national power by influential CCP-affiliated economist Hu Angang (胡鞍钢, *1953) and strategic studies expert Men Honghua (門洪華, *1969). In 2004, they wrote:

> [. . .] we hold that the objectives of China's grand strategy should be to make the people rich and the country strong, that is, constantly raising the percentage of its CNP in the world's total so as to become a big world power in the middle of this century. The basic objectives of China's grand strategy in the future 20 years should be to including [*sic*] 6 goals: 'high growth, great national power, affluent people, national security, improvement of international competitiveness and sustainable development' (Hu and Men 2004: 30–31).

According to the bluntly worded, non-official, but quite representative position taken here by two leading intellectuals, China's rise to great power status in every sense of the word—political, economic, and military-strategic—is definitely a strategic goal, along with the country's continued economic development and the regime's survival (cf. Wacker 2006: 55). The navy's build-up is critical for attaining these goals, both materially and symbolically.

In order to better understand China's national and military power sources as well as the challenges it faces in its further economic and political development in relation to the issue of sea power, it is necessary to review China's domestic and foreign policy imperatives—what China dubs its national "core interests." I will then return to the issue of naval development.

2.2.1 The Historical Meaning of China's Economic Growth Miracle

Viewed from a broader historical perspective, China only recently emerged from an era of weakness and trouble that lasted roughly 150 years. The world's most wealthy and advanced country and the self-proclaimed center of civilization in the eighteenth century, China faced a dramatic downturn towards the middle of the nineteenth century, when it was degraded to semi-colonial status in the face of technologically superior forces. China's ultimate humiliation came from the sea when a small fleet of British gun boats inflicted heavy damage on Chinese war junks during the so-called Opium Wars (1839–1842 and 1856–1860). The twentieth century, however, would prove to be China's most difficult century. Two revolutions in 1911 and 1949, the Japanese occupation and a brutal civil war between Communists and Nationalists had already wreaked havoc in the country by the time

---

[12] A grand strategy, according to Wacker (2006: 55), is a "vision and internal logic connecting a country's paramount goals to its diplomatic, security, economic, and military strategies for deploying means and resources, taking into account the possible responses of other actors." It need not necessarily be explicitly stated, but can also be inferred by observers as an "external attribution [. . .] or an *ex post* rationalization of foreign policy actions" [italics in the original].

Mao's failed economic policies ruined most of the progress that had been made in the few years after 1949. Altogether, the communist takeover, the "Great Leap Forward" (1957–1961) and the class struggle mayhem of Mao's "Cultural Revolution" (1966–1976) claimed at least 50 million lives and left the economy in shambles.[13] This era took its toll on all areas of industrial development as well as on human education and welfare. It also inhibited overall progress in naval modernization, even though some military breakthroughs such as development of the atomic and hydrogen bombs and nuclear-powered submarines were achieved due to high-level backing for projects like these of extreme strategic importance.[14]

When the reform era began in the late 1970s, China's post-Mao leadership inherited a country devastated by natural and man-made calamities. With starvation still widespread after 30 years of communist rule, China was technologically backward and internationally isolated. Apart from the unresolved conflict with its so-called "renegade province" Taiwan, China was on very cool terms with its former ally, the Soviet Union. It had also made the dubious choice of supporting the genocidal Khmer Rouge regime in Cambodia and had recently been at war with India. The People's Liberation Army (PLA), although numerically strong, was so badly equipped and trained that it suffered humiliating losses against the combat-proven forces of Vietnam in 1979.[15]

Starting from a very low pedestal, the development path taken by the Chinese government since then has ushered in a period of accelerated economic growth. Ever vigilant towards threats against its monopoly on power, the CCP leadership chose a gradualist reform approach that was designed to let market forces stimulate productivity without endangering the Communist Party's exclusive control over the political system.[16]

Once the CCP allowed growth to occur in the early 1980s by relaxing growth-stifling policies, China started catching up. Economic progress was fairly steady

[13] According to CCP Central Committee figures cited by Domes (1980: 66–68), 100 million rural Chinese were still affected by hunger in late 1978, with per capita grain production levels lower than in 1957.

[14] Cole (2010: 7–18) provides a concise overview of naval development during China's communist era until today.

[15] For a detailed description of this war and the Chinese military's perception of it, cf. Li (2009: 252–263). He writes: "Many of the PLA's commanding officers were shocked by the poor discipline, low morale, combat ineffectiveness, and high casualties [...]. During the nineteen days of the first two phases, the PLA suffered 26,000 casualties, about 1,350 per day" (2009: 258).

[16] Notwithstanding attempts by some observers to find a trend towards political pluralization in today's China, the political system must still be described as structurally Leninist. This means that the ruling CCP insists on its right to control the three vital pillars of state power: The propaganda apparatus, the armed forces and the Party's control over leadership personnel appointments within all meaningful organizations in China's state administration, economy, and society. Cf. Oksenberg (2002: 193–194), McGregor (2010), Li (2010), Shambaugh (2010: 81–86) and Guo (2013: 23–25).

throughout the entire reform era, reaching an average level of quarterly GDP growth of ca. 9 % during 1989–2010.[17] Systemic shocks such as the 1989 Tian'anmen incident, the Asian Financial Crisis of 1997/1998, or the Global Financial Crisis of 2008 have not had a severe impact on China's growth, at least as of the time of this writing. The Tian'anmen incident however had grave repercussions on naval modernization, for it was the cause of the still-ongoing arms embargo by the U.S. and the EU that forced China to abandon co-operative projects begun between 1979 and 1989 and to orient itself towards Russian and Ukrainian naval technology. These developments continue to severely limit China's technological cooperation options to this day.

By 2011, Deng Xiaoping's reform strategy had finally made China an economic superpower.[18] Through its hugely successful foreign trade policy, China has been able to assemble the world's largest reserves of foreign currency. This not only bolstered China's ability to withstand external financial shocks such as the 2008 crisis, but also generated the cash necessary for large-scale military procurement and industrial infrastructure development. Military spending has since picked up speed, even though a recent study by Liff and Erickson (2013) has shown convincingly that the increase is still fairly in line with government spending growth overall (cf. Sect. 3.6). Moreover, China's budget for internal security since 2011 has been found to be even larger than the defense budget, pointing to the myriad challenges to regime stability the CCP leadership has to face each day. This also puts the military build-up into perspective as a state priority (Guo 2013: 225).

2.2.2 The Imperative of Regime Survival

Despite China's economic reforms, it is not possible to speak of any softening of authoritarian rule, which modernization theorists often expect to automatically follow from economic liberalization. In addition, the closely interwoven relationship between the Communist Party and the military—a "symbiosis," as Shambaugh calls it—remains more or less intact (Shambaugh 2004: 11–12). Following the revolutionary credo that political power flows from the barrel of a gun, Party control over the PLA—originally the military wing of the Communist Party—is considered to be one of the three pillars of communist rule, next to media and leadership personnel control. This view was expressed in a famous dictum by Mao Zedong, who wrote in 1938: "Our principle is that the Party commands the gun, and the gun must never be allowed to command the Party" (Mao 1965 [1938]: 224).

---

[17] Data from *Trading Economics* online, available at http://www.tradingeconomics.com/Econom ics/ GDP-Growth.aspx?Symbol=CNY (accessed March 4, 2011).

[18] In 2010, China became the world's largest exporter for the first time. As of 2013, it was the world's second largest economy after the U.S. in both nominal and PPP terms and held the world's largest foreign currency reserves of US$ 3.82 trillion. China's public debt stood at a mere 31.7 % of GDP (CIA 2014).

The CCP's dependence on support from the armed forces has been vital in times of increasingly numerous popular uprisings, especially in hot spots like Tibet and Xinjiang. Repeated evidence of this was shown in the years since the PLA ended Cultural Revolution outrages by force. The Party, however, has had to take active measures to ensure PLA loyalty to the Party line, e.g. by adhering to the Leninist principle of adding Party-appointed "political commissars" to each command level and requiring higher-level commanding officers to be Party members, and by establishing about 90,000 Party cells among the troops, or roughly one for every 25 soldiers (McGregor 2010: 110). Nevertheless, in 1989, the Party leadership encountered strong resistance among PLA commanders ordered to take part in the suppression of the Tian'anmen demonstrations, leading to a widespread purge in its aftermath among both the armed forces and within the civilian Party bureaucracy.[19]

It is clear, however, that for the time being and recent overhauls notwithstanding, the PLA remains a Party's army rather than a people's army. The amount of political-ideological training at cadre schools, previously amounting to as much as 30 % of total training, has greatly declined in recent years and given way to a greater emphasis on professional military training.[20] Rising professionalization and less emphasis on ideological indoctrination, however, have given rise to doubts about the military's continued loyalty towards the CCP, especially since the younger top leaders in the politburo usually have few military credentials and connections of their own compared to earlier veteran heroes of the revolution such as Deng Xiaoping. A 2006 article in the CCP's principal theory magazine *Qiushi* ("Seeking Truth") entitled "For the implementation of our armed forces' historic mission, enforce the ideological and political construction within our Navy" by Hu Yanlin (胡彦林, *1943), the then PLA Navy's Political Commissar, was notable for strongly and repeatedly emphasizing the need to adhere to the "basic rule of Party leadership":

> In the process of fulfilling the historic mission of our armed forces, the severity of the political challenges encountered by the naval forces becomes ever more obvious. Therefore we have to adhere strictly to the basic principle of obeying the Party's commands (Hu 2006: 45, my transl.).[21]

Interestingly, the same text continues by blaming destructive "outside influences" for discipline problems inside China's armed forces:

> While preparing for military conflict and finding themselves at the front of Reform and Opening, the naval forces not only encounter the severe challenge posed by Western enemy forces who are employing a strategy of "Westernization" and "splitting" against us; they also face the difficult experience of manifold new contradictions and problems brought

[19] For detailed statistics on these developments see Heilmann and Kirchberger (2000).

[20] See the description and analysis in Shambaugh (2004: 53–55, 133–134 and 11–13).

[21] I am grateful to Shi Ming for drawing my attention to a highly interesting series of *Qiushi* articles on military affairs published during 2004–2006 that in sum point to an intense internal definition phase regarding military affairs during that time.

> about by our country's reform and development entering a decisive phase, and moreover face all kinds of challenges and influences from social discourse. We must not underestimate the danger posed by infiltration and subversive activities of Western enemy forces, by the encroachment of wrong thinking such as 'separation of the armed forces from the Party and their de-politicization' (军队非党化, 非政治化) or their 'becoming state-led' (国家化), by the influence and challenge to the armed forces brought about by society's pluralist value tendencies, or by a decayed ideological culture to the thinking and morale of cadres and soldiers (Hu 2006: 45, my transl.).

The appellative character of this particular article (which for all its length is remarkable for its lack of substance and its repetitive use of identical phraseology centering around the "need to obey the Party Central" while "withstanding" unnamed "enemy forces") is only one example among many similar texts that harken back to a bygone era, specifically 1976/1977 when would-be Mao heir Hua Guofeng issued similarly structured and worded articles to rally support for his claim to leadership—largely without success.[22] At that time, at least, such rhetoric was indicative of weakness more than strength. Whether these much later attempts to boost morale are in fact indications of an "erosion of values" within the armed forces, however, is hard to assess from the outside. They almost certainly allude to the endemic problem of corruption and hedonism among military officers that was targeted by a new anti-corruption campaign starting from late 2012 (cf. Mulvenon 2013b). Evoking the threat of enemy subversion has been a long-standing feature of CCP rhetoric from the beginning, right down to the CCP constitution's preamble itself, which contains similar references to "enemies" through all of its revisions to this day.[23]

A perceived slackening of Party control over the PLA in recent years would at least explain why Xi Jinping, right after taking over the helm from Hu Jintao as the new "core" (核心) of the "collective leadership" (集体领导) in 2012, seemed to cater to the military by embarking on a hardliner course against Japan in the Senkaku/Diaoyutai Islands issue. China's establishment of an "East China Sea Air Defense Identification Zone" (ADIZ) covering the disputed area in November 2013 in an unusually abrupt manner was a source of consternation in the region as well as abroad. Xi, it seems, was bent on proving to observers both at home and abroad that the "core interests" of China are not open to debate. Xi, the son of revolutionary hero Xi Zhongxun, is also the first leader since Deng Xiaoping to have first-hand experience with the military, having served as an assistant (秘书) to General Geng Biao, the Secretary-General of the Central Military Commission (CMC) during the late 1970s when China was at war with Vietnam (Mulvenon 2013a: 1–2; Hu and Liang 2012: 137–138). His position as personal assistant to

[22] It is not without irony that critical support for Deng Xiaoping's challenge against Hua during that time came, of all places, mostly from the military (cf. Kirchberger 2004: 132–135).

[23] The CCP's Party Constitution was last amended in November 2012 and can be read in Chinese at http://dangshi.people.com.cn/n/2012/1119/c234123-19618241.html and in its official English translation at http://www.china.org.cn/china/18th_cpc_congress/2012-11/16/content_27138030.htm (accessed January 21, 2014).

such a high-level military leader must have given Xi unparalleled insight into the workings of the CMC.[24] Nonetheless, his actual standing within the armed forces—an elusive factor that has often been decisive in phases of leadership transition—is difficult to assess.

Even though the exact state of Party control over the PLA is currently unknown, the CCP's dependency on military support for its rule, as well as the actions of recent leaders (as citied in the examples above) suggest that the Party leadership is heeding the military's modernization needs and policy preferences more than the governments of liberal democracies typically would.

2.2.3 *Domestic Threats to Political Stability*

Despite widespread inner dissent within the Chinese populace, as reflected in rising numbers of public protest incidents recorded in China year after year, an unprecedented budget for internal security surpassing even the defense budget in 2012, and extreme measures taken by the leadership to control popular discourse through tight Internet censorship and heightened propaganda efforts, there have been no major country-wide uprisings in China's core areas since the violent suppression of peaceful demonstrations in Beijing in 1989.[25] To many observers it appears that the rulers and the ruled have tacitly agreed on a truce: In effect, as long as the ruling party keeps delivering on its promise of progress, the urban population majority will refrain from engaging in political protests. That said, if the economic engine ever grinds to a halt and large portions of the urban population, which soon will make up more than half of China's total population,[26] are put under pressure, this mutual understanding might no longer hold, with incalculable consequences for regime survival. China watchers and Chinese intellectuals alike for the most part agree that regime legitimacy among the populace now rests primarily on the country's economic performance, although quality-of-life issues such as food safety, pollution control, social security, and healthcare seem more pressing of late.[27] Although firmly entrenched and without any rival organization to challenge its power, the Communist Party still has no choice but to deliver on its promise of

[24] On the functions and positions of various types of *mishu* within the Chinese leadership system cf. Li and Pye (1992).

[25] According to official Chinese figures, instances of public unrest have increased 10-fold between 1993 (8,700 recorded cases) and 2005 (87,000 recorded cases). This means "that each province experienced an average of nearly eight incidents per day" (Chung et al. 2006: 6). For an insightful analysis of the different types and triggers of such incidents and the various types of regime responses see Göbel and Ong (2012).

[26] Cf. the urban/rural population growth projections published in STRATFOR (2007).

[27] According to Qinghua University Professor Yan Xuetong in an Interview with Richard McGregor, this is the mainstream view in China, much to his personal chagrin (cf. McGregor 2010: 107).

progress. Otherwise the leadership fears it could meet the demise of its Soviet sister party—a momentous event that has perhaps been the most extensively studied problem in China's political circles since 1991 (cf. Shambaugh 2010: 53–81).[28] The Romanian military's arrest and execution in late 1989 of the country's leader Ceaușescu, who happened to be a personal friend of Deng Xiaoping, was supposedly shown on videotape to CCP politburo members and Party elders and reportedly horrified China's leadership (Yang 1998: 256–257).

The goal of avoiding the pitfalls that lead to the Soviet Union's collapse has already demanded painful choices. The CCP, once the self-proclaimed vanguard of the working class before officially re-casting itself in the role of a "Marxist governing party" in 2012, presided already in the late 1990s over the large-scale scrapping of inefficient state-owned enterprises which were instrumental for providing urban employees with the so-called "iron rice-bowl" system of social security. Millions of jobs were lost without adequate social security in place to alleviate the hardships. At the same time, the military was stripped of its business empire—a fairly controversial decision at the time that created a great deal of resentment within the armed forces (Cheung 2009: 118–129). Today, about 200 million migrant workers flock to the urban growth centers, attesting to the ever-increasing wealth gap between impoverished rural and flourishing urban areas. They also often only benefit from minimal work safety standards and endure the living conditions of second-class citizens. Such a level of social inequality would have been unthinkable in the egalitarian, though universally impoverished, society of the Mao era. For all its bloodshed, this period is again being idealized by disillusioned segments of the populace and even within some quarters of the ruling CCP, the so-called "Old" and "New Leftists" (老左派/新左派; cf. Zhao 2011: 222–224). The inequality is also hard to justify under the umbrella of a communist state ideology, contributing to a widespread loss of values and an ideological vacuum that creates legitimacy problems for a leadership increasingly perceived as hopelessly corrupt (Holbig 2009: 42).

2.2.4 Creating National Unity Through Nationalist Propaganda

In the face of rising regional wealth disparities, China's leaders have increasingly resorted to nationalist propaganda for engendering a sense of unity and identity—often with quite astonishing success. Outside observers have been appalled by violent manifestations of popular outrage against Japanese companies, as tensions

[28] Incidentally, a TV documentary series on this subject entitled "Alert to danger while dwelling in safety" (居安思危) that was prepared by Party intellectuals a few years ago became mandatory viewing for all CCP cadres (cf. the extensive analysis of this series in Müller 2013: 133–176).

over the Japanese-controlled Senkaku/Diaoyutai Islands issue have been heightened in recent years. Toward this end, the Party-controlled media apparatus is being employed with great efficiency and is now also offering special TV programing on weapons development catering to a youthful urban audience, apparently with the aim of boosting confidence in the leadership and enhancing national pride and self-identification (Cheung 2011: 135–136). As Cheung writes: "Programming portrays a positive and strongly patriotic image of the military as the defenders of the country's sovereignty and ideological purity" (Cheung 2011: 141).

China's government has also succeeded in making a great majority of China's populace uncritically accept a number of tenets that serve the end of "national unity" and "patriotic thinking." Its methods include censorship of school history textbooks, tight media control, strict Internet censorship and an unending injection of repetitive phraseology into the public discourse. As a result, many Chinese firmly believe that Taiwan's reunification with the motherland is inevitable, that it is what the Taiwanese truly desire, and that it must be brought about by the use of force if necessary. Moreover, they are convinced that the South China Sea issue—like the Senkaku/Diaoyutai Islands issue—is a "core interest" of China worth fighting a major war over, and that there is a consensus among the industrialized powers including Japan that China should be denied its rightful place in the world (cf. Swaine 2013: 4).[29] Such positions have been widely echoed in nationalist bestsellers published by members of the "New Left" movement such as *China Can Say No* (中国可以说不, Song et al. 1996) and *China Is Unhappy* (中国不高兴, Song et al. 2009), and their success reflects a general climate of disillusionment with China's current position in the world order.

By placing the blame for domestic problems on "hostile outside forces" and keeping alive the memory of wartime atrocities committed by Japan, the CCP leadership ensures that a number of skeletons of a disturbing past will remain firmly locked up in the closet of the Party's history. Calamities like the horrific Great Leap starvations and the 1989 Tian'anmen massacre in fact remain unknown to many younger Chinese. Meanwhile, the CCP successfully claims sole responsibility for China's post-1978 growth miracle, without however acknowledging that during the preceding three decades of CCP rule very little economic progress was achieved, while some 50 million Chinese lost their lives as collateral damage of government utopian experiments or due to violent "class struggle" and downright political persecution. This failure to deal with the darker aspects of CCP history is slowly being addressed within progressive circles among the Party's 83-million membership, which is evident in a multitude of commemorative and autobiographical articles published in outspoken media such as the history magazine *Yanhuang*

[29] See the interesting analysis of official and unofficial Chinese publications on the islands conflict issue with Japan by Swaine. He found that "no major difference seems to exist between the views of civilian and military sources, whether authoritative or otherwise" (Swaine 2013: 11).

chunqiu (炎黄春秋). This intellectual movement—initiated by reform-minded Party elders—is an indication that China still awaits a moment of historical reappraisal for coming to terms with its national history comparable to those already conducted in other post-totalitarian states. If this were officially endorsed at some point in the future, it could bring about a significant shift in the perception of Chinese national identity, and perhaps even influence the Chinese leadership's foreign policy behavior.

2.2.5 Necessary Conditions for China's Further Development

So far, China's vibrant growth has kept all potential dangers to regime stability in check. What, then, are the major factors that have contributed to this economic boom? Experts in the field of China's economic development tend to agree that high global demand for China's manufactured goods, bolstered by the still (as of 2014) relatively low value of China's currency, plus a proportionally high influx of Foreign Direct Investments (FDI) with their associated technology transfers, are at the root of China's current economic success. Domestic demand however is still relatively underdeveloped. Political stability (domestic and regional) thus furthers a positive investment climate, while internal unrest or conflict with regional neighbors would likely endanger it.

There is a second dimension to the imperative of keeping a positive investment climate that is often overlooked. The systemic problem of widespread official corruption and rent-seeking—which is common to most post-socialist economies and which has already resulted in the misappropriation (sometimes euphemistically dubbed "informal privatization") of large portions of state wealth by corrupt cadres—has so far not impacted as severely on China's economy as it did on the Soviet Union in its final phase. The major reason for this, according to relevant studies, seems to be China's dynamic investment climate, which ensures that most of the misappropriated funds never actually leave the country, but are reinvested in China itself, even if privately. This means they still contribute in some way or another to overall growth and infrastructure development. Late Soviet cadres-turned-oligarchs in contrast preferred to transfer their misappropriated funds overseas, effectively draining the Soviet Union's national economy of resources (cf. Dickson 1997: 197; Heilmann 2000: 232, 241). Lately, there have been massive indications that the tide may be turning in China, with increasing numbers of high-ranking cadres transferring both family members and funds to secure overseas destinations (Boehler 2014). If left unchecked, this alarming tendency could represent a real threat with regard to regime stability. In any case, it should be clear from the discussion above that China is even more dependent on maintaining domestic and regional peace for its "peaceful development" or "rejuvenation" strategy[30] than many observers tend to realize.

It was thus no exaggeration when veteran China watcher Susan Shirk remarked in her 2007 study of China's domestic and international constraints that China, because of its multitude of unresolved domestic problems that could yet endanger its rise, can be considered a "fragile superpower" at best. Addressing a U.S. audience, Shirk writes: "Seeing China objectively by recognizing the domestic fragility of this emerging superpower can help us avoid mistakes that could provoke a confrontation" (2007: 269). Similar points have recently been made by David Shambaugh, an expert noted for his long-term experience in analyzing the complexities of China's party state and military. Shambaugh argues that "China is a confused and conflicted rising power undergoing an identity crisis of significant proportions" and, what's more, that "it is not so much an aggressive or threatening China with which the world should be concerned, but rather an insecure, confused, frustrated, angry, dissatisfied, selfish, truculent, and lonely power" (Shambaugh 2013: 316–317). These observations nicely capture the complexities and pitfalls of China's transitory international identity.

2.2.6 China's Integration into the World Trade System

As the graph in Fig. 2.1 shows, China's share of global trade volume as of 2012 was already the second largest in the world and almost on par with the United States. If Hong Kong's share were added, which is calculated separately by the World Trade Organisation (WTO), China would actually surpass the U.S. The figure also shows that world-trade participation is highly uneven. Twenty-nine countries (18 % of all WTO members) altogether represent a share of more than 80 % of total world trade; the remaining 130 WTO member states, or 82 % of trading countries, represent a share in world-trade volume of less than 1 % each. Only a small number of countries therefore can be considered highly or well integrated into the world trade system, while the majority of countries are still left on the sidelines.

One effect of China's high degree of integration into the world trade system has been the fact that China and its major trading partners have entered into quasi-symbiotic relationships. This is especially noteworthy in the case of the U.S., China's main trading partner, which is ironically also its most powerful potential adversary in a possible military conflict over Taiwan. The U.S. also leads a quasi-military alliance in the Asia Pacific composed of U.S.-bilateral security partnerships with Japan, South Korea, Australia, and several other strategically positioned

[30] China's leaders have struggled to put a fitting name tag on their national development strategy for some time. Before 2004, "Peaceful Rise" (和平崛起) was advocated, and later abandoned for the less ambitious-sounding "Peaceful Development" (和平发展). "Rejuvenation" (复兴) is another term expressing the desire to put China back to its rightful place in the world. It was originally coined by late-Qing reformer Liang Qichao (1873–1929).

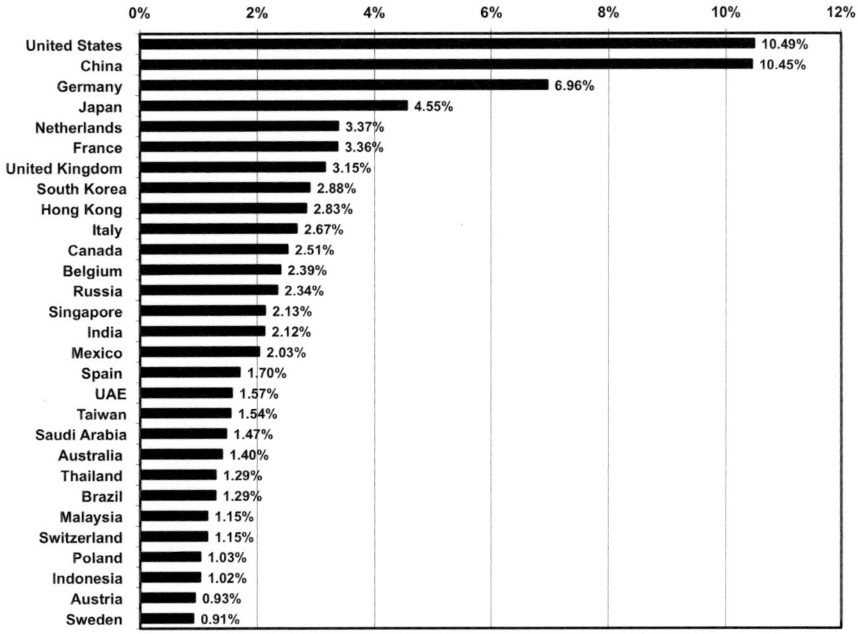

Source: Shares computed from the WTO International Trade Statistics Database (WTO 2013).

Fig. 2.1 World trade shares, 2012 (%)

countries, and maintains numerous costly military bases in these and other countries in the region. China meanwhile finances a large portion of the foreign-owned U.S. state debt. The economic interdependency between China and the U.S. is however mutual and, again, fraught with irony. China, for instance, needs its major export customers, foremost the U.S., to fuel export growth, which, in turn, is necessary to generate the vast financial resources China needs for its military modernization, internal security, as well as other costly projects such as infrastructure development, social security, and pollution control.

China, as Mark Leonard puts it, is on "a quest for markets, natural resources and political support" for its continued development (Leonard 2008: 87). Such a grand strategy however effectively precludes belligerent behavior, notwithstanding the hawkish rhetoric sometimes voiced by Chinese military officials.[31] For Western observers, this means that those arguing in favor of "containment" against China, or those adding fuel to the fire of mutual threat perceptions by publishing texts

[31] Examples in recent years have included the various publications by air force Colonel Dai Xu (戴旭, *1964), who has written a book named "Totem of the Sea" (海图腾) on China's aircraft carrier programme (Dai 2009) and another one on China's so-called "C-shaped encirclement" (C-形包围) by the U.S. and other Asian powers (Dai 2010).

implying an inevitable military conflict with China,[32] would perhaps do well to think through the myriad implications of China's complex role in the world trade system alone before advocating any such strategy.

2.3 Maritime Geography as a Factor for China's Naval Development

2.3.1 Dependence on Maritime Trade Routes

Bearing in mind the above-mentioned factors necessary for China's continued development, what is their relation to naval power? One obvious concern is the need for uninterrupted access to crucial materials (steel, oil, other raw materials, machinery) that fuel China's growth and that rely heavily on maritime transport. China also needs secure sea lanes for the transport of its export merchandise. Already in 2007, a report stated that China's oil reserves would last for only 7 days. In 2013, China became the world's largest importer of oil from the OPEC countries after the United States. Its dependence on maritime transport routes are such a source of concern for China that it actively supports the construction of the planned Thai Canal at the Isthmus of Kra, which would offer an alternative to the piracy-infested Strait of Malacca. China is also building pipelines across Pakistan, and, as of 2013, it has taken steps to open up the Northern Sea Route across the Arctic for commercial shipping. China invests in oil refinery installations in Singapore and Sri Lanka, and is heavily involved in building the port of Gwadar in Pakistan, which could eventually be used by China as a naval base (cf. Ruppik 2007: 27). China's need for food imports could soon become another critical issue. In sum, these few points illustrate why maritime security in the global trade choke points (shown as red circles in the map in Fig. 2.2) is as much a strategic concern to China as to any other nation largely dependent on maritime trade (cf. Holmes and Yoshihara 2008: 4–5). Western critics of China's "hunger for resources" however often fail to acknowledge the fact that China has the same legitimate concerns in this regard as other industrialized nations. As a relative newcomer, however, it finds

[32] Examples of articles sure to fuel mutual distrust include Kaplan's (2005) piece in *The Atlantic* entitled "How We Would Fight China" that was featured on the magazine cover together with a photograph of a grim-looking Chinese seaman; or Kraska's (2010) article entitled "How the United States Lost the Naval War of 2015" depicting a hypothetical scenario of a successful Chinese attack on a U.S. aircraft carrier. This tendency is nothing new. Yang and Liu, in their highly interesting quantitative analysis of "China Threat" content in U.S. print media between 1992 and 2006, have found that since 1995, the military threat theme was the dominant strain, before economic/trade threat and political/ideological threat, and has made up almost 70 % of all "China Threat" related media content over a 15 year period (Yang and Liu 2012: 706). For an outline of the original "China Threat" argument see Roy (2009).

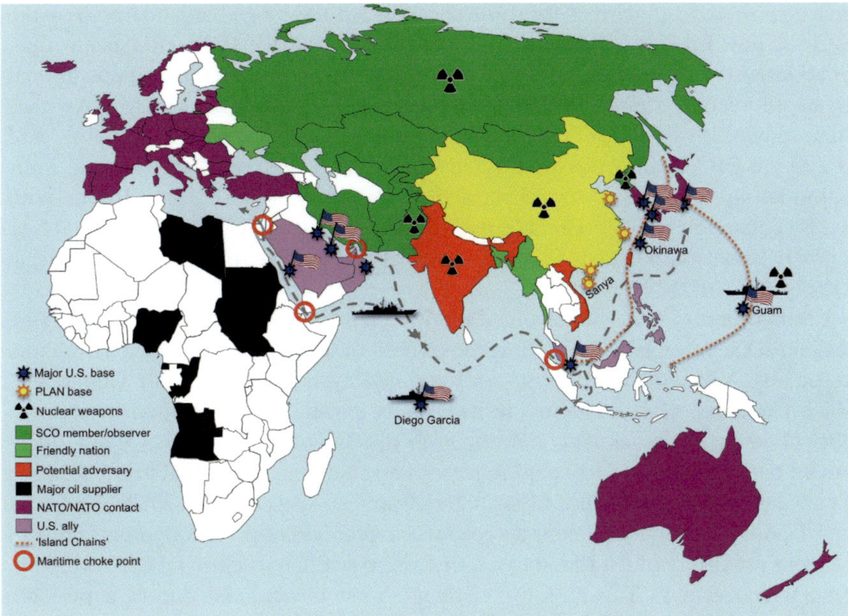

Source: Map concept adapted from Le Monde, Atlas der Globalisierung (2006) with additional data from various news reports.

Fig. 2.2 China's geostrategic situation from Beijing's point of view

itself at a disadvantage compared to the established powers when it comes to accommodating its needs on the global market (cf. Gu and Mayer 2007: 1–6).

The systemic danger inherent in China's economic vulnerability to blockades and embargoes, the so-called "Malacca Dilemma" (cf. Chen 2010), has been clearly noted by Chinese naval strategists such as Zhang Wenmu, who wrote:

> [. . .] China is almost helpless to protect its overseas oil import routes. This is an Achilles heel to contemporary China, as it has forced China to entrust its fate (stable markets and access to resources) to others. Therefore, it is imperative that China, as a nation, pay attention to its maritime security and the means to defend its interests through sea power (a critical capability in which China currently lags behind) (Zhang 2006: 20).

Zhang also states that, in an age of economic globalization, "national defense must go where a nation's economic interests lie," thus making a case for extended out-of-area operations (Zhang 2006: 20).

Apart from this problem, however, China has several traditional maritime security concerns as well. The need for keeping up credible nuclear deterrence vis-à-vis the five de-facto nuclear-armed powers in its immediate vicinity—India, Pakistan, Russia, North Korea, and the U.S. Navy's 7th fleet—is one factor. At the moment, though, it seems that its first submarine-borne nuclear ballistic missile system still awaits completion, even though China has operated indigenously built nuclear-powered submarines for some decades. A long-range ballistic missile—the *JL-2*—is now finally in the testing phase (Wertheim 2013: 105). This means that, for all intents and purposes, China probably does not at present have an operational sea-based second strike capability, although this could soon change. Another Chinese concern is the safeguarding and policing of China's large coastline and its 200 nm EEZ against maritime threats such as smuggling, illegal fishing, infiltration by spies, maritime terrorism, and piracy, to name just a few. In line with overall economic growth, China's maritime economy grew steadily by ca. 11 % annually between 2001 and 2005, adding another incentive for enhancing maritime security (Li 2010: 291–292).

China furthermore has unresolved maritime claims that conflict with Japan (the Senkaku/Diaoyutai Islands) and with several of its Southeast Asian neighbors (especially Vietnam, the Philippines, and Malaysia) in the South China Sea, which China has claimed almost in its entirety since 1947. To demonstrate to the other claimant countries its commitment to these offshore territories, China must remain prepared to regularly deploy patrol vessels there. And, last but not least, in the absence of a formal peace treaty, China's unresolved conflict with Taiwan over the island's sovereignty remains a geostrategic problem of prime importance due to the close proximity of the island to the Chinese coastline, its close relations with the U.S., which acts as Taiwan's effective guardian power, and due to a possible U.S. involvement in a military conflict.

2.3.2 Military Alliances in China's Neighborhood

The main military-strategic problem from China's point of view is the presence in its close vicinity of a powerful informal alliance of countries led by the U.S. composed of bilateral U.S. security partnerships. This coalition has for all practical purposes achieved a high level of interoperability through shared technology, extensive military contacts, and regular joint large-scale maneuvers. In fact, the U.S. can be dubbed a "resident power" in the Asia Pacific region given the nature of its permanent forward presence and vested interests, which it pursues through a combination of bilateral trade and security partnerships as well as through arms trade relations (Sakhuja 2011: 58). Without a doubt, U.S. forces in Asia are the most powerful military power in the entire region, with more than 120,000 troops, 6 aircraft carrier battle groups carrying up to 500 fighter aircraft, 31 nuclear-powered attack submarines, 12 heavily armed cruisers, 29 destroyers, and 12 frigates deployed there in 2008, among them *AEGIS*-equipped surface vessels armed with cruise missiles, surface-to-air missiles, and anti-ballistic missiles (Holslag 2010: 31). The so-called "first island chain," as Chinese strategists call the line of islands ranging from the Korean peninsula Southwards over Japan, Okinawa, the Senkaku/Diaoyutai Islands and Taiwan, is literally dotted with U.S. military installations and listening stations that are positioned close to Chinese shores. Altogether, as of 2013, the U.S. has officially deployed more than 57,340 military personnel to

23 countries within the Asia Pacific region, excluding the large force at Hawaii. The vast majority, about 54,400, are stationed in Japan.[33] As of early 2014, six of the ten currently serviceable U.S. aircraft carriers have been deployed to the region, as have 60 % of the nuclear-powered U.S. submarines, among them eight SSBN (cf. Park 2014).

In sum, the high number of U.S. troops stationed permanently in South Korea, Japan, the Philippines, Singapore, Australia, and on U.S. bases such as Guam, Hawaii, and Diego Garcia, as well as intensifying U.S. political and arms trade linkages with historic opponents of China such as India and Vietnam, present a formidable challenge to China's own regional aspirations. U.S. relations with its former war enemy Vietnam, China's primary contender for control over the South China Sea, were elevated to the status of a defense agreement in 2011 (which included negotiations for the use of ports) and were intensified in 2013 with an agreement on the transfer of nuclear technology from the U.S. to Vietnam.

The fact that, in the field of naval technology, more and more of the region's U.S. allies—South Korea, Japan and Australia, to be exact—have chosen to adopt the *AEGIS* combat system for their latest surface vessels to increase interoperability and intelligence sharing with U.S. forces is also worrisome from China's viewpoint, not least because its only military ally in East Asia is currently North Korea.

One major security benefit enjoyed by all members of the informal U.S.-led alliance is a kind of "extended nuclear deterrence" provided by U.S. forces to its non-nuclear allies both in Asia and in Europe, and to NATO as a whole:

> [...] US nuclear weapons provide critical reassurance against conventional and nuclear threats from regional adversaries. The United States continues to station nuclear weapons in five European countries—each of whom is a non-nuclear weapon state under the NPT—and strongly reaffirmed the extension of its nuclear umbrella to Japan and South Korea in the wake of North Korea's nuclear tests in 2006 and 2009 (Frühling and O'Neil 2012: 86).

For the above-mentioned reasons, it is useful from a strategic viewpoint to conceive of the U.S., South Korea, Japan, and Australia as members of a hypothetical strategic alliance of powers that China would be pitted against more or less alone in the unlikely case of military conflict—for North Korea is hardly a reliable partner—and has no hope of ever joining. As General Yao Yunzhu (姚云竹, *1954) of the Chinese Academy of Military Science observed, "the international military order is U.S.-led—NATO and Asian bilateral alliances—there is nothing like the WTO for China to get into" (quoted in 'China's military rise' 2012). Japan, China's primary contender for control over the Senkaku/Diaoyutai Islands has, on the other hand, decided to issue a Joint Political Declaration with NATO in April 2013 that is intended to further mutual cooperation and formalize Japan's relations with the organization. Developments such as these are directly related to a growing sense of a "China Threat" within the region, while the alliance-building resulting from this,

[33] Deployment data as published by the U.S. Defense Manpower Data Center (DMDC) on September 19, 2013, available at https://www.dmdc.osd.mil/appj/dwp/getfile.do?fileNm=SIAD_309_Report_P1307.xlsx&filePathNm=milRegionCountry (accessed January 22, 2014).

in turn, creates a feeling of exclusion among China's leaders that fuels distrust and gives rise to a perception of being "trapped" or "encircled" (cf. He and Feng 2011: 20).

Moreover, apart from the main members of the informal U.S.-led alliance, other countries within the "Asian Security Complex" such as Taiwan and the Philippines are directly dependent upon U.S. aid, arms exports, and interventions for their security and therefore have close political and military ties with the U.S., even in the absence of formal security treaties. As a result, they lean heavily on this alliance politically, economically and culturally.

During the past 10 years, news media in Asian and Western countries have repeatedly drawn attention to China's attempts to forge closer military ties with some of the lesser powers in the region, notably Bangladesh, Bhutan, Nepal, Maldives, Myanmar, Pakistan, and Sri Lanka. China's strategy of building, refurbishing, and operating ports in some of these countries has been referred to as a "string of pearls"[34] strategy aimed at the eventual establishment of a Chinese overseas base network (cf. Blazevic 2009: 63–64). So far, this has yet to materialize. Within the above-mentioned group of countries, Pakistan must be considered the most dependable and influential bilateral partner, and mutual security relations have been enhanced by China's involvement in building the port of Gwadar and numerous joint arms procurement projects also in the naval shipbuilding sector. China has officially been asked by Pakistan to build a naval facility next to the commercial port of Gwadar, but Chinese officials have so far publicly declined the offer (Mahadzir and Hardy 2011). Gwadar, therefore, is the most likely candidate for eventually becoming a true PLAN forward base. However, Pakistan is also a long time U.S. ally, mutual irritations notwithstanding. Therefore I concur with Moss who wrote in 2011:

> [...] the PLAN's steps into the wider world have so far been tentative. Chinese-built ports in Bangladesh, Myanmar, Pakistan and Sri Lanka certainly have strategic potential, but there is no evidence yet to suggest that the PLAN intends to use these facilities as forward naval bases, although it may ultimately do so (Moss 2011).

Holmes and Yoshihara present a similarly balanced view of the issue in their 2011 paper on the "string of pearls."

Irrespective of the true intentions of China's leadership regarding such a base-building strategy, it is a useful exercise to compare the current situation of China's tentative out-of-area bases with the firmly entrenched network of military bases operated by U.S. forces worldwide for assessing the strategic balance. Analysts have had surprising difficulty when trying to establish a reliable figure of worldwide

[34] Apart from the Pakistani ports Gwadar and Karachi, other Chinese-built or refurbished ports often mentioned in connection with the "string of pearls" include Hambantota on Sri Lanka and ports on Small and Great Coco Island, Sittwe, Kyuakpu, Mergui and Hainggyi Island (Myanmar). As Blazevic notes, "In spite of the impressive number of projects, few appear to have notable military resources. Moreover, even if China does intend to use the pearls for military purposes, its access to them in a time of conflict is doubtful" (Blazevic 2009: 63–64).

U.S. military bases due to perhaps intentional lack of official transparency. Nevertheless, according to estimations by several critical analysts, the U.S. maintained between 700 and 800 bases in about 63 countries around the world 10 years ago, and has more than 1,000 today (cf. Dufour 2007; Turse 2011; Vine 2012). Many of the newer ones seem to belong to a new type of small, unobtrusive installation dubbed a "lily pad" that is designed to have a low profile.[35] In addition to the bases on land, the 10 currently active aircraft carrier task forces are, in effect, large floating air bases. The level of military integration with host countries is supported by about 170 exercises and about 250 U.S. Navy port visits each year. The total cost of maintaining the vast network of overseas bases has been estimated to amount to about $250 billion annually. This huge investment may be justified from a U.S. point of view because, as Vine has observed:

> Beyond their military utility, the lily pads and other forms of power projection are also political and economic tools used to build and maintain alliances and provide privileged U.S. access to overseas markets, resources, and investment opportunities (Vine 2012).

All in all, when viewing the issue disinterestedly, it would seem that worrying about a hypothetical "string of pearls" strategy of China as a challenge to the U.S. is somewhat disingenuous, or at the very least, premature.

2.3.3 China's Difficult Maritime Geography

In sum, it must be conceded that China's geostrategic situation, as illustrated by the map in Fig. 2.2, is indeed unusually complex. China has land borders with 14 countries, and it has been at war with two of these since the 1960s. Four of its immediate neighbors are nuclear powers. Several potentially hostile powers are not only allied with the U.S., the sole remaining superpower, but also located unnervingly close to China's maritime borders. In the case of Taiwan, the shortest distance is only 100 nm (or 185 km). Douglas MacArthur's famous phrase that Taiwan is an "unsinkable aircraft carrier," which is moreover located next to vital sea lines, has resonated widely in Chinese strategic thinking, partly explaining the near-obsessive quality of China's preoccupation with the Taiwan question (cf. Holmes and Yoshihara 2008: 54–62).[36]

China's challenging maritime geography—its so-called "littoral dilemma"—has been extensively analyzed in Mahanian terms by Homes and Yoshihara (2008:

[35] They are in officially called "lily pads" in reference to the way water lily pads are used by a frog for "jumping across a pond toward its prey." They are "small, secretive, inaccessible facilities with limited numbers of troops, spartan amenities, and prepositioned weaponry and supplies." Vine (2012) estimates that "the Pentagon has probably built upwards of 50 lily pads and other small bases since around 2000, while exploring the construction of dozens more."

[36] Interviews conducted by me in 2001 with Chinese analysts in a Beijing think tank support this view.

48–70). The characteristic of the "first island chain" as a natural barrier separating the "Chinese Caribbean" from the Oceans has been summed up by them as follows:

> For Chinese Mahanians, Taiwan, Japan, the Korean Peninsula, and the Southeast Asian Archipelago represent an exceedingly long natural geographic barrier to Chinese naval power projection. [...] Proximity, especially along the southeastern and northeastern shores, is a problematic characteristic of the island chain for Beijing. At its closest point, Taiwan is less than 100 miles from China's shoreline, the Korean Peninsula only 50 miles farther. To the north, commercial and naval shipping entering or exiting the Yellow and East China seas confronts something akin to layered defense. As Chinese vessels, both commercial and military, transit past the Korean Peninsula, they quickly encounter the Japanese archipelago. The governments of these territories are treaty-based allies of the United States (Holmes and Yoshihara 2008: 69).

The combined effects of the geographic proximity of numerous foreign military installations and listening stations plus the existence of U.S.-dominated military alliances mean that China is under much heavier surveillance by powerful strategic rivals than is customary or would even be possible in the case of most other rising powers. This situation renders the free movement of Chinese naval vessels into deeper Pacific waters without prior detection difficult.

The resulting threat perception in China occasionally finds expression in the form of writings such as PLA Air Force Colonel Dai Xu's *The C-shaped Encirclement*, a book from 2010 that deplores the perceived "shutting in" of China by U.S. partners in the Asia Pacific, Southeast Asia, and Central Asia. Chinese fears of "encirclement" should be taken note of by outside observers, and it should be granted that this perception is not unfounded. Polemical writings like Colonel Dai's capitalize on such fears to make an impact on the mass-market audience, but they are also a reflection of deeper concerns within the Chinese military regarding the country's actual strategic vulnerability.

To illustrate the effect upon the thinking of Chinese strategists, the tables can be turned and a hypothetical situation envisioned where numerous Chinese military installations are located close to U.S. borders, on e.g. the Caribbean Islands, in Central America, and Southern Canada. The question may be asked, along with Vine (2012): "[H]ow would the United States respond if China, Russia, or Iran were to build even a single lily-pad base of its own in Mexico or the Caribbean?" When contemplating the possible range of U.S. reactions to such a development, the Cuban missile crisis of 1962 certainly comes to mind as a historical precedent, even if the analogy is arguably flawed due to a much-different world political climate today. Nevertheless, it is helpful to realize that Chinese strategists may conceive of the waters inside the first island chain, and of the South China Sea, as the "Chinese caribbean", and of the Malacca Strait as a counterpart to the Panama Canal (cf. Crisp 2010: 205).

A more sober, but no less dire assessment of the Chinese surveillance situation can be found in a recent article on China's electronic warfare challenges by three academics from Wuhan's Naval University of Engineering (Liu et al. 2013). This article describes the problem confronting the Chinese Navy as follows:

> We are facing a full-fledged surveillance with communication electronics covering all bands. [...] The surveillance range extends all over the Asia-Pacific region. Using Japan as a forward base, the U.S. forces perform multi-faceted, full-fledged, and three-dimensional communication surveillance operations against our navy using ground stations, surface vessels, electronic surveillance systems, and reconnaissance satellites. Advanced surveillance installations and specialized personnel are active 24 hours a day, watching our navy's every move. Electronic surveillance ships and aircraft in all parts of the world, as well as reconnaissance satellites in space, also perform monitoring tasks. Meanwhile, the U.S. has established reconnaissance intelligence sharing agreements with Japan, South Korea, and other countries (Liu et al. 2013: 197, my transl.).

The authors continue by describing in more detail the comprehensive nature of surveillance measures directed towards China:

> The surveillance basically extends to all radio bands. The U.S. military listening station at Misawa on the Japanese Island Honshu, Yokohama's Kamiseya Communications Station and Okinawa's Sobe Communication Site [decommissioned in 2006, S.K.], as well as the listening stations at Osan and Kunsan Airbases in South Korea simultaneously listen to all kinds of shortwave communication in Russia, China, and North Korea and also locate and decipher the signals. In Australia, the [Defence Satellite Communications] Station established at Geraldton and the [Joint Defence] Facility Pine Gap have Antennas capable of intercepting all kinds of signals from satellite dishes, and intercept primarily the shortwave communication of all the Asia-Pacific countries. The [Boeing] RC-135v/w, [Lockheed] EP-3E and [Boeing] EA-18G electronic reconnaissance aircraft deployed at Misawa Air Base, Kadena Air Base and Naval Air Facility Atsugi also conduct comprehensive surveillance and intercept our military's shortwave and VHF communications signals. The U.S. military "Magnum" class electronic reconnaissance satellites with their large antennas of more than 90 m in diameter can intercept a very wide frequency range of radio communication signals (Liu et al. 2013: 197, my transl.).

The authors conclude that "the wide range of surveillance is able to cover our entire sea area." More research on electronic countermeasures, the authors point out, could effectively prove to be "the Chinese Navy's life insurance in the event of a maritime conflict" (Liu et al. 2013: 197–198).

From the above example it seems clear that as a result of China's sudden establishment of the new "East China Sea Air Defense Identification Zone" in late 2013, conflicts with the U.S. about surveillance flights over Chinese-claimed territory are almost certain to get more intense.

China's geostrategic situation becomes even more challenging when taking underwater geography into account. From the point of view of naval development, China's surrounding waters have two major characteristics: The extreme proximity of foreign military installations discussed above, and a striking shallowness of average water depth (see the map in Fig. 2.3). Almost three-fourths of the East China Sea is less than 200 m deep; the average depth is 350 m (cf. 'East China Sea' 2014). The situation in the Yellow Sea further to the North is even more extreme: With a maximum depth of 152 m and a mean depth of only 44 m, it is difficult terrain, especially for submarine operations. In effect, it "[...] forms a flat, shallow, and partly enclosed marine embayment. Most of the sea, which is deeper than the Bo Hai, consists of an oval-shaped basin with depths of about 200 to 260 feet (60 to 80 metres)" ('Yellow Sea' 2014). The South China Sea, which is claimed by

Source: Reprinted with permission from Encyclopædia Britannica, © 2013 by Encyclopædia Britannica, Inc.

Fig. 2.3 East China Sea: depth contours and submarine features

China almost in its entirety, is much farther removed from the Chinese mainland, and with an average depth of about 1,210 m it is the only sea area significantly deeper than China's directly accessible territorial waters (cf. Cole 2010: xxi). Its deepest portion is the China Sea Basin, which "[...] has a maximum depth of 16,457 feet (5,016 metres) and an abyssal plain with a mean depth of some 14,100 feet (4,300 metres)" ('South China Sea' 2014). This body of water "[...] contains the most heavily used SLOCs between the Indian and Pacific oceans. It opens to several strategically vital seaways, none more important than the Singapore and Malacca straits" (Cole 2010: xxi). To reach it, Chinese vessels from the Yellow and East China Seas must first transit the Taiwan Strait, which is only ca. 100 nm wide and, according to Cole, who cites interviews with U.S. meteorological experts, "is potentially one of the most dangerous bodies of water in the world" to navigate. Its mean depth is just under 70 m, and it is

> [...] marked by even shallower areas, wide-ranging tides, unpredictable currents, and the potential for sudden bad weather. Its waters are also extremely 'noisy' due to the extensive ship traffic, fishing boats, shipwrecks, and biological activity, making the use of both passive and active acoustic sensors (sonar and sonobuoys) difficult for submarine and antisubmarine warfare operations (Cole 2010: xxi).

This peculiar, as well as completely inescapable fact of maritime geography is surprisingly little discussed with regard to China's naval development. Nonetheless, it has grave strategic consequences. The problems resulting from China's narrow and shallow territorial waters are most relevant for submarine operations but also affect naval surface vessels, at least to some degree. The defining characteristic of the submarine as a naval weapon system is, after all, its ability to hide from enemy view. This ability is compromised in shallow waters, at least for transiting submarines, because "[t]he deeper the submarine can go the bigger the volume of sea it can hide in" (Canadian Forces n.y.: 49). Shallow waters are also difficult terrain to navigate safely, especially for larger submarines such as China's giant new *Jin* class SSBN. At over 130 m in length, its hull diameter must be large enough to accommodate the *JL-2* missile, which is about 13 m long. Shallow waters therefore not only make such vessels vulnerable to detection, but also to accidental grounding.[37]

The question of using smaller submarines in shallow coastal waters gets mixed reviews, as there can also be some strategic benefits. An older Canadian training manual for the *Oberon* class submarine sums it up this way:

> The submarine is vulnerable in shallow water, and many writers consider that the use of submarines for cloak and dagger or pinpricking operations is unnecessarily hazarding a valuable major war vessel. Nevertheless, there have been many examples of submarines landing agents, special parties and V.I.P.s, landing saboteurs and commandos, attacking targets ashore, watching enemy coast lines and harbors, acting as navigational beacons (Canadian Forces n.y.: 48).

[37] The table of recorded submarine accidents in Sect. 5.3.3 shows quite a number of navigational errors in shallow areas and crowded sea lanes as the main cause of such accidents within the U.S. Navy.

Smaller diesel submarines would be better able to cope with shallow waters. Moreover, since several large rivers empty into the Yellow and East China Seas, causing varying levels of salinity and carrying large amounts of sediment, anti-submarine warfare is also heavily compromised, not least because sonar pulses reflect off the sea bottom as well. This means that the maritime territory in question is probably best suited to the "classic" submarine tactic in shallow water: "to lie on the bottom, with engines stopped, to minimise noise" (Friedman 1984: 175). In addition, diesel submarines can be used to covertly mine an area with ground mines (Scott 2011).

The confined and shallow nature of China's territorial waters, which are comparable in this regard with Europe's Baltic Sea, effectively means that Chinese surface vessels can hardly leave shore unnoticed by foreign surveillance, and they have a hard time hiding during transit. It further means that submarines and surface vessels have more limited options for tests and training (as certain minimal water depths are necessary for both submarines and surface vessels conducting tests and trials as well as for a variety of training missions), and, finally, that the terrain is uncommonly dangerous and difficult to navigate. If water depth levels around Hainan island are taken into account, China's decision to secretly build a large submarine base on the Southern tip of Hainan by drilling deep tunnels into the rock for launching submarines directly into deeper water to avoid surveillance becomes much more understandable (cf. 'Secret Sanya—China's new nuclear naval base revealed' 2008).

Some observers nevertheless have difficulty conceding this point to China. Cole writes:

> PLAN strategists have written much about China's lack of access to the open Pacific due to the presence of the 'first island chain.' This is cited as one reason why China must control Taiwan to 'break' this restriction. As a practical matter, and considering the geography, it is difficult to accept an assertion that Taiwan or any other entity significantly limits China's access to the Philippine Sea or other parts of the Pacific Ocean, any more than Cuba could be seen as blocking U.S. access to open Ocean (Cole 2010: 36–37).

The problem, however, is not so much that the island chain in any way "blocks" China's access to the open ocean, but that bathymetry and narrow distances between China and the first island chain provide easy conditions to China's neighbors for electronic and other forms of surveillance. This consequently makes some kinds of naval training and testing operations difficult to accomplish, putting China's vessels at a strategic disadvantage compared to all the other nations in the region who have direct access to deep oceanic waters at least from some parts of their coastline. The same cannot be said of the U.S., India, Brazil, South Korea, Taiwan, or Japan, because all these countries have ready access to deep waters more or less directly from their shores.

This situation would of course be remedied to some extent if Taiwan were brought back under Chinese control, for the waters to the East of Taiwan are extremely deep and Taiwan could thus serve as a springboard for, among other things, China's nuclear submarines. This author therefore tends to agree with Holmes and Yoshihara who think that

> [...] the Chinese leadership almost certainly conceives of Taiwan not only in the nationalist terms that are the stock-in-trade of Western China watchers, but as a barrier to the nation's maritime destiny (Holmes and Yoshihara 2008: 56).

This line of reasoning is supported by statements made by Chinese maritime strategist Zhang Wenmu, who wrote:

> In the near to medium term, unifying Taiwan with the motherland and recovering China's sovereign islands is both the great historical mission that the Chinese government must shoulder and a necessary foundation for China to safeguard its national sea rights. [...] Whether these goals are realized peacefully or otherwise, the Chinese navy's future military role in unifying the country will be of great importance. In this sense, and only within the scope of national sovereignty, the expansion of China's sea power is unlimited (Zhang 2006: 25).

Maritime geography can also go a long way towards explaining China's curious insistence on its "historical" sovereignty rights over the Spratly Islands, which are essentially nothing more than largely uninhabitable reefs and shoals that harbor rich fishing grounds and possibly also rich deposits in hydrocarbons, but little else. China's claim seems untenable at first glance because these rocks and reefs are much farther removed from its Southern coastline than they are, for instance, from Vietnam's. Nonetheless, as the map in Fig. 2.3 shows, the South China Sea is the only maritime area directly accessible from China's shores that is markedly deeper than China's average coastal waters. Controlling these islands and claiming a 200 nm Exclusive Economic Zone around them would give China effective control over much deeper waters and thus help remedy the strategic weakness posed by its enclosed and shallow coastal waters described above.

Be that as it may, this critical disadvantage, as difficult as it may be for Chinese naval strategists such as Zhang Wenmu to accept, is a fact of life and China will have to come to terms with what is an accident of history. It is no different for other states that find their geographical surroundings challenging. Thus, while China's interest in Taiwan, the Spratly Archipelago, and the Senkaku/Diaoyutai Islands may seem understandable from the geostrategic point of view, this of course does not give China any natural right to enlarge its territory by military means and violate other countries' legitimate interests. If this were not the case, then many other nations with foreign-held territories close to their own shores could make a plausible argument for war. Therefore, Zhang Wenmu's assertion that "[i]f Taiwan and other islands are not within China's control, China will not be able to guarantee the border security of commercial centers such as Shanghai, Guangzhou and Shenzhen" (2006: 25) sounds eerily akin to statements made in the 1930s by invaders of peaceful neighboring countries such as Stalin and Hitler, who justified their wars of aggression against Poland, Finland, Estonia, Lithuania, and Latvia with similar "Lebensraum"-type arguments.[38] In fact, as Hughes notes, it is quite strange that in Zhang's as well as other Chinese writings discussing Mahanian notions of sea power,

> nowhere in this theory can there be found any awareness that it may be controversial to base a theory of China's rise on a set of ideas that contributed to the naval arms race that led up to World War I and had a decisive impact on the architects of naval power in rising Germany and Japan, feeding into the *geopolitik* thinking that drove the policies of those rising powers (Hughes 2011: 603, italics in the original).

Viewed from a European historical perspective this line of reasoning, in combination with an unhesitant use of problematic terms such as "Lebensraum" (生存空间) in Chinese strategic writings is worrisome. It is therefore reassuring to note that in the above-mentioned government-commissioned *Rise of Great Powers* study (Tang 2006), Germany was explicitly cited as an example of a rising power that made the mistake of trying to upset the balance of power by force, leading to disaster. The path of peaceful development taken up by Germany after its crushing defeat in 1945, on the other hand, is credited in the same analysis with ultimately making it the leading economic power in Europe, and thus an example worth emulating (cf. Tang 2006: 276–277).

References

Bimber, Bruce A. and Steven W. Popper. 1994. *What is a Critical Technology?* Santa Monica: RAND.

Blazevic, Jason J. 2009. Defensive Realism in the Indian Ocean: Oil, Sea Lanes and the Security Dilemma. *China Security* Vol 5, No. 3: 59-71.

Boehler, Patrick. 2014. China's elite hiding billions overseas, US report says. *South China Morning Post* online. January 23, 2014. http://www.scmp.com/news/china/article/1411335/chinas-elite-hiding-billions-overseas-us-report-says. Accessed January 23, 2014.

Canadian Forces. n.y. *C.F. 'O' Class Submarines - Submarine Principles Training Notebook.* Electronic version available at Historic Naval Ships Association Website. http://www.hnsa.org/doc/oberon/principles/index.htm. Accessed January 22, 2014.

Chen, Shaofeng. 2010. China's Self-Extrication from the 'Malacca Dilemma' and Its Implications. *International Journal of China Studies* Vol. 1, No. 1 (January 2010): 1-24.

Cheung, Tai Ming. 2009. *Fortifying China. The Struggle to Build a Modern Defense Economy.* Ithaca and London: Cornell University Press.

Cheung, Tai Ming. 2011. Engineering Human Souls: The Development of Chinese Military Journalism and the Emerging Defense Media Market. In *Changing Media, Changing China,* ed. Susan Shirk, 128-149. Oxford: Oxford University Press.

[38] Arguments with explicit mentioning of the Chinese term 生存空间 ("Lebensraum") are also made in *China is Unhappy* (Song et al. 2009, p. 164). Stalin cited the supposed threat to the security of Leningrad as the major reason for invading an almost demilitarized Finland in 1939, while Hitler claimed a supposed need for more "Lebensraum" as his justification for invading Poland. Another example of aggressive expansionism justified in a similar manner was the Japanese occupation in China from 1937 to 1945.

'China's Military Rise. The dragon's new teeth: A rare look inside the world's biggest military expansion'. 2012. *The Economist* (el. ed.), April 07, 2012. http://www.economist.com/node/21552193. Accessed April 18, 2014.

Chung, Jae Ho, Hongyi Lai and Ming Xia. 2006. Mounting Challenges to Governance in China: Surveying Collective Protestors, Religious Sects and Criminal Organizations. *The China Journal* No. 56, July 2006: 1-31.

CIA. 2014. *CIA World Factbook* https://www.cia.gov/library/publications/the-world-factbook/. Accessed March 23, 2014.

Cole, Bernard D. 2010. *The Great Wall at Sea. China's Navy in the Twenty-First Century*. 2nd ed. Annapolis: Naval Institute Press.

Crisp, Michael. 2010. The Great Chinese Sea Power Debate: A Review Essay. *Journal of Contemporary China* 19(63), January 2010: 201-212.

Crosston, Matthew. 2014. Phreak the Speak: The Flawed Communications within Cyber Intelligentsia. *Cyberspace and International Relations: Theory, Prospects and Challenges,* eds. Jan-Frederik Kremer and Benedikt Müller, 253-267. Berlin et al: Springer

Dai, Xu 戴旭. 2009. *Hai tuteng: Zhongguo hangmu* (海图腾: 中国航母 | Totem of the Sea: The Chinese aircraft carrier). Beijing: Huawen chubanshe.

Dai, Xu 戴旭. 2010. *C-xing baowei* (C形包围 | C-Shaped Encirclement). Shanghai: Wenhui chubanshe.

Dickson, Bruce. 1997. *Democratization in China and Taiwan: The Adaptability of Leninist Parties*. Oxford: Clarendon Press.

Domes, Jürgen. 1980. *Politische Soziologie der Volksrepublik China* (Political Sociology of the People's Republic of China). Wiebaden: Akademische Verlagsgesellschaft.

Dufour, Jules. 2007. The Global Deployment of US Military Personnel. *Global Research,* July 1, 2007. http://www.globalresearch.ca/the-worldwide-network-of-us-military-bases/5564. Accessed January 22, 2014

'East China Sea'. 2014. *Encyclopædia Britannica Online*. http://www.britannica.com/EBchecked/topic/176576/East-China-Sea. Accessed January 23, 2014.

Erickson, Andrew S. and Lyle J. Goldstein. 2009. China Studies the Rise of Great Powers. In *China Goes to Sea. Maritime Transformation in Comparative Historical Perspective,* eds. A. S. Erickson, L. J. Goldstein and C. Lord, 401-425. Annapolis: Naval Institute Press.

Friedman, Norman. 1984. *Submarine Design and Development*. London: Conway Maritime Press.

Frühling, Stephan and Andrew O'Neil. 2012. Nuclear Weapons and Power in the 21st Century. In *Power in the 21st century. International Security and International Political Economy in a Changing World,* eds. Enrico Fels, Jan-Frederik Kremer and Katharina Kronenberg, 81-95. Berlin et al.: Springer.

Göbel, Christian and Lynette Ong. 2012. Social Unrest in China. ECRAN Paper. http://www.euecran.eu/publications-2011/social-unrest-in-china-1. Accessed December 16, 2012.

Grebe, Jan. 2011. The Global Militarization Index (GMI) – A tool for evaluating development orientation of states as well as regional developments. BICC Occasional Paper, February 2011. Bonn: BICC, http://www.bicc.de/old-site/uploads/pdf/publications/papers/occ_paper_07/occasional_paper_VII_02_11_eng.pdf. Accessed September 7, 2013.

Gu, Xuewu and Maximilian Mayer. 2007. Chinas Energiehunger: Mythos oder Realität? (China's hunger for energy: myth or reality?). München et al.: Oldenbourg Wissenschaftsverlag.

Guo, Sujian. 2013. *Chinese Politics and Government: Power, Ideology, and Organization*. London and New York: Routledge.

He, Kai and Huiyun Feng 2011. 'Why is there no NATO in Asia?' revisited: Prospect theory, balance of threat, and US alliance strategies. *European Journal of International Relations* 2011 (published online January 27, 2011): 1-24. http://ejt.sagepub.com/content/early/2011/01/26/1354066110377124.abstract. Accessed January 22, 2014.

Heilmann, Sebastian. 2000. *Die Politik der Wirtschaftsreformen in China und Russland* (The Politics of Economic Reform in China and Russia). Hamburg: Institut für Asienkunde.

Heilmann, Sebastian and Sarah Kirchberger. 2000. The Chinese Nomenklatura in Transition. A Study Based on Internal Cadre Statistics of the Central Organization Department of the Chinese Communist Party. *China Analysis* No. 1 (June 2000), Center for East Asian and Pacific Studies, Trier University. http://www.chinapolitik.de/resources/analysis1.pdf. Accessed December 1, 2013.

Holbig, Heike. 2009. Remaking the CCP's Ideology: Determinants, Progress, and Limits Under Hu Jintao. *Journal of Current Chinese Affairs* 3 (2009): 35-62.

Holmes, James R. and Toshi Yoshihara. 2008. *Chinese Naval Strategy in the 21st Century. The Turn to Mahan,* London/New York: Routledge.

Holslag, Jonathan, 2010. *Trapped Giant: China's Military Rise.* The International Institute of Strategic Studies (IISS), London: Routledge.

Hoyt, Timothy D. 2007. *Military Industry and Regional Defense Policy: India, Iraq, and Israel.* London and New York: Routledge.

Hu, Yanlin 胡彦林. 2006. Weirao lüxing wojun lishi shiming jiaqiang haijun sixiang zhengzhi jianshe (围绕履行我军历史使命 加强海军思想政治建设 | For the implementation of our armed forces' historic mission, enforce the ideological and political construction within our Navy). *Qiushi* 求是 2/2006: 44-46.

Hu, Lili 胡麗麗 and Jian Liang 梁劍. 2012. *Xi Jinping zhuanji* (習近平傳記 | A Biography of Xi Jinping). Taibei: Lingxiu chubanshe.

Hu, Angang 胡鞍钢 and Honghua Men 门洪华. 2004. The Rising of Modern China: Comprehensive National Power and Grand Strategy. Paper presented at the International Conference of the Korea Institute of International Economic Policy (KIEP), 19–20 March 2004, in Seoul, Korea. http://irchina.org/en/xueren/china/pdf/mhh3.pdf. Accessed March 2, 2011.

Hughes, Christopher. 2011. Reclassifying Chinese Nationalism: The *Geopolitik* Turn. *Journal of Contemporary China* (2011), 20(71), September: 601-620.

Kaplan, Robert D. 2005. How We Would Fight China. *The Atlantic,* June 2005. http://www.theatlantic.com/magazine/archive/2005/06/how-we-would-fight-china/303959/. Accessed December 6, 2013.

Kirchberger, Sarah. 2004. *Informelle Regeln der Politik in China und Taiwan* (Informal Rules in Chinese and Taiwanese Politics). Ph.D. dissertation, University of Hamburg. Hamburg: Kovac.

Kraska, James. 2010. How the United States Lost the Naval War of 2015. *Orbis* Winter 2010: 35-45.

Lampton, David M. 2014. *Following the Leader: Ruling China, from Deng Xiaoping to Xi Jinping.* Berkeley et al.: University of California Press.

Leonard, Mark. 2008. *What Does China Think?* London: Fourth Estate.

Li, Cheng. 2010. China's Communist Party-State: The Structure and Dynamics of Power. In *Politics in China: An Introduction,* ed. William A. Joseph, 165-191. Oxford: Oxford University Press.

Li, Xiaobing. 2009. *A History of the Modern Chinese Army.* Lexington: The University Press of Kentucky.

Li, Mingjiang. 2010. China and Maritime Cooperation in East Asia: Recent Developments and Future Prospects. *Journal of Contemporary China* 19(64), March 2010: 291-310.

Li, Wei and Lucian Pye. 1992. The Ubiquitous Role of the Mishu in Chinese Politics. *The China Quarterly* Vol. 132 (December 1992): 913-936.

Liff, Adam P. and Andrew S. Erickson. 2013. Demystifying China's Defence Spending: Less Mysterious in the Aggregate. *The China Quarterly.* doi:http://dx.doi.org/10.1017/S0305741013000295.

Liu, Zhipeng 刘志鹏, Wei Hao 郝威 and Hongjun Zheng 郑红俊. 2013. Woguo haijun dangqian mianlin de tongxin dianzi zhan xingshi ji duice yanjiu (我国海军当前面临的通信电子战形势及对策研究 | The current communications electronic warfare situation confronting our navy and analysis of countermeasures). *Xinxi tongxin* 信息通信 (Information & Communication) No. 124, 2/2013: 197-198.

Mahadzir, Dzirhan and James Hardy. 2011. Chinese Defence Minister Dismisses Pakistan's Port Plans. *Jane's Defence Weekly* (el. ed.), June 8, 2011. http://www.janes.com.

Mahan, Alfred Thayer. 1987 [1890]. *The Influence of Sea Power Upon History 1660-1783*. Reprint of the 5th ed. (1894). New York: Dover Publications.

Mao, Tse-tung (a.k.a. Mao Zedong 毛泽东). 1965. Problems of War and Strategy (Nov. 6, 1938). In *Selected Works of Mao Tse-tung: Vol. II*. Mao Tse-tung. Beijing: Foreign Languages Press. https://www.marxists.org/reference/archive/mao/selected-works/volume-2/mswv2_12.htm. Accessed February 14, 2015.

Mayer, Maximilian. 2012. Exploring China's rise as knowledge power. In *Power in the 21st century. International Security and International Political Economy in a Changing World*. Ed. Enrico Fels, Jan-Frederik Kremer and Katharina Kronenberg, 287-311. Berlin et al.: Springer.

McGregor, Richard. 2010. *The Party. The Secret World of China's Communist Rulers*. New York: Harper Collins.

Moss, Trefor. 2011. Briefing: PLAN of Action. *Jane's Defence Weekly* (el. ed.), February 4, 2011. http://www.janes.com.

Müller, Gotelind. 2013. *Documentary, World History, and National Power in the PRC: Global Rise in Chinese Eyes*. New York: Routledge.

Mulvenon, James. 2013a. The New Central Military Commission. *China Leadership Monitor* No. 40, January 14, 2013: 1-10. http://media.hoover.org/sites/default/files/documents/CLM40JM.pdf. Accessed January 20, 2014.

Mulvenon, James. 2013b. Comrade, Where's My Military Car? Xi Jinping's Throwback Mass-Line Campaign to Curb PLA Corruption. *China Leadership Monitor* No. 42 (07 Oct 2013): 1-5. http://www.hoover.org/publications/china-leadership-monitor/article/158546. Accessed March 23, 2014.

Oksenberg, Michel. 2002. China's Political System: Challenges of the Twenty-First Century. In *The Nature of Chinese Politics: From Mao to Jiang*, ed. Jonathan Unger, 193-208. Armonk and London: M.E. Sharpe.

Park, Hyun. 2014. US stationing more aircraft carriers and troops in Asia-Pacific region. *The Hankyoreh* online, January 17, 2014. http://www.hani.co.kr/arti/english_edition/e_international/620232.html. Accessed January 22, 2014.

Pfetsch, Frank R. 1994. *Internationale Politik* (International Politics). Stuttgart et al.: Kohlhammer.

Pillsbury, Michael. 2000. *China Debates the Future Security Environment*. Washington D.C.: National Defense University Press.

Potgieter, Thean. 2009. When Maritime Security is Absent. In *Maritime Security in Southern African Waters*, eds. Thean Potgieter and Reiner Pommerin, 5-22. Centre for Military Studies, Stellenbosch University. Stellenbosch: Sun Press.

Pugh, Philip. 1986. *The Cost of Seapower: The Influence of Money on Naval Affairs from 1815 to the Present Day*. London: Conway Maritime Press.

Rice, Susan E. and Stewart Patrick. 2008. Index of State Weakness In the Developing World. Washington, D.C.: The Brookings Institution. http://www.brookings.edu/~/media/files/rc/reports/2008/02_weak_states_index/02_weak_states_index.pdf. Accessed March 2, 2011.

Roy, Denny. 2009 (1996). The 'China Threat' Issue: Major Arguments. In *China and Southeast Asia, Vol. VI: The People's Republic of China and Southeast Asia*, ed. Geoff Wade, 188-200. London and New York: Routledge.

Ruppik, Dirk. 2007. China unterstützt Umgehung der Straße von Malakka (China supports Malacca Strait bypass). *Schiff & Hafen*, Jan. 2007: 27.

Sakhuja, Vijay. 2011. *Asian Maritime Power in the 21st Century: Strategic Transactions – China, India and Southeast Asia*. Singapore: Pentagon Press.

Scott, Richard. 2011. Conventional wisdom. *Jane's Defence Weekly* (el. ed.), April 7, 2011. http://www.janes.com.

'Secret Sanya - China's new nuclear naval base revealed.' 2008. *Jane's Intelligence Review* (el. ed.), April 21, 2008. http://www.janes.com.

Shambaugh, David. 2004. *Modernizing China's Military. Progress, Problems, and Prospects*. Paperback ed. Berkeley et al.: University of California Press.

Shambaugh, David. 2010. *China's Communist Party. Atrophy and Adaptation*. 2nd ed. Berkeley et al.: University of California Press.

Shambaugh, David. 2013. *China Goes Global. The Partial Power*. Oxford: Oxford University Press.

Shirk, Susan. L. 2007. *China: Fragile Superpower*. Oxford: Oxford University Press.

Song, Qiang 宋强 et al. 1996. *Zhongguo keyi shuo bu* (中国可以说不 | China can say no). Beijing: Zhonghua gongshang lianhe chubanshe.

Song, Xiaojun 宋晓军 et al. 2009. *Zhongguo bu gaoxing* (中国不高兴 | China is unhappy). Beijing: Jiangsu renmin chubanshe.

'South China Sea'. 2014. *Encyclopædia Britannica Online*. http://www.britannica.com/EBchecked/topic/556146/South-China-Sea. Accessed January 23, 2014.

STRATFOR. 2007. China: Rural Migration and Plugging the Rural-Urban Gap. STRATFOR Global Intelligence, 2 April 2007. http://www.stratfor.com/memberships/29016/china_rural_migration_and_plugging_rural_urban_gap. Accessed February 25, 2011.

Swaine, Michael D. 2013. Chinese Views Regarding the Senkaku/Diaoyu Islands Dispute. *China Leadership Monitor* No. 41 (6 June 2013): 1-27. http://media.hoover.org/sites/default/files/documents/CLM41MS.pdf. Accessed January 20, 2014.

Tang, Jin 唐晋. 2006. *Daguo jueqi. Yi lishi de yanguang he quanqiu de shiye jiedu 15 shiji yilai 9 ge shijiexing daguo jueqi de lishi* (大国崛起: 以历史的眼光和全球的视野解读15世纪以来9个世界性大国崛起的历史 | The Rise of Great Powers: Decoding the history of 9 global rising powers since the 15th century from a historical perspective and with a global point of view). Beijing: Renmin chubanshe.

Tellis, Ashley J. et al. 2000. *Measuring National Power in the Postindustrial Age*. Santa Monica: RAND.

Till, Geoffrey. 2009. *Seapower: A Guide for the Twenty-First Century*. 2nd ed. London and New York: Routledge.

Turse, Nick. 2011. Does the Pentagon Really Have 1,180 Foreign Bases? *The World Post*, January 10, 2011. http://www.huffingtonpost.com/nick-turse/does-the-pentagon-really-_b_806817.html. Accessed March 23, 2014.

Vine, David. 2012. The Lily-Pad Strategy: How the Pentagon Is Quietly Transforming Its Overseas Base Empire and Creating a Dangerous New Way of War. TomDispatch.com, July 15, 2012 and *The Asia Times Online*, July 19, 2012. http://www.tomdispatch.com/blog/175568/. Accessed January 22, 2014.

Wacker, Gudrun. 2006. China's Grand Strategy. In China's Rise: The Return of Geopolitics?, ed. Gudrun Wacker, 55-60. SWP Research Paper No. 1, Berlin: SWP. http://www.swp-berlin.org/en/publications/swp-research-papers/swp-research-paper-detail/article/china_rueckkehr_der_geopolitik.html. Accessed February 16, 2015.

Wertheim, Eric. 2013. *The Naval Institute Guide to Combat Fleets of the World*. 16th ed. Annapolis: Naval Institute Press.

WTO 2013. WTO International Trade Statistics Database. http://www.wto.org/english/res_e/statis_e/its2013_e/its13_world_trade_dev_e.htm. Accessed January 21, 2014.

Yang, Benjamin. 1998. *Deng: A Political Biography*. Armonk and London: M.E. Sharpe.

Yang, Yi Edward and Xinsheng Liu. 2012. The 'China Threat' through the Lens of the US Print Media: 1992-2006. *Journal of Contemporary China* (2012), 21(76): 695-711.

'Yellow Sea'. 2014. *Encyclopædia Britannica Online*. http://www.britannica.com/EBchecked/topic/652686/Yellow-Sea. Accessed January 23, 2014.

Zhang, Wenmu 张文木. 2006: Sea Power and China's Strategic Choices. *China Security* Summer 2006: 17 – 31.

Zhao, Yuezhi. 2011. Sustaining and Contesting Revolutionary Legacies in Media and Ideology. In *Mao's Invisible Hand: The Political Foundations of Adaptive Governance in China*, eds. Sebastian Heilmann and Elizabeth Perry, 201-236. Cambridge and London: Harvard University Press.

Chapter 3
Comparing Naval Capability and Estimating the Cost Impact

> 夫兵者，不祥之器，天道惡之。
> *"Weapons are ominous instruments, and the Dao of Heaven abhors them."*
> —*Huang Shi Gong san lüe*, sect. 3. Transl. by Lo Ping-cheung (2012: 410).

This chapter describes the major factors to consider when evaluating naval power from the point of view of military hardware. This encompasses not only naval weapon systems and vessel types, but also their strategic functions, life-cycle issues, and cost influences. Although these interrelated factors have a strong impact on the actual naval capability level of all countries, they are not sufficiently taken into consideration as the material basis that defines the upper limit of what is achievable through naval strategy. Just as a chain is only as strong as its weakest link, the performance of a navy relies upon *all* the contributing factors of naval power. While state ideology and geostrategic orientation may shape a country's world outlook and determine the range of possible actions perceived by its leaders, it is the available technology that largely defines what is practicable. Therefore, this part of the study will attempt to demonstrate how different types of naval weapon systems can be systematically evaluated from the point of view of their strategic value when put to use.

The factors discussed so far raise an important question: the problem of finding criteria for evaluating the current developmental status of various naval forces from a total capability level viewpoint that transcends the mere list of naval inventory. It should be clear from the preceding chapters that the approach of counting vessels is not revealing in itself, because it neither takes into account the extent of a country's homeland defense commitment nor its mission competency. As Tellis et al. have pointed out, "[...] military threats, geography, and alliances [...] help shape a country's force architecture and, ultimately, its effective military capabilities" (Tellis et al. 2000: 135). This factor explains why ostensibly widely divergent forces can nevertheless be classified as fulfilling roughly the same capability criteria. The militaries of Israel and New Zealand are good examples of this phenomenon:

S. Kirchberger, *Assessing China's Naval Power*, Global Power Shift,
DOI 10.1007/978-3-662-47127-2_3

> Both have modern economies, well-educated populations, access to world markets and modern technologies, and freely elected governments. Yet their armed forces are quite different. [...] Israel is directly threatened by nearby neighbors and defends itself without formal allies. New Zealand is an island nation, faces no apparent external threats, and is allied with Australia and the United States (Tellis et al. 2000: 135).

Under fortunate geostrategic conditions, vastly inferior forces can thus maintain an adequate level of militarization compared with countries in high-threat environments. Fleet age structures and fleet sizes should therefore always be evaluated in relation to homeland defense commitment, compared with the other navies in the "Regional Security Complex," weighing training and maintenance infrastructures. Moreover, as Tsouras has noted, the institutional culture must not be underestimated, as it is ultimately what defines a navy's "character":

> A navy is not just a collection of ships, men, and deadly weapons—the pieces in a war game. It is a human institution that follows patterns set in its developmental stages by diverse factors of history, culture, and geography. The distillation of these intangible experiences form a navy's character, and this character determines the navy's conduct and success under the ultimate stress of war (Tsouras 1986: 3).

Several scholars who have tried to find a solution to the above-mentioned methodological problem have come up with more or less compatible ranking systems for classifying entire navies according to their rough overall capability. These systems use a combination of factors which are qualitatively evaluated and then combined for assigning an overall capability rank. Although it may seem crude to the specialist, when applied to a large number of cases a framework such as this can serve to determine the range of the "normal" with regard to naval militarization and thus help to identify "anomalous" cases of naval build-ups.

The system of naval combat proficiency published by Tellis et al. (2000)—ranking naval missions from the most basic to the most sophisticated level—will thus be described and used as a heuristic tool. It will be shown that hardware, logistics and training requirements clearly demarcate the boundaries between a total of nine successive levels of mission difficulty. This system will serve as the basis for describing in more detail a few key types of naval hardware that, by their very existence or absence from the fleet, define the maximum level of a country's naval proficiency. Likewise, a few major components (or sub-systems) will be considered in order to assess the maximum capability of different ships belonging to the same vessel types. This will give the reader a better grasp of the meaning of "critical" technological advances in China's naval shipbuilding and the strategic consequences discussed in more detail in Chaps. 5 and 6.

In order to determine the trajectory and long-term prospects of ongoing build-ups, cost implications of naval development as a portion of state spending priorities must also then be addressed using a comparative intra- and interregional yardstick. This appraisal will be undertaken in Sect. 3.5.

3.1 The Task of Classifying Navies According to Capability

The key criteria used by various analysts for evaluations of total naval development are typically *force structure* (i.e. type, age, and number of vessels); *sustainability* (i.e. the ability to keep naval forces on station), *function and capability* (the navy's primary task, maximum mission capability and geographic reach), *flexibility* (i.e. the variety and number of missions that can be undertaken concurrently as determined by surplus forces or redundancy), and *access to state-of-the-art technology* (cf. e.g. Till 2009: 114–115). In accord with the system of naval combat proficiency levels introduced in Sect. 3.2, it is possible to construct a fairly consistent ranking system for classifying any navy's overall developmental status.

The existing classification system of ten naval ranks employed by Todd and Lindberg seemed useful for the present study because their classification criteria are explicit and consistent, and because they provide a complete ranking of all navies according to their respective status as of 1996 (cf. Todd and Lindberg 1996: 53–64). This starting point allows for easy adjustments of changed force levels using current data, and moreover allows for the upward or downward tracking of changing candidates such as China.[1] The key aspects of the system are depicted in Fig. 3.1, with examples in the right column reflecting 2013 data.

In order to newly classify the ca. 149 currently existing naval forces (navies, coastguards or riverine forces) into this system according to 2013 data, the qualitative summary evaluations provided in the latest edition of *Jane's World Navies* were used as a basis for making the overall classifications. These were then checked for consistency with the above-mentioned vessel inventory characteristics and the other given criteria using standard warship catalogues (*Jane's Fighting Ships* 2014; Wertheim 2013). As a caveat, it is important to keep in mind that this classification system is only a heuristic tool for structuring reality. Of course, any classification system will inevitably be found to have minor or major flaws if checked carefully on a case-by-case basis, and each naval rank represents a rather broad spectrum itself. Therefore, instead of providing the complete list of navies for all the ten ranks as Todd and Lindberg have done, only a few examples are mentioned in the right column of the table in Fig. 3.1 for the sake of illustration. The purpose of this exercise, after all, is not to degrade any one country, but rather to make the extremely complex world of naval development more easily accessible to research.

One striking difference to the older ranking provided by Todd and Lindberg lies in a marked overall rise in naval capability in the developing world. While in 1996, 58 naval forces were still included in the lowest Rank, my new survey found that this group has shrunk to no more than 14 naval forces as of 2013, reflecting efforts

[1] Till (2009: 114) cites a slightly different yet somewhat compatible classification system of nine ranks which is based on previous work by Eric Grove and Michael Morris. The main difference to Todd's and Lindberg's system consists of a further rank inserted between their Ranks 1 and 2, while their Ranks 7 and 8 are lumped together and their Rank 9 is left out completely.

	Rank	Designation	Typical Inventory	Defining Capabilities	Current Examples
BLUE WATER NAVIES	1	Global-reach power-projection	All larger ship types in high numbers	Multiple, regular, sustained power projection missions globally	USA.
	2	Limited global-reach power-projection	CVN, other aviation-capable ships, many SSN/SSK, support ships	At least one major power-projection operation globally in addition to homeland defense	France (↓), UK (↓).
	3	Multi-regional power-projection	CVL, other aviation-capable ships, SSN/SSK, enough support ships	Power-projection missions in regions beyond own EEZ in addition to homeland defense	India, Russia, Italy, Spain, Brazil.
	4	Regional power-projection	Aviation-capable ships (DD, FF), submarines, some support ships	No at-sea fleet air support other than organic helis, thus limited to area of land-based aircraft range for power-projection missions	China (↑), Japan, Australia, South Korea, Pakistan, Singapore, Taiwan, Germany (+12).
NON - BLUE WATER NAVIES	5	Regional offshore coastal defense	Smaller ships (FF, Corvettes), no underway replenishment	Coastal defense operations at least in own EEZ and slightly beyond	Thailand, Malaysia, Indonesia, Bangladesh, Vietnam, Israel (+11).
	6	Inshore coastal defense	Only smaller ships (Corvettes, FAC)	Confined to inner reaches of own EEZ	North Korea, Myanmar, Brunei, Sri Lanka (+17).
	7	Regional offshore constabulary	Lightly armed OPV, PB, PC for CG–type duties	Geographic reach like Rank 5, but maritime policing instead of defense	Iceland, Ireland, Tunisia, Estonia, Iraq (+14).
	8	Inshore constabulary	Only Patrol Boats and Patrol Craft	Confined to missions well within own EEZ	Cambodia, Cyprus, Philippines (+30).
	9	Inland waterway	Patrol Craft	Waterborne riverine defense of landlocked states	Bolivia, Paraguay, Burundi (+14).
	10	Token	Often only one or two small craft	Only very basic constabulary capabilities, if any	Ca. 14 examples worldwide.

Source: Compiled from Todd and Lindberg (1996: 53–64)
Examples originally given by Todd and Lindberg reflected 1996 capability levels and were adjusted to reflect 2014 capabilities. Arrows indicate upward or downward development tendency.

Fig. 3.1 World naval hierarchy according to the Todd/Lindberg classification system with data from 2013

in many developing countries at strengthening maritime security. Relatedly, the group of Rank 8 forces has markedly grown from only 13 in 1996 to ca. 33 in 2013.

3.1.1 The World's Naval Hierarchy and China's Current Status

Depending on the criteria used, about 28 navies, or roughly 19 % of the world's ca. 150 naval forces, can currently be categorized into capability Ranks 1–4 and therefore constitute the world's true "blue-water" navies. Thirty-eight navies (or 26 %) belong to the still relatively powerful Ranks 5 and 6, meaning they have respectable naval forces capable of effectively defending their maritime borders. Ca. 52 naval forces (35 %) can be considered Rank 7–8 navies and thus basically have only coast guard-type capabilities, while 31, or 21 %, belong to the least powerful Ranks 9 and 10. How should the major Asia Pacific naval forces be classified according to the same categories? An overview of an approximate ranking was attempted with the results shown in Fig. 3.2.

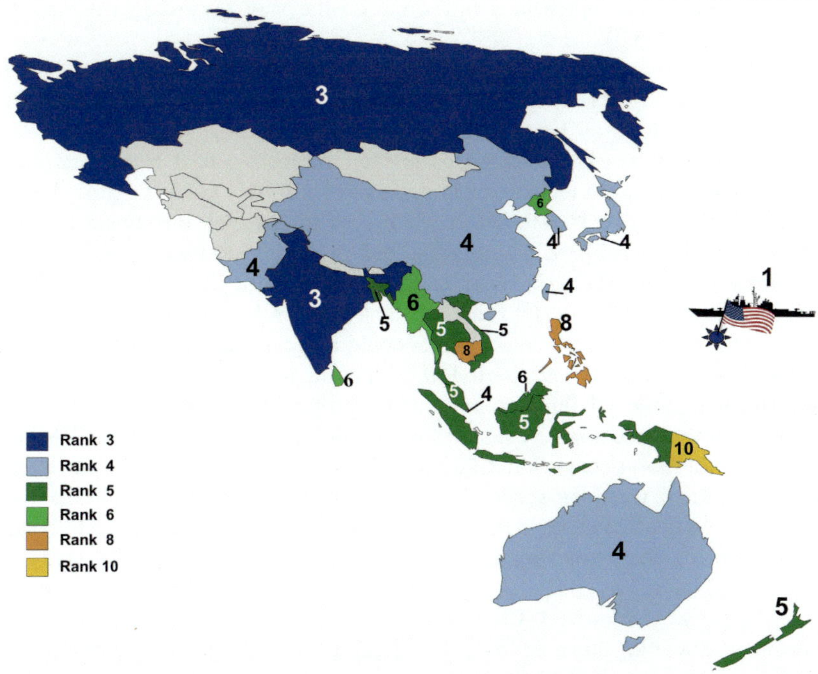

Source: Navies ranked according to the Todd and Lindberg (1996) classification system using current data from Wertheim (2007, 2013), *Jane's Fighting Ships*(2014) and *Jane's World Navies* (2013)

Fig. 3.2 Approximate ranking of Asia Pacific naval forces according to Todd-Lindberg capability levels, 2013

Figure 3.2 shows that China, as of 2013, was still ranked in Rank 4, or the lowest of the so-called "blue-water"[2] ranks, according to overall operational capability. This is mainly due to a lack of fleet air support resulting from the still-nascent carrier capability. Should China succeed in introducing a true and full-fledged carrier capability to its fleet, which it is currently in the process of doing, it would automatically rise to Rank 3. Without the air wing in place, the commissioning of the *Liaoning* (ex-*Varyag*), China's first operational aircraft carrier, was not sufficient in itself to achieve this benchmark. In fact, it seems probable—as Chinese leaders as well as independent naval analysts have repeatedly stated—that the vintage Soviet-era carrier will likely only serve as a test and training platform to prepare the field for China's first indigenous carrier fleet, which is currently under development and could encompass three ships. The *Liaoning's* actual combat capability and strategic value as a weapon system is therefore in doubt, even though

[2] Note that Todd and Lindberg's definition of the term "blue-water navy" differs from its common usage, which often implies an existing carrier capability.

its commissioning did represent a milestone on the way towards a true carrier capability (cf. Sect. 5.3).

According to the criteria of the naval ranking system employed here, China's overall naval capability is considered to be roughly on par with such seemingly disparate naval powers as Japan, Australia, South Korea, Pakistan, and Taiwan, notwithstanding the marked differences in their respective fleet sizes and force structures (and downward tendencies in some cases). In this system, fellow BRIC states India, Russia and Brazil rank even higher than China due to an already operational carrier capability, which is the defining criterion that separates Rank 3 from Rank 4. While Malaysia, Thailand, Indonesia, Vietnam and Bangladesh all wield respectable non-blue-water forces of Rank 5, other regional navies fall short.[3]

Currently, it seems that the most China could achieve within the medium term would be the already envisaged rise to Rank 3, putting it on par with the other BRIC states. This would mean having at least one operational carrier plus its air wing in place. In order to then compete with the sole remaining Rank 2 powers (the UK and France) within a short time period, China would not only need quick access to state-of-the-art high-tech weapons, sensor systems, and CDS technology, but also have to embark on a costly offshore base-building strategy. Neither scenario seems likely as long as the arms embargo persists and China is limited to less advanced Russian and Ukrainian technology to compensate for its own shortcomings. Nonetheless, if the arms embargo were removed, China could conceivably rise very quickly if the leadership were willing and able to steadily invest the necessary financial resources, build a regional network of bases, while at the same time significantly enhancing naval training. Be that as it may, even if this scenario were to soon materialize, China would still be far from competing with the only truly global naval power, the U.S. Navy, and its vast network of bases and allies all over the world.

3.1.2 What Is an "Adequate" Level of Naval Capability?

In order to determine whether or not a country's naval modernization plans are within the realm of the reasonable given its economic size and geostrategic situation, and also within the regional context, it is a useful exercise to correlate naval capability rank with the level of integration into the world trade system as shown in Fig. 2.1. The country's percentage share in the world trade volume can be used as a proxy for estimating the country's perceived need of protecting its Sea Lines of Communication (SLOCs).[4] It was a central tenet of Mahan's classic treatment of

[3] The Royal Thai Navy is an anomaly, because although it operates a small carrier, it is not included in the Rank 3 navies. The reason for this is that Thailand does not fulfil the other criteria necessary for Rank 3, and the carrier is mainly used for ceremonial purposes.

[4] Other possible proxies that could be used for the present purpose include the country's degree of connectivity to global shipping networks as measured by UNCTAD's *Liner Shipping Connectivity Index* (UNCTAD 2014) or by container port traffic (World Bank 2014). Though there are

sea power already in 1894 that "the necessity of a navy [. . .] springs [. . .] from the existence of a peaceful shipping, and disappears with it, except in the case of a nation which has aggressive tendencies" (Mahan 1987 [1894]: 26; note the qualification).

The reasons for a tendency to invest large sums in naval modernization as a result of rising trade volumes are not difficult to understand. Active integration into world trade usually corresponds with relative wealth, which is a necessary precondition for the maintenance of powerful naval forces. On the other hand, sizeable overseas investments, large merchant marines, and dependence on uninterrupted trade routes usually associated with the large trading countries require adequate and active maritime protection. In short, world-trade leaders have both the need as well as the means to maintain powerful navies.

It seems reasonable to assume, therefore, that the level of integration into world trade is a relatively reliable indicator of a country's propensity for building highly capable naval forces. This idea was already advocated in the late nineteenth century, when British organizations such as the London Chamber of Commerce held that a country's naval strength and share of world trade should be directly proportional, and that maintenance costs for a navy were "akin to an insurance premium on the trade to be protected" (Pugh 1986: 21).

This study finds that the expected correlation indeed continues to hold to this day. Figure 3.3 illustrates the close relationship between a high level of integration into world trade and the maintenance of a so-called "blue-water navy" in the Todd/Lindberg system (capability Ranks 1–4). The indicated 29 top trading nations constitute only ca. 18 % of the world's total 159 WTO member states. These 29 states (or 18 %), however, control more than 80 % of total world trade, with the remaining 130 (or 82 %) trading nations having only a tiny share of less than 0.9 % each. While all naval forces classified in the "token navy" category belong to countries with low levels of trade integration, the world's most capable navies of Ranks 1–3 are without exception found within the small group of the top 25 trading nations.

Except for Switzerland and Austria, the only two landlocked countries appearing in the list, and Hong Kong, which is a Special Administrative Region of the PRC, almost all the 29 top-trading nations maintain comparatively powerful navies. 18 (or 62 %) maintain navies of the "blue-water" types (Ranks 1–4). A further eight (or 27.6 %) still have quite powerful Rank 5 or 6 navies.

Based solely on the above empirical observations, and given China's prominent share in the present world trade system—it ranks second to the USA only because Hong Kong's share is calculated separately—China's desire to rise to blue-water navy Rank 3 or 2, which is on par with other trading nations such as France, the UK, Spain, Brazil, or Italy (countries with a far smaller world trade share than China) does not seem unnatural.

---

variations in position, the above named indices yield strikingly similar lists as the world trade share distribution shown in Figs. 2.1 and 3.3. I am indebted to Michael Funke for pointing this out.

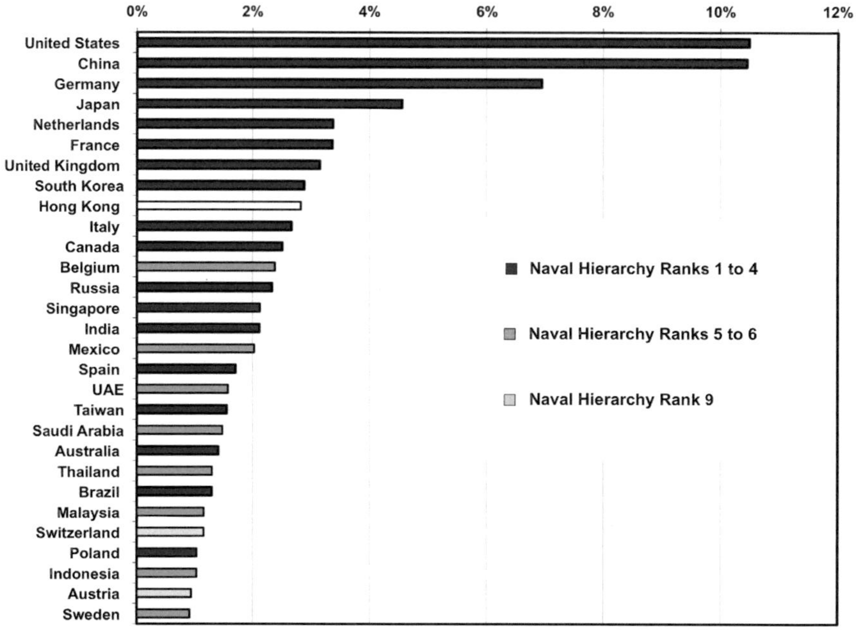

Source: World trade shares computed from WTO 2013 data. Navies ranked according to the Todd and Lindberg (1996) classification system using current data from Wertheim (2007, 2013) and *Jane's World Navies* (2013)

Fig. 3.3 Share of world trade volume, 2012 (%) and rank in the "world naval hierarchy"

3.2 Principal Naval Weapon Systems and Their Strategic Functions

Naval forces are first and foremost responsible for safeguarding maritime security near a country's coastline. This can be considered the most basic function of any navy. How far from shore a navy can safely operate and what kind of additional missions it can accomplish is determined in no small measure by the available naval hardware, as well as by the level of training and logistics.

In order to assess the relative difficulty of various naval mission types, it is necessary to first recognize the fact that the sea poses unique challenges as a strategic environment. For one, the sea is not partitioned, but, as Till writes, "all joined-up," making it difficult to clearly demarcate boundaries and frontier lines. The sea is also "largely un-owned and un-ownable," as well as impossible to occupy continuously. For this reason, "[p]ossession of the sea is not generally an object of maritime operations"—a major difference to land operations. The most decisive characteristic of the sea as a strategic environment, however, is its multi-dimensionality. More precisely, it has a sub-surface, surface, and air dimension that "may interface at any time, presenting threats and challenges from all azimuths." On top of its particular strategic challenges, the sea is "large, opaque, varied and often hostile—finding the enemy is often the main problem [...]. The sea itself indeed can be an enemy" (Till 2009: 43–44). This sentiment is reflected in a medieval proverb that those who cannot pray should go to sea, while those who cannot sleep should go to church. So far, technological progress has neither managed to make seafaring on the high seas completely safe nor much more comfortable than previously. There is still no technological remedy available against tsunamis, hurricanes, or rogue waves. Moments of intense fear, discomfort, and sea sickness are thus just as bound to confront seafarers today as in the past.

3.2.1 Naval Mission Types and Related Hardware Requirements

In contrast, for example, to combat airplanes which are only airborne for relatively short durations and typically spend the better part of their service lives in maintenance, naval vessels must be able to withstand the elements for weeks and months on end, survive in hostile environments and under a multitude of climatic conditions, give shelter to a large crew over extended periods, and navigate safely at any hour and in any weather. At the same time, they are required to perform a wide variety of difficult and dangerous tasks and missions, from search-and-rescue to actual combat against surface, sub-surface, and airborne combatants.

Naval operations can be roughly classified according to one of four major types:

(a) Law-enforcement missions (i.e. coast guard-type tasks such as anti-piracy, maritime patrol, counter terrorism, or anti-smuggling)
(b) Sea-control missions (i.e. SLOC protection, protection of territorial integrity, embargo operations)
(c) Power-projection missions (i.e. support for land attack or special operations missions, carrier operations)
(d) Humanitarian missions (i.e. disaster relief, evacuation, search and rescue).

This functional classification according to the naval mission's main objective does not yield any explicit information on the relative mission difficulty, or the technological requirements necessary for such tasks. It should, however, be intuitively clear that the farther a vessel operates from homeland shores, the larger and more seaworthy it has to be for good survivability and seakeeping, and the better the communication and sensor equipment has to be to navigate safely. In contrast, operations within confined and shallow waters or very near to a coastline require smaller vessels with a shallower draft, for large surface vessels and submarines find it difficult to navigate safely here.

In order to show the relative complexity of naval operation types, I will use a classification system of naval combat proficiency levels originally developed by

Level	Designation	Defining Capabilities	Required Hardware	HR / Training Requirements	Logistics, Maintenance, Organizational Infrastructure
I	Coastal defense and mining	Basic naval competency: independent patrolling of coastal waters, mine-laying for defensive operations	Small craft (patrol craft ≥21m), small-caliber guns	Limited (commercial) ship-operating skills, limited weapons proficiency	Almost none; spare parts, elementary repairs
II	Coastal anti-surface warfare	Ships operate independently close to shore; LOS targeting of surface vessels; limited VHF communications	Smaller ships with converted army weapons (grenades, .50 caliber machine guns, shoulder-launched rockets); land-based missile batteries	Limited (commercial) ship-operating skills, limited naval training for weapons proficiency	
III	Anti-surface and anti-air warfare with surface ships	Ship vs. ship warfare; ships act independently or form small surface action groups; cannot operate far from coast for extended periods; limited air/surface search, LOS targeting of low-tech missiles, naval gun fire support	Corvettes, older frigates, destroyers, minesweepers; medium caliber guns, elementary radars, elementary naval CM, ASM, SAM	Elementary naval ship-handling skills at dedicated naval school or training base; elem. training for damage and fire control and AAW	Elementary logistics and maintenance; elementary naval shipyard infra-structure with trained personnel
IV	Anti-surface warfare with submarines	Targeting of military and civilian shipping traffic at moderate distances; subs acting independently, stationing themselves at choke points	Small elementary diesel submarines; elementary sonars and torpedoes	Complexity and challenges of submarine ops require dedicated training at submarine schools including improved weapons and sensor training	Logistics and maintenance also for submarine operations
V	Anti-submarine warfare with surface ships	Combined anti-submarine operations with ships and helicopters acting in coordination	Dedicated surface combatants with capable sonars and embarked helicopters; elementary data links for tactical coordination; target submarines for fleet training	Dedicated ship manning, naval aviation and maintenance training, also tactical coordination, sonar use, oceanography; elementary fleet exercises	Tactical coordination and integration between ships and helos; elementary intelligence support
VI	Anti-submarine warfare with submarines	High integrative demands on fleet due to inherent difficulty of subsurface ASW targeting	Quiet, advanced submarines (nuclear or modern diesel); advanced passive sonar and fire control; basic torpedoes	Training for passive sonar operations and advanced fire control, advanced ship-handling skills	Basic level of skilled maintenance and effective logistics support, basic infrastructure
VII	Naval strike and limited air control	Important threshold which signals the ability to project power ashore; forming of carrier battle groups sharing moderate amounts of tactical data	Basic carrier with embarked aircraft, other large aviation-capable ships for carrier protection, basic radar and naval CM, ASM, SAM; satellite imagery or land-based, long-range maritime patrol aircraft, basic data links, secure comms	Dedicated carrier operations and naval aviation training; regular basic fleet exercises	Basic intelligence support; extremely high level of integrative efficiency required
VIII	Multi-mission air control, limited sea control, and deep strike	True "blue water" capability; deep strike capability against both land and sea targets, forward presence (power projection)	Advanced carriers (CVN) capable of launching CTOL aircraft; advanced high-speed data transfer and comm systems, sophisticated offensive and defensive systems, advanced cruise missiles, advanced radar	Advanced training and support for a multitude of complex systems and operations	Advanced large-scale fleet exercises, substantial joint exercises, underway replenishment, advanced shore establishment for maintenance and logistics, advanced intelligence support
IX	Comprehensive sea control	True network-centric forms of warfare that enable a force to successfully interdict an adversary's assets in any operating medium (the naval warfare ideal)	OTH reconnaissance, sur-veillance, target acquisition systems, real-time processing, pervasive communications, adv. satellites, adv. tactical displays	Advanced training and support, technically competent manpower, complete computer literacy at all levels, training emphasis on info ops	

Source: Compiled from Tellis et al. (2000: 164 –176). System originally developed at RAND by Jeffrey A. Isaacson et al.

Fig. 3.4 Hierarchy of naval combat proficiency levels

Jeffrey A. Isaacson et al. at RAND and described by Tellis et al. (2000: 164–176). It is helpful for illustrating the interrelatedness of hardware, training, and logistics when assessing naval capability.

Figure 3.4 shows nine successive levels of complexity, starting with Level I—"Coastal Defense and Mining"—as the most basic task. It cites the main characteristics, required hardware, training facilities, and necessary logistics infrastructure to unfold each level of capability. The system works like a pyramid in that every succeeding combat proficiency level automatically encompasses all the proficiencies of the lower levels. Moreover, the number of navies that reach a certain level grows consistently smaller towards the top. Level IX—"Comprehensive Sea Control"—marks a theoretically possible level of capability that is nonetheless unreachable by any force, including the U.S. Navy, for the foreseeable future.

Translated into the common color-coding scheme often employed to differentiate between types of navies with regard to the respective distances from the shoreline, mission levels I and II could be referred to as "brown water," Levels III through VI as "green water," and Levels VII to IX as "blue water."

As shown in Fig. 3.4, the first two levels of combat proficiency—designated "Coastal defense and mining" and "Coastal anti-surface warfare," respectively—require little in terms of specialized training and use naval hardware that mostly consists of converted civilian shipping equipment with weaponry adapted from ground forces. Level III—"Anti-surface and anti-air warfare with surface ships"—thus marks an important capability threshold characterized by the ability to employ some elementary radar systems and low-tech naval missiles and countermeasures as well as mines. Since larger surface ship types such as basic frigates, destroyers, and mine warfare vessels tend to be in use in such navies, and the weapon systems are much more difficult to handle than the converted army guns of Levels I and II, a dedicated ship-handling training infrastructure, and usually a naval academy, is already a necessity at this level, along with a basic naval shipyard infrastructure able to perform elementary maintenance and repairs.

The next important technology threshold in Level IV—"Anti-surface warfare with submarines"—is marked by the presence of an elementary submarine capability. The integration of this special naval vessel type into the force poses dedicated training and logistics requirements. Submarine operations are extremely hazardous compared to surface missions, and the necessary sensors—mainly sonars—require intensive training to be operated, as does torpedo warfare. Submarines, moreover, require a different maintenance infrastructure than surface vessels. The introduction of a submarine capability of even the most basic kind therefore has far-reaching, formidable, and far more costly consequences than the mere purchase price of the submarines themselves suggests. Some navies in the Asia Pacific are currently planning to break this capability threshold, i.e. Bangladesh, Vietnam, Thailand, and even the Philippines, but the total investment volume needed for the related infrastructures in addition to the procurement price has often been difficult to accommodate within the defense budgets of the respective countries.[5] Strategic considerations can outweigh the disadvantages, however, since enemy submarines pose a grave strategic threat to all the surface ships in the inventories of navies at Levels I through Level III that are not capable of anti-submarine warfare. In order to conduct and train for effective anti-submarine

[5] Cf. the discussion in Hardy and Pape (2013) and Grevatt (2011).

warfare operations, a naval force needs a submarine capability of its own. Submarine hunting on the other hand is much more effective using ship-borne ASW helicopters than ship-borne bow-mounted sonar systems, suggesting yet another technology threshold.[6]

The next technology and training threshold that defines Level V—"Anti-submarine warfare with surface ships"—is, accordingly, the existence of a naval aviation capability. Integrating ships and ship-borne helicopters requires dedicated data links. Embarking helicopters in hangars also has massive consequences for ship design, not only because the hull design must accommodate a hangar aft, but because the flight deck needs to be stable enough for safe starting and landing operations. It goes without saying that naval aviation, similar to submarine operations, is a hazardous and difficult activity, especially at night and in bad weather. It therefore provides its own dedicated training and logistics requirements in addition to all those already mentioned. Furthermore, the enhanced intelligence requirements, using oceanography, satellite imagery and other data sources, need to be met for these kinds of complex operations integrating different types of surface vessels, submarines and helicopters.

If anti-submarine warfare is to be conducted with submarines—the most effective way of hunting submarines—another technology gap appears to define the border between capability Levels V and VI, which is called "Anti-Submarine Warfare with Submarines." For this task, modern quiet submarines with high endurance are required, either nuclear-powered or modern diesel-electric with Air-Independent Propulsion (AIP). This dimension of naval capability adds considerably to the complexity of fleet operations that now integrate surface ships, shipborne aviation, and highly advanced submarine operations. Here, the quality and technological sophistication level of the submarines is an important differentiating criterion. Only very few countries are in fact currently able to produce submarines that meet the necessary technological standards.[7] What's more, not all of these are willing to export such weapon systems, effectively excluding some

[6] On a sober note, it must be stated that in recent years modern diesel-electric submarines have on various occasions successfully evaded detection by even the most advanced naval forces during numerous exercises, and it is quite possible that this strategic problem could be beyond remedy, rendering even the most advanced large surface vessels inherently vulnerable to modern subs (various shipbuilding and navy sources interviewed by the author of this study during 2013).

[7] The established manufacturers as of 2014 are the U.S. and Britain (nuclear powered), France, Russia and China (nuclear and diesel-electric), Germany, Sweden, Japan and Spain (only diesel-electric), with modern German and French AIP designs reaching performance levels comparable with nuclear submarines in terms of quietness and endurance (cf. Scott 2011). India, Italy and South Korea have built submarines under license agreements for their own use (Phipps and Hardy 2014). Germany, France, Spain, Russia and Sweden have so far been willing to export diesel submarines, while nuclear-powered submarines are generally not available on the export market, even though Russia has agreed to lease one to India, and France has offered "assistance" to Brazil for building the nuclear power-plant. India and Brazil are currently developing a capability for building nuclear powered submarines.

interested countries (e.g. Taiwan) from this capability level if they cannot produce the necessary boats themselves (Phipps and Hardy 2014).

Whereas anti-submarine warfare can by itself be seen as defensive, to the extent at least that there might be an actual submarine threat in the vicinity, the next capability Level VII—"Naval strike and limited air control" must be seen as being oriented towards offensive power-projection operations, as such capabilities are not strictly necessary for purely defensive purposes. At issue here is what is commonly called a true "blue water" capability, i.e., the capability of a navy to operate in the high seas far from its home shores and conduct limited offensive strike operations. China seems to be aiming for precisely this level of competency at the moment. It further entails dedicated technological, logistics, and training requirements not present in the earlier levels, the most visible being aircraft carriers and the requisite ships and aircraft for forming complete battle groups. Few navies currently possess this high level of capability. Apart from the U.S., only Russia, France, the United Kingdom, India, Italy, Spain and Brazil can at least theoretically conduct such complex operations, with China still in the preliminary phase. While Thailand owns a small carrier, it has not developed the necessary capability to operate it in the way described here. Japan and Australia, on the other hand, while officially not operating aircraft carriers, likely could develop this capability relatively quickly due to their existing (Japan) and prospective (Australia) helicopter carrier capabilities and highly advanced naval aviation infrastructures.

Apart from the necessary hardware in terms of either conventionally or nuclear-powered carriers, their respective air wings, redundant numbers of large surface ships and submarines for carrier protection, as well as the presence of sophisticated sensors, weapon systems, and data links, the training and logistics component becomes even more crucial here. Carrier flight operations pose a heightened level of difficulty compared with shipborne helicopter operations, both in training and technology, and fleet exercises must regularly be conducted to establish and maintain the skills needed for such complex operations under realistic conditions. The general education standard of fleet personnel is higher than in the preceding capability levels, and the sheer size and heterogeneity of the fleet poses integrative demands that are quite difficult to meet. Only large-scale and regular training exercises can ensure that an accomplished proficiency level is being maintained, while a breakdown in just one of the contributing component capabilities means that the overall capability of Level VII will be compromised at least temporarily.

Level VIII—"Multi-mission air control, limited sea control, and deep strike"—differs only in degree from Level VII. Here, more than one type of the related missions can be simultaneously conducted, while the geographical reach of the navy is also larger. The main technological and training difference to the preceding level is the presence of network-centric warfare, which poses all kinds of additional technological requirements regarding data links and fleet integration, because

> [...] sensors, communication systems, and weapon systems are linked into an interconnected grid that provides an integrated picture of the battlefield to all levels of command and control down to the individual soldier. This concept of full connectivity depends on a host of sophisticated electronic and information technologies, as well as

specialized services and technical support that have generated a sizable demand from both the civil defense and military sectors (Neuman 2009: 63).

This concept lies at the center of the strategic goal of transformation, which "aims to expand communication among military units and to achieve greater real-time target acquisition, longer-range projection capabilities, increased unit mobility, and improved weapon accuracy and lethality" (Neuman 2009: 63).

In order to handle the sizeable information flows, vessels, aircraft, sensors, and weapons must be more state-of-the-art than in Level VII. To provide enhanced geographical reach, a network of bases is needed to ensure access to offshore replenishment facilities and logistic support. Currently, only the U.S. Navy can be said to possess this level of capability, and the initial investment and constant funding necessary to establish and maintain a comparable global base network would likely prove prohibitive to even the most ambitious rising power with a mind to replicate this success. It should not be forgotten that the current status and worldwide presence of the U.S. Navy is a by-product of very peculiar wartime conditions that are fortunately unlikely to occur again in the foreseeable future. It seems far-fetched to assume that any country would be willing or able to compromise other fiscal priorities to the degree necessary to build a comparable force level during the current world political climate.

The final level described in this system—Level IX "Comprehensive sea control"—would require such a high degree of true network-centric warfare capability that it seems fairly utopian today and is currently not within reach for any navy.

To sum it up, notwithstanding the complexities of various navies' inventories with a myriad of ship types, there are in fact only a few major types of naval hardware that are of critical importance for the analyst when trying to determine a navy's maximum level of mission complexity. These are submarines (diesel-electric and/or nuclear-powered), large surface ships with embarked helicopters such as destroyers and frigates, and offensive power-projection vessels such as aircraft and helicopter carriers, each of course including the necessary sensors, weapons and data links to integrate them into the fleet and provide network-centric warfare capability.

However, since real capability hinges not only on the existence of pieces of hardware, but on the actual skill levels and operational experience of personnel, a record of regular exercises and live tests, and a suitable training infrastructure are just as relevant. Equally critical is the necessary logistics infrastructure, including maintenance, repair and replenishment facilities, and access to offshore bases.

I will describe the major naval vessel types in more detail in the next subchapter in order to draw attention to the technology areas where innovation yields the most strategically relevant results. This will also shed light on those areas where innovation is the most difficult to achieve, namely in the transformation of navies towards the above-described ideal of "network-centric warfare."

3.2.2 *Major Naval Vessel Types and Their Role Within the Fleet*

Just like entire navies, the naval combatants of today are best conceptualized as complex systems (or, as the slogan goes, "systems of systems") because they almost resemble living organisms rather than being homogenous units or mere assemblages of unrelated pieces of technology. This quality of naval vessels—that they are in effect comparable to quasi-organisms with computer networks and data links serving as brains and nervous systems, with effectors as action organs, various sensor equipment as sense organs, and propulsion systems to ensure their mobility—is important to recognize, and means that each naval weapon system is only as capable as its weakest component part. Sometimes this is the human crew if training levels do not match the requirement for operating the systems safely. This is most visible in submarine operations where extreme cautionary measures are taken to ensure that orders are correctly understood, such as when the entire crew repeats aloud commands from the commanding officer, and potential submariners are screened more carefully than other personnel for mental instability because of its grave implications for a mission (cf. Canadian Forces n.y.: 30). As a training manual for the Canadian *Oberon* class submarines stated during the 1970s:

> The best weapon system in the world is no use in the hands of an untrained team. The submarine is part of the system, and if you open the wrong valve, lose the trim, press the wrong switch or just do not know the right thing to do, all may come to naught. Basic training, continuation training, team training, exercises, all lead up to one thing—an ability to use the weapon system to the full extent of its capability (Canadian Forces n.y.: 49).

The human crew's mental stability, motivation and training quality become essential factors for success especially in the case of nuclear-powered submarines, which can remain submerged for months at a time and thus put extreme demands on the human psyche. Due to the fact that submarines as launch platforms for ballistic missiles are "[...]inherently stealthy and, while a potential enemy might be able to locate one or two at a time, the chances of being able to locate 10 or more [are] extremely remote", nuclear-powered ballistic-missile submarines (SSBN) to this day remain a crucial component of credible deterrence strategies in post-Cold War times even though their overall number has decreased (Miller 2002: 402–404). In fact, "the United States, Russia, France and Britain have moved significant numbers of their nuclear arsenal onto submarines" (Sakhuja 2011: 67).

Due to the advances in weapon design and the extreme velocities reached by modern missiles, however, fully automated responses to such threats have become the norm. Naval vessels capable of automatically detecting and countering approaching threats in a matter of milliseconds independently of any human operator are thus common. Missile development is the factor that ultimately drives much naval technological progress today. Fully automated self-defense systems such as the U.S. *Phalanx,* European *Goalkeeper* or Russian *Kashtan* close-in weapon systems (CIWS) or the U.S. *Rolling Airframe Missile System (RAM)* became a necessity because of the threat posed by supersonic missiles that leave little to no reaction time from detection to impact. The sinking of the British destroyer *HMS Sheffield* in 1982 during the Falklands War as well as a more recent missile hit taken by the Israeli corvette *INS Hanit* in 2006 confirmed the catastrophic effect of these weapon systems. Countering new developments in missile technology is therefore one of the key areas in naval technology research. The development of effective countermeasures has been possible only due to the advent of advanced information technology aboard vessels, and represents the greatest departure from traditional forms of naval warfare up until the immediate post-World War II era.

Figure 3.5 gives an overview of the most important vessel types arranged according to technical complexity. The approximate length and displacement ranges assigned to the ship types only provide a rough idea and do not necessarily correspond with official usage in all the world's navies. In fact, some navies consciously choose awkward or inappropriate designations for their vessels as a result of political pressures.[8] In this study, I have chosen to employ functional criteria such as vessel size and armament for determining vessel types that are in accord with those used by most naval analysts regardless of the respective navy's own system of reckoning (cf. Wertheim 2013: xxi–xxii).

If ranked according to technical complexity, the high end of the warship spectrum is currently occupied by nuclear-powered vessels such as aircraft carriers, cruisers and submarines due to the complexity of the nuclear propulsion system. This is followed by large surface combatants such as conventionally-powered cruisers (which only a few navies now operate), destroyers, and frigates—i.e. ships that are "extraordinarily dense with large engines and high-technology systems which are manned by large crews" (Birkler et al. 2005: 27). The most technologically complex among these ship types are anti-air-warfare vessels equipped with vertical-launch systems (VLS) for surface-air-missiles (SAM) that need extremely powerful radar systems that can keep track of dozens of targets at a time.

Smaller, less heavily armed surface combatants such as corvettes and Fast Attack Craft (FAC) are much less difficult to build, and at the lower end of the spectrum. Auxiliaries such as commodity carriers like oilers and tankers differ very little from their commercial counterparts, as they are basically "large steel boxes with relatively small engines at the rear [...] manned by one- or two-dozen mariners" (Birkler et al. 2005: 28).[9] A ship type that has recently become popular with several navies in the Asia Pacific is the Amphibious Assault type of large

[8] The criteria for assigning vessel types to ships are less than clear-cut, especially with the larger surface combatants. Examples for downplaying actual capability would include the under-classification of the German *F-124* and the Franco/Italian *FREMM* as frigates instead of destroyers, as well as the initial under-classification of the South African *Valour*-class frigates as "corvettes." Examples of over-classification can also be found, e.g. in the Argentine Navy.

[9] Even relatively advanced naval shipbuilding nations, however, such as Brazil and Poland, have experienced surprising difficulties with indigenous corvette programs during the last 10 years, even though in Poland's case a basic design was purchased in the export market.

Vessel type	Major function/role and defining characteristics	Approx. full load displ. (t)	Approx. unit price (US$)
Aircraft carriers (CV, CVL, CVN)	Major offensive power-projection platforms that enable various aircraft to land and take off from. Nuclear-powered (CVN) or using gas or steam turbines for propulsion (CVL). *Subtypes* include CTOL carriers (steam catapults and brake lines for conventional takeoff and landing aircraft) in service with the USA, France, and Brazil; and STOVL carriers (ski jump for short take-offs and vertical landing of STOVL aircraft) in service with the UK, India, Italy, Russia, Spain, Thailand and China.	From 11,000	Up to several billion US$
Amphibious assault ships or helicopter carriers (LPD, LHD)	Havc helideck for several helicopters, car deck, heavy duty goods lifts davits for landing craft and rigid hull inflatable boats and a dock for landing craft. Missions: transport of troops and vehicles, provision of command facilities, contingency services, evacuation, hospital services (examples: *Mistral* class, *Dokdo*).	From ca. 16,000	From ca. 300 million
Cruisers (C, CG)	Very large, heavily armed surface combatants capable of operating independently in multi-threat environments. Mainly employed by the U.S. Navy as part of aircraft carrier battle groups (example: *Ticonderoga* class).	>10,000 t	From ca. 2 billion
Destroyers (DD, DDG)	Very capable ships of > 5,000 t full load displacement carrying high level sensors and advanced weapons systems to operate independently in high-threat environments and provide air-defense for larger numbers of smaller vessels. Modern destroyers (DDG) typically carry vertical launch guided missiles and 1_2 embarked helicopters (example: *Arleigh Burke* class, *Type 45*).	5,000 – 10,000	From ca. 700 million
Frigates (FF, FFG)	The smallest surface combatants able to operate independently in a high-threat environment, often limited to one specific role. Able to defend a complete squadron against air raids. Can conduct air defense, submarines warfare or surface combat missions, but are usually more role-specific than destroyers. Carry extensive sensor and communication equipment and relatively heavy armament. Typically carry 1 embarked helicopter (examples: FFG 7/*Perry* class, *MEKO 200*).	2,500 – 5,000	From ca. 300 million
Corvettes (FS)	Small, fast, well-armed combatants for regional missions equipped for medium threat environments. Similar in size to Offshore Patrol Vessels (OPV), but much more heavily armed, with higher survivability and greater top speed but less endurance (example: *K130*).	1,000 – 2,500	From ca. 150 million
Offshore patrol vessels (PV, OPV)	Vessels of up to corvette size, but much less heavily armed. Used for coast guard type duties such as maritime patrol, crime prevention, terrorism fighting, anti-piracy, search and rescue, logistic support, emergency service, humanitarian aid, maritime interception, evacuations, etc. Often built to commercial standards. Usually one embarked helicopter. Smaller patrol vessels are called Patrol Boat or -Craft (example: Fassmer *OPV 80*).	>700 up to 3,000	From ca. 30 million

Fig. 3.5 (continued)

LHDs/LPDs. These "helicopter dock landing ships," sometimes dubbed "helicopter carriers," have the size of small aircraft carriers and are filled with installations for a great variety of missions ranging from troop transport, landing support, hospital service, evacuation, and disaster relief. In some navies, the idea has been to use these ships as "surrogate aircraft carriers in lower-threat areas" (Friedman 2006: ix).

Vessel type	Major function/role and defining characteristics	Approx. full load displ. (t)	Approx. unit price 09/:2
Fast attack craft (FAC)	Smaller than corvettes, very fast (from >25 up to 50 knots speed), and heavily armed with missiles for hit-and-run strikes close to the shore (within ca. 100nm). Little endurance, limited sensors and defenses (examples: *Osa* class, *Combattante* class).	<700	From ca. 60 million
Nuclear-powered ballistic missile submarines (SSBN)	The most powerful warships in existence, and also the largest (length 120_170_m, displacement comparable to destroyers). The most expensive submarines. Armed with ballistic missiles. Can remain submerged for months, e.g. for missions under Arctic ice. Essential asset for a credible nuclear deterrent. In service only with the U.S., UK, France, Russia, China, with India in the process of building one (examples: *Ohio* class, *Vanguard* class).	Up to 19,000 submerged	Up to several billion
Nuclear attack submarines (SSN)	Slightly smaller nuclear powered subs (length 70_150_m) armed with torpedoes against submarines and surface vessels, underwater to surface missiles, and cruise missiles for land attack. In service only with the above-named navies, plus India (examples: *Barracuda* class, *Los Angeles* class, *Akula* class).	Up to 13,000 submerged	From ca. 1 billion
Conventional (diesel-electric) submarines (SS)	Small to medium-sized submarines with torpedoes and USMs. Powered by large accumulators. To charge the batteries, subs have to sail at periscope/snorkel depth regularly. Modern developments include air-independent propulsion (AIP) systems that make the subs extremely quiet and capable, with maximum submerged time extended to several weeks (examples: HDW *764* class, *Lada* class, *Scorpène* class).	1,000 – 3,500	From ca. 400 million
Mine countermeasure vessels (MCMV)	Vessels designed specifically to locate, destroy, or lay naval mines (example: *Katanpää* class).	ca. 500t	From ca. 100 million
Auxiliaries	At the low end both technically and regarding unit price, this category encompasses various ship types used for logistic support such as tankers, oilers, tugs, and the like which closely resemble their civilian counterparts. More complex examples include Joint Support Ships (JSS) or Combat Replenishment Vessels (or EGV).	All kinds of sizes	Various

Source: Descriptions assembled from Dokkum (2011: 60–61), Wertheim (2013), and various reports from *Jane's Defence Weekly* and *Jane's Navy International*, and information from own interviews during 2007–2013

Fig. 3.5 Principal warship types of modern navies

Australia, Japan, South Korea, and China, as well as the U.S., all operate different types of such ships.

In the subsurface realm, huge and costly nuclear-powered submarines represent the high end both technologically as well as financially. According to a U.S. study cited in Lewis (2008), the construction of a single SSBN involves about a million separate parts. The then CEO of French shipbuilding group DCNS, Jean-Marie Poimboeuf, was quoted in an interview as follows:

> When you count the number of hours required to design and build it, and the amount of technology involved—nuclear, space, communications, information systems and missiles—the boat is 30 times more complicated to make than a fighter aircraft (quoted in Lewis 2008).

Nuclear-powered submarines are therefore operated by just a handful of navies, while much smaller diesel-electric submarines of various quality levels are far more affordable also to smaller navies. Unlike nuclear-powered submarines, they can also at least theoretically be purchased on the export market. Modern conventional submarines are nevertheless technically complex systems and their smaller size, which gives rise to the necessity of integrating many sensitive systems within a very small pressure hull, becomes a challenge of its own:

> The submarine must provide living facilities and stores to enable the crew to stay on patrol for long periods and similarly carry adequate supplies of fuel, equipment spares, etc. It must be able to discharge its weapons on the surface, at periscope depth or deep, and operate to the maximum possible depth whenever necessary. To do this effectively it needs communications, sonar, navigational and weapons systems all crammed into a hull which must be as small as possible (Canadian Forces n.y.: 6).

Building and operating conventional submarines is therefore a challenge not to be underestimated, even for countries already disposed to building and operating nuclear-powered subs. This explains why the U.S. has so far not been able to honor its pledge to help Taiwan acquire new diesel submarines once all currently active manufacturers stalled for political reasons (cf. Sect. 6.5).

To illustrate the relative differences in technological complexity and cost, Fig. 3.6 shows an approximate graphic distribution pattern of these ship types along the two dimensions of technical complexity and average unit price.

The systemic differences between commercial and military shipbuilding are clearly reflected in the average unit prices of vessels. According to a 2005 survey, worldwide military shipbuilding cost is on average 100 times the commercial cost per unit of volume (GRT). While commercial shipbuilding has seen a worldwide

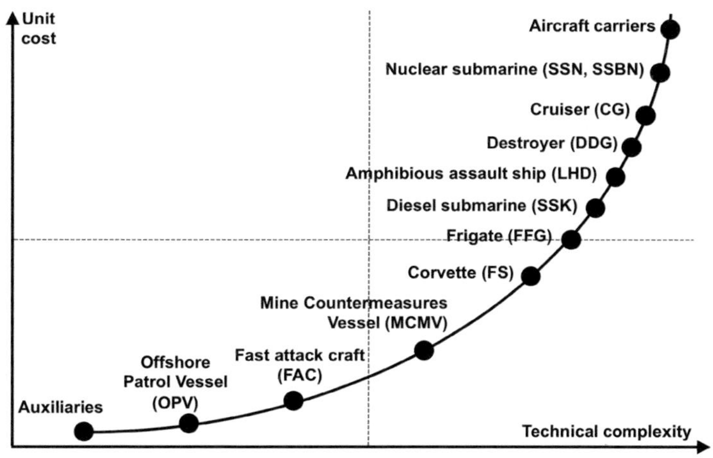

Source: Estimation based on various reports in the *Defense News*, *Jane's Navy International*, *Jane's Defense Weekly* over several years, and interviews with various naval shipbuilding representatives, March 2013

Fig. 3.6 Relative technical complexity and cost of major naval vessel types

price decline of 30–50 % during the last 20 years, military shipbuilding in contrast saw an average annual cost increase of almost 10 % within the same time period. This is mainly due to ever more expensive sensor and weapon systems and the related system integration demands, i.e., to their higher complexity per weight unit (cf. Birkler et al. 2005: 30–32).[10]

3.2.3 China's Warship Inventory

Looking at China's current inventory of naval vessels, it is noteworthy that with the commissioning of China's first functioning carrier hull, the *Liaoning*, into the fleet as a training platform, all major vessel types shown in Fig. 3.5 are now present, with the exception of nuclear-powered carriers and cruisers. Therefore, most naval mission types are theoretically within reach for the Chinese navy, at least from the hardware point of view. Strategic capabilities of warships belonging to the same overall type, however, can still vary considerably depending on age, weapons and sensor outfit, shipbuilding quality, hull type, propulsion plant, quality of maintenance, etc. This is why technical issues and deficiencies are of particular relevance when assessing any country's—not just China's—actual level of naval power. What is needed in addition to a purely numerical analysis of the fleet inventory is therefore a qualitative evaluation of capability levels that differentiates between vessels of the same general type. Obsolete sensor, communication, or weapons technology clearly has a degrading effect upon the respective naval weapon system's overall strategic value. It is therefore useful to conceptualize several "generations" of technical standards within each major vessel type. These generations can be determined by their weapon outfits, sensor technologies, propulsion systems, and hull form characteristics such as stealth and survivability features, which are usually shared by all the ships of the same flight or class.

From a strategic point of view, the most important technological watersheds with regard to the above-mentioned hierarchy of naval mission types have been the introduction of shipborne naval aviation, developments in missile technology, and the employment of advanced information technology aboard vessels resulting in a tendency towards network-centric warfare. As of this writing, only the most modern Chinese ship classes can be assumed to have the necessary sensor technology to perform the more difficult types of integrated missions. These factors will be discussed in more detail in Chap. 5.

[10] For an informative and balanced account of the various reasons for cost increases in large-scale arms projects worldwide, see 'The Cost of Weapons: Defence Spending in a Time of Austerity' (2010).

3.3 Service Lives, Fleet Age Structures, and Procurement Cycles

One characteristic of naval vessels that has significant strategic consequences is their relative longevity compared with other types of military technology. Naval vessels are typically in service for decades. The shipbuilding industry sources I interviewed insisted that 30 years—the service lifespan expected by the U.S. Navy for its vessels—is the safe maximum both for steel surface vessels and submarines because of rapidly degrading hull strength towards the end of its duration, and empirical evidence from navies around the globe supported the finding that the more capable navies do indeed usually retire their vessels after this interval. Some smaller navies, however, take the risk of keeping vessels in service longer than this, sometimes up to 40 years and even longer (cf. Arena et al. 2006: 67). This study, therefore, generally assumes a maximum lifespan of 40 years for surface vessels but only 30 years for submarines, as they are even more at risk from hull degradation than surface vessels and are therefore seldom left in service for more than 30 years (Taiwan being a notable exception due to its inability for political reasons to renew its submarine fleet).

It is consequently useful to conceive of warships as having a definite time-dimension, that is, a service life with clearly demarcated stages that are by no means uniform regarding capability and operational readiness.

Another important factor is the fact that the procurement and production processes are quite lengthy, for the full life cycle of a naval vessel is often longer than 50 years all in all from the definition phase to the scrapping phase. This means that vessel designs, which are inevitably products of the prevailing global political climate at the time of conception, may have to be used for extended periods, even if a dramatic shift in the security environment renders such vessels unsuitable. One example of this is the post-Cold War era of the early 1990s: Many of the recently built giant anti-air warfare ships designed for a full-fledged nuclear war scenario now appear somewhat obsolete with respect to new asymmetric threats such as terrorism and piracy, which call for greater numbers of smaller, more flexible and less heavily armed vessels. According to Friedman, the three main new developments in naval warfare since the end of the Cold War thus include "the rise of what is now called network-centric warfare," "the impact of the global war on terror," and "the rise of new developers of sophisticated naval systems, particularly command and control systems" (Friedman 2006: vii).

Because naval vessel building programs represent such an enormous investment of national financial resources and take so long in their planning and execution, it is necessary, and yet also difficult, for defense planners to anticipate global strategic changes. For the industries involved, this problem can also produce difficulties in sustaining a profitable business base in-between projects, which in turn contributes to cost increases. As Arena et al. have found in their 2006 study:

> Many shipyards have a monopsony relationship with the government—that is, the government is their main, if not only, customer. At the same time, fluctuating ship orders from the Navy, with initially forecast orders typically exceeding what is ultimately purchased, discourage shipyards from making investments that could ultimately reduce the cost of ships. More importantly, an unstable business base causes fluctuations in the demand for skilled labor that are expensive and difficult to manage. The unstable business base also prevents contractors from leveraging purchases (long-term contracts) from subcontractors and suppliers that might result in more stable pricing. [...] Finally, the unstable business base makes it difficult for the shipbuilders and suppliers to manage their workforce (Arena et al. 2006: xvii).

This is an observable phenomenon in most countries with shipbuilding industries catering solely to their national governments, and the global trend in the industry has been shipyard concentration. In China, however, the unique Leninist power structure of the political economy yields power tools to the leadership unavailable to typical Western governments. The combination of rich financial resources freely available to the Chinese government, a lack of public oversight over the spending of such funds, and effective Communist Party control over all the strategically viable state-owned industries (among them the 11 leading arms-producing companies)[11] means that China's leaders can partially offset the above-mentioned negative effects for naval shipbuilding industries and allocate funds strategically without being subject to public scrutiny. Japan, although for different reasons, is another example of a country that has taken active measures against these problems by means of a shrewd industrial policy aimed at creating more stable conditions in naval procurement. Ships are kept in use for shorter time spans (only 20 years for submarines), which enhances the turnover rate and creates more stable build cycles (cf. Fig. 3.9). One byproduct of this is that Japan could easily take advantage of the established industrial production procedures and quickly enlarge its submarine fleet by simply retiring older boats later than planned.

Figure 3.7 illustrates the various stages—and approximate duration spans—of a typical naval shipbuilding project such as a destroyer or a conventionally powered submarine. Different time spans can be observed in technologically more complex projects such as aircraft carriers and nuclear submarines, where the concept, design, and build phases for prototypes can sometimes stretch to decades.

As can be seen in the example, the design phase can take up to several years even in a "regular" naval project. According to a naval architect interviewed by me in 2012 and 2013, in the case of a mid-sized surface combatant it typically takes 7–10 years from contract award before a naval vessel is actually commissioned into service. The product development phase can easily take up a total workload of several million hours before construction actually begins. System integration problems of different degrees seem universal to naval shipbuilding programs beyond rudimentary complexity, even in the most advanced shipbuilding countries, and

[11] For a discussion of the industrial influence and economic standing of these companies, see Surry (2007), Lague and Zhu (2012) and Grevatt (2013).

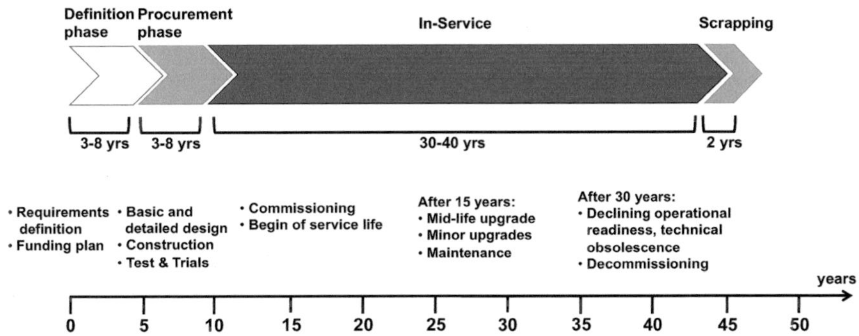

Source: Information collected from interviews with naval shipbuilding executives in Hamburg, Germany, March 2013

Fig. 3.7 Lifespan of a naval vessel

tend to make military shipbuilding an increasingly expensive and time-consuming endeavor with actual delivery times of vessels sometimes hard to predict:

> It can take two years or more to design a military ship compared with six months for a commercial ship. Predesign work on a frigate or submarine can amount to 10 times that needed for a tanker (Birkler et al. 2005: 35).

This is a problem noted also by Chinese analysts, as shown in a comparative discussion of technical risk factors in some major Western (mainly U.S.) shipbuilding programs published by researchers at Wuhan's Naval University of Engineering in 2008:

> According to our analysis, technical risk originates primarily from the fact that in warship weapons outfitting and manufacture, new technologies, new materials and new work processes are used. During the concrete warship development process, technical risk is the result of newly developed systems and equipment being simultaneously developed and manufactured (Lü et al. 2008: 34, my transl.).

A ship's service life itself does not begin directly after it has been constructed and outfitted, but only after somewhat lengthy tests and trials during which malfunctions and deficiencies are typically discovered and addressed. This phase can take up several years, especially in the case of a first-of-class prototype vessel, if graver technological problems occur. The vessel can only be officially commissioned and begin its service life after this stage is completed. Nevertheless, even if the latest available state-of-the-art weapons and electronics equipment have been selected and built into the vessel during the definition phase, some of the electronics will already be outdated at the time of commissioning, and after about half the service life has passed it is inevitable that some of the installed systems will be obsolete enough to justify a costly mid-life upgrade. Sometimes, far-reaching changes are implemented that occasionally even require hull modifications, putting the vessel out of service for a considerable time span. Afterwards, the already aging vessel can complete its service life, regular maintenance and repair work notwithstanding, until approximately 30 years of material degradation to the steel hull finally make decommissioning inevitable, depending on the number of days per year actually spent at sea. Some of the more durable installed equipment may be salvaged before the vessel is scrapped, and installed on other, newer-generation vessels, or sold on the export market. In some cases, instead of being refitted or scrapped, obsolete or near-obsolete vessels are transferred to lesser-developed navies that cannot afford newly built vessels.

Figure 3.7 also illustrates the fact that naval vessels, just like any other item of defense equipment in service, automatically incur costs at all stages of their use, not just in their initial procurement phase. These so-called "life-cycle costs" are incurred during the ship's entire service life, from pre-design to scrapping. This means that "much of the budget for any year is bespoken by decisions made many years earlier to procure this or that equipment now in service" (Pugh 1986: 121). Contrary to public perception, the initial procurement cost of naval weapon systems does not by a long shot make up the major part of total life-cycle cost. As Fig. 3.8 shows, operating cost is by far the largest expenditure, while the initial procurement cost (with maintenance closely behind) makes up only about one quarter of the expense.

When discussing fleet renewals, procurement programs and cost implications, it is easy to forget the above-mentioned time dimension of naval shipbuilding programs and the long-term financial commitment associated with all the prior

Fig. 3.8 Typical life-cycle cost breakdown for warships

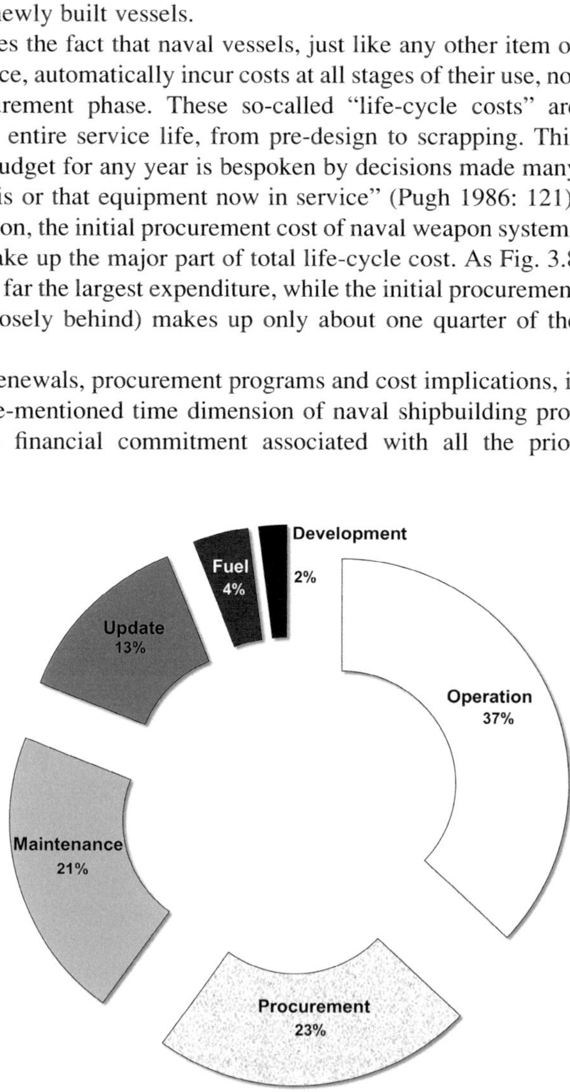

Approximate life -cycle cost shares assuming a class of six and 25 years of service life (as of 1986)
Source: Graph drawn using data from Pugh (1986: 124)

procurement decisions. One result of this is a definite and characteristic age structure and technology distribution pattern within fleets. This is just as important to consider as the vessel type distribution pattern.

Fleet age structure has several major strategic implications: For one, because of the fixed nature of a vessel's lifespan, the latest point in time at which a vessel, or in some cases entire classes of ships, will need to retire is predetermined. Moreover, because the procurement phase of new building projects can be drawn out, programs must be initiated while the older vessels are still about 10 years from being decommissioned lest there appear a capability gap within the fleet. Consequently, when commenting on the supposed "expansion plans" of a fleet, observers sometimes overlook the fact that in most cases one ship class is merely being replaced with newer vessels in regular fashion through so-called replacement procurement and no expansion is actually taking place. Only when smaller numbers of older vessels are replaced with larger numbers of new vessels, or when an entirely new vessel type is added to the fleet, does a fleet actually grow. This, however, only happens rarely.

It is important to keep this point in mind with regard to China because it clears up misconceptions and makes more transparent the areas in which the Chinese fleet actually expands and changes structure, and those where it merely continues at the same level.

Because of the lengthy service life of naval vessels spanning several decades, it is an economic and strategic imperative that the newest available technology be integrated into each new vessel class so that its strategic value is not degraded ahead of time. As a result, highly innovative, usually extremely expensive and sometimes experimental or untested state-of-the-art technologies are destined for integration into the newest vessel designs, sometimes years before they actually come into existence. This uncertainty not only drives up the cost, but can also cause significant delays if the technology in question cannot be developed on time or suffers unplanned setbacks and malfunctions. Newly built ships are therefore almost always more costly and more problem-ridden at the beginning of their service lives than their predecessors. This is inevitable given the nature of the technologies being integrated, and therefore a problem all major navies share. It is only reasonable to assume that China's navy has to deal with such problems, even if they are not publicly advertised. This also explains why some navies opt for "tried and proven" designs instead of state-of-the-art technology. China's relatively recent purchase of Soviet-era *Sovremenny* class destroyers and *Kilo* class submarines from Russia as complete weapon systems would be an example for this type of risk-minimization strategy.

On the other hand, since different generations of vessels within a fleet usually carry highly divergent equipment, they can effectively be grouped into different levels of combat capability. Describing the U.S. Navy, Arena et al. write:

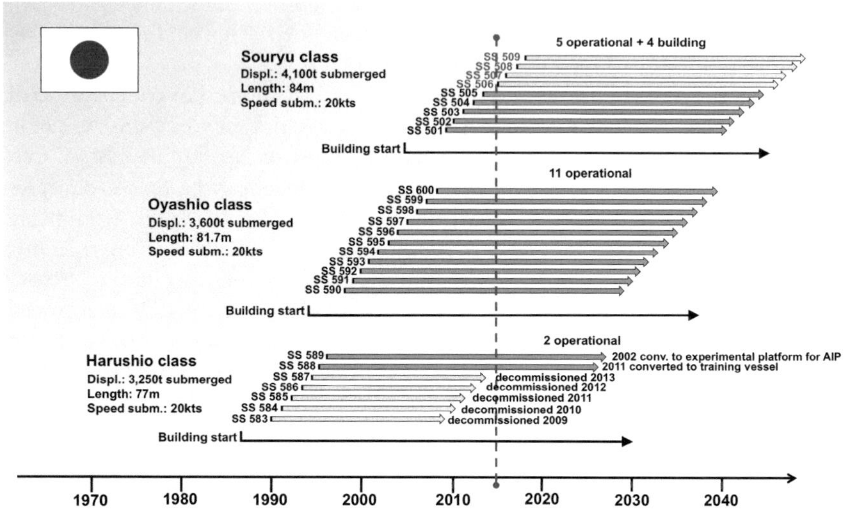

Source: Drawn using data from *Jane's Fighting Ships* (2014) and Wertheim (2013)

Fig. 3.9 Fleet age structure of the JMSDF Submarine Fleet, 2015

> With each new class or flight,[12] weapon systems on combatants have improved with each newer version. For example, ships are progressively able to track more targets at a given time. Their detection range has also increased. The ability to strike inland has increased in terms of range and accuracy. [...] Other changes have also improved ship performance in various operational dimensions. Propulsion plants now use gas turbines rather than steam engines. Higher-strength steel alloys have helped to reduce weight. Better hull designs have reduced radar cross-section. Better machinery and improved isolation systems have reduced radiated noise (Arena et al. 2006: 40).

A system differentiating between capability levels of vessel generations from a strategic viewpoint is therefore needed to complement the analytic tools presented so far.

This study uses a simple graphic device for indicating fleet age structures which will be used in Chaps. 5 and 6 (see Fig. 3.9 using the Japanese submarine fleet as an example). In the figures showing fleet structures, each vessel is represented by an arrow on a timeline, with the year of commissioning as the starting point, and is identified by its pennant number. The length of the arrow indicates the theoretical maximum lifespan of the vessel, i.e. ca. 40 years for surface ships and ca. 30 years for submarines. Retired vessels are drawn in light gray and active vessels in darker colors. All vessels of the same shipbuilding program (called a "class" of ships) are

[12] While a "class" of ships means the group of vessels built to the same design initially, a further "batch" of vessels is sometime built later on to the same or a slightly modified design. Such follow-on vessels are usually then designated as "x class flight II" (or batch II) vessels.

grouped together, and all classes of the same major vessel type (i.e., destroyers) are drawn together into one picture. This kind of graphic representation allows for a quick overview of the age structure within a vessel type. Moreover, by drawing a vertical line through any of the years on the timeline, the approximate vessel count at that point in time can be easily deduced, thus making it possible to determine whether newly built vessels are in fact replacement procurement or actual fleet expansion programs. As can be seen from the figure, Japan currently does not make use of its submarines' total possible service life span, but chooses to retire and scrap submarines far ahead of time in order to create more stable build cycles.

A detailed case study of China's fleet age structure using this analytic tool will be conducted in Chap. 5, with comparisons to other Asia Pacific and BRIC navies drawn on a case-to-case basis in Chap. 6.

3.4 The Impact of Network-Centric Warfare on Naval Shipbuilding

Overall, the ongoing development of militaries transforming towards "network-centric warfare" has had a massive influence on naval shipbuilding worldwide.[13] This development is the result of two factors: First, the advances in weapons technology that have rendered human reaction times to approaching missiles insufficient, thus giving rise to a need for fully automated responses towards incoming threats; and second, the progress in computer technology that first provided the necessary data processing capability and storing and displaying infrastructures needed for such a revolutionary development. Data links between ships, aircraft and shore headquarters then enabled the sharing of situational awareness between diverse units, thus creating a so-called "joint battlespace" in which various offensive or sensor capabilities provided by some can be effectively shared among the entire group. Replacing earlier concepts of "point defense," where each unit is charged with countering the attacks directed at it, very few units could now provide so-called "area defense" for e.g. an entire flotilla, or provide target data for offensive operations to other units. The degree of interoperability actually in reach depends heavily on the available sensor and data link technology.

As a result, the larger types of naval vessels such as destroyers, frigates, and submarines of newer generations have largely become giant floating computer systems. Their respective Combat Direction Systems (CDS, alternatively called Combat Management Systems/CMS) effectively function as the ship's brains by connecting the data input from various sensors (radars, sonars, infrared, optronic, etc.) with a multitude of effectors and countermeasures that constitute the "action organs" of the ship. The *AEGIS* combat system in use aboard the U.S. Navy's

[13] For a very accessible treatment of this subject, see Friedman (2009).

Arleigh Burke class destroyers, for example, is credited with the ability of detecting and tracking flying objects within 1,000 km, of simultaneously tracking up to 900 targets within 500 km, and intercepting any target within 170 km.[14] So far, the system has been exported to five U.S. allies.

The critical components of a vessel's combat system (CS) or integrated warfare system (IWS) are its main sensor, usually a phased-array air search/fire control radar system such as the *AEGIS* system's *AN/SPY-1* radar in the case of large surface combatants; a vertical-launch system (VLS) such as the *Mk 41* for surface-to-air missiles (SAM) such as the U.S. *SM-2 MR "Standard Missile"*; an automated ship self-defense system such as the *Phalanx* close-in weapon system (CIWS); and an associated computerized data processing system (the CDS) with data links such as the U.S. *Link 16* and *Link 11* to other units (e.g. helicopters, aircraft, other vessels, or shore headquarters), to which various types of other compatible sensors and effectors (e.g. sonar, cruise missiles, torpedoes) can be connected. The level of complexity of modern combat system architecture is such that it can easily take several years for a new prototype combat system to begin functioning properly after first being installed. The potential for malfunctions seems incredibly high, while the list of naval combat system suppliers active on the global export market remains short. As Neuman (2009: 63) explains:

> "[...] given the sophistication and complex manufacturing techniques involved, few defense industries are capable of producing them. In East Asia, for example, Japan, South Korea, Singapore, Taiwan, and China to a lesser extent, have integrated advanced computers, communications equipment, and sensors with precision-guided weapons in an effort to transform their militaries."

It is striking how limited the range of state-of-the-art combat system suppliers actually is. The short list given in Fig. 3.10 is almost exhaustive, showing the main Western producers of modern naval CDS, their countries of origin, their main products, their amount of defense-related revenue in the year 2011, and their rank within the *Defense News Top 100* Defense and Aerospace Companies (2012).

Due to the huge development and procurement cost of these systems, it is not surprising that almost all the leading suppliers of naval CS are found within the biggest 11 defense companies listed by the *Defense News* database, and that their combined share of defense-related revenue amounted to one third of the total worldwide defense-related revenue volume in 2012. Due to unavailability of official data, however, China's share is not included in the aggregate volume cited.

Apart from supplying their own "parent navies," the leading CS-producing companies also dominate the worldwide naval CS export market. Lockheed Martin's *AN/SPY-1* radar system is the centerpiece of the tried and proven, if not exactly state-of-the-art, *AEGIS* combat system which has been selected by the Japanese, South Korean, Australian, Norwegian, and Spanish navies for their large surface

[14] Cf. *Jane's World Navies* (2013), el. ed., South Korea chapter.

Company	HQ	Major combat systems (CDS | main radar)	Rank among top 100 defense companies (2012)	Defense-related revenue 2011 (US$ bn)	% share of total defense-related revenue of top 100[a]
Lockheed Martin	USA	AEGIS | AN/SPY-1	1	43.978	10.6
BAE Systems	UK	PAAMS | Sampson ("Sea Viper")	3	29.130	7.0
Raytheon	USA	SSDS	5	23.055	5.6
EADS (Cassidian and Atlas Elektronik)	NL	ANCS | TRS-4D	7	16.092	3.9
Finmeccanica (Selex)	Italy	IPN-10, IPN-20; PAAMS | EMPAR	8	14.584	3.5
Thales / Thales NL	France	TAVITAC TACTICOS | Héraclès SEWACO | APAR	11	9.492	2.3
Almaz-Antey	Russia	S-300F / S-300FM / Ryf-M a.k.a. "SA-N-6", SA-N-20" | "Tombstone"	21	3552	0.9
SAAB	Sweden	9LV | Sea GIRAFFE	27	3.259	0.8
Total				143.142	34.6

[a]100 / =US$414.304 billion
Data source: Defense News Top 100 database (2012). Own calculations

Fig. 3.10 Major naval combat systems manufacturers, 2012

combatants as well. The interoperability of their vessels with U.S. naval forces has undoubtedly played a large part in their selection processes, giving manufacturer Lockheed Martin an effective hold over the U.S.-oriented naval export market share. The wave of recent admissions of former Warsaw Pact countries to NATO has also had the effect of promoting American technical standards in naval modernization in their navies, as the countries in question have in many cases decided to supplant obsolete Soviet-era technologies with American systems in order to ensure NATO interoperability (cf. Mommsen 2005: xxv). Nonetheless, this has merely opened up new markets for the U.S. based market leaders. Already during the 1980s, Todd (1985: 310) stated that "the American pattern of procurement has, of course, influenced the naval programs of its allies, especially the NATO countries" (Todd 1985: 310). This was certainly helped by the employment of financial incentives such as the U.S. system of Foreign Military Sales (FMS) that financially supports the acquisition of U.S. military technology by its allies. It would not be an exaggeration to suggest that there are aspects of dependence and hegemonialism within this power structure, not least because the purchaser of a complex military technology essentially remains dependent on the source company for spare parts, technical support and training for the whole duration of its use. If vital support is withheld e.g. for political reasons, the whole weapon system may become useless, making recipient countries vulnerable to political pressures by the exporting country.

Apart from U.S. market leader Lockheed Martin, a few European system companies have remained firmly entrenched in their respective home markets, though in some cases they have also gained some export customers, often in parts of the world with traditional ties to the exporting country.[15]

In naval electronics and combat system design, much depends on the quality of the radar systems and signal processing systems that work together:

> Sensor technologies employ microelectronic devices to monitor and/or observe changes in their environment, while signal processing technologies transform the sensors' electrical signals into usable information and transmit it to users. Together these two technologies enable automated systems to interact with the external world. The quality of sensor capabilities can be measured by the accuracy, reliability, and responsiveness of the sensors. Redundancy is also a key quality in a sensor array, since it may correct for localized flaws in the first three areas. The quality of signal processing is indicated by the system's ability to discern between false and true signals, eliminate irrelevant noise, and produce accurate readings (Tellis et al. 2000: 59).

It should be clear from the discussion above that it is very difficult, if not altogether futile, to combine core components from different combat systems to form a coherent whole because of the level of signal precision that is needed. The pitfalls of automated friend-foe-recognition systems were demonstrated by a tragic event from the *AEGIS* system's early days, when in 1988 the U.S. cruiser *USS Vincennes* mistakenly shot down a civilian Iranian airplane, resulting in 290 casualties. Another drastic example illustrating the problems of system integration is mentioned by Bussert (2003):

> Among the factors causing the loss of the Royal Navy's HMS *Sheffield* in the Falklands could be radars that should have tracked incoming Exocets but were shut down because

[15] Airbus Defence and Space will supply the German Navy's new frigate *F-125* with the *TRS-3D NR* radar while TKMS and Airbus Defence and Space subsidiary Atlas Elektronik supplies the *ANCS* CDS to this project. Atlas has also supplied CDS to many German submarine projects, both domestic and export. The French partially state-owned Thales Group's *SETIS*, *TACTICOS* and *SEWACO* combat systems have so far been in use in many surface vessels worldwide, with *TACTICOS* alone installed on hundreds of vessels. Italian Finmeccanica's subsidiary Selex combines its *PAAMS* combat system and *EMPAR* Radar with *Aster 15/30* SAM on the Italian variant of the pan-European *FREMM* frigate project (which, like the German *F-124* and *F-125*, is a destroyer rather than a frigate by the standards used in this book). British defense giant BAE Systems combines a different radar (*SAMPSON*) with a variant of the Italian *PAAMS* combat system, while Raytheon's *SSDS* is the combat system in use aboard the US Navy's nuclear-powered aircraft carriers (which are never exported), LPDs, and LHDs.

> they interfered with communications frequencies. The inability to conduct vital communications and radar operations simultaneously should have been noted during integration tests (italics in the original).

Due to the extreme procurement cost and political export restrictions imposed by Western suppliers, there has been a recent trend for more indigenous combat system development in a number of countries. Friedman writes:

> A striking feature [. . .] is the number of countries that are now developing their own combat direction systems, or are now abandoning foreign models for entirely local developments. Examples are the new distributed system [sic] currently being advertised in China—and probably equipping the new frigates—and the Turkish Genesis program. Another interesting development is the spread of national-level command systems similar in role to the U.S. GCCS-M or the British JOCS (Friedman 2006: x).

China was heavily afflicted by the 1989 embargo especially in the field of naval electronics. During the short period in the 1980s when advanced Western combat systems became available to China, the French *TAVITAC* and the Italian *IPN-10* CDS became influential models for China's indigenous combat system development. Based on thorough analyses of publicized photographs of consoles observable on various Chinese warships, Friedman even goes so far to assume that

> [a]t least until recently, Chinese destroyer and frigate combat systems were probably descended from two types purchased abroad in the 1980s, the French TAVITAC (two sold) and the Italian IPN-10. [. . .] Elements of TAVITAC were, presumably, delivered with the Crotale naval [SAM] system. The Italians supplied electronic warfare equipment and the Aspide missile, and probably also the A-244 torpedo. Note that there was almost certainly a third supplier, Signaal [now Thales NL], the main evidence being the Ramses EW system. There have never been claims that the Chinese received SEWACO in any form, however (Friedman 2006: 61).

In addition to this Western European technological legacy, Russian naval export technology has had a very strong impact on China's naval development. Russian military technology is still rather dominant in some parts of the world, notably in China and India, some Middle Eastern, Maghreb, and African countries, some former Soviet Union and Warsaw Pact members, as well as in countries with political ties to Russia such as Vietnam. Such countries usually buy complete "combat suites"—i.e. sensor and weapons equipment—from this source. Despite some problems experienced within the post-Soviet Russian navy—due not least to the fact that most of the Soviet Union's major naval shipyards were "inherited" by Ukraine, not Russia—it is important to note that

> Russian weapons and systems [. . .] remain very important, partly because two major navies, those of India and of China, are supplied mainly with Russian systems. The Russian navy itself is still quite powerful, despite its post-Cold War problems (Friedman 2006: x).

Russian technology differs widely from Western state-of-the-art systems in several respects, making direct comparisons between vessel types with similar designations difficult. As frustrating as this may be to the analyst, this core technological difference is not necessarily a detrimental factor from the point of view of smaller recipient navies. In 2003, a Chinese analysis came to the conclusion that in direct comparison with Russian naval technology, U.S. capabilities in anti-air warfare (AAW) and anti-submarine warfare (ASW) were still clearly superior. In anti-surface warfare (ASuW), however, U.S. and Russian capabilities were judged to be roughly comparable, with unique strengths on both sides (cf. Chen 2003: 21). Moreover, many Russian vessels are built from proven and reliable designs and are available at very competitive prices compared with the usually much more expensive Western systems. Countries used to Soviet or Russian weapon systems have often build up related training and maintenance infrastructures, which helps to minimize integration challenges and makes further acquisitions of Russian military technology attractive. However, as Bussert (2003) notes, there are many clues that Soviet vessels have experienced ample system integration problems on par with Western navies over the years, even if such problems are rarely publicized.

China has long been Russia's most important export customer for weapon systems. This includes some recent transfers of naval technology—the above-mentioned *Sovremenny* class destroyers and *Kilo* class submarines, which were imported as complete weapon systems likely in order to minimize system integration challenges—even though China is aiming at more indigenous development. It seems likely that due to the traditional Chinese dependence upon Soviet-era vessel designs, and also the recent acquisitions of Russian technology, China's indigenous vessel designs will be at least partly modeled on and thus heavily influenced by Soviet-era technology. The implications of this factor are discussed in more detail in Chaps. 4 and 5.

3.5 Cost Implications of Naval Power

When a navy endeavors to extend its reach beyond the confines of coastal defense and aims at securing the country's Sea Lines of Communication (SLOC) and other overseas interests, this invariably becomes a heavy burden to even relatively powerful navies' budgets. One reason for this is that related duties can only be performed if there is enough redundancy in the fleet to conduct extensive out-of-area patrols in addition to homeland defense. Moreover, vessels needed for operations far away from the home shores must be larger, more seaworthy, and have better endurance than those suitable for coastal defense. More ships means more sailors who need to be paid, fed, clothed, equipped, trained, and paid pensions after retirement. Another cost-intensive necessity is an infrastructure of offshore bases or ports where maintenance and supplies can be obtained—something China so far is still lacking. Any significant naval build-up surpassing traditional coastal defense capabilities therefore requires the constant and long-term input of financial resources far surpassing the budget needed to simply build more ships. Moreover, maritime weapon platforms are generally characterized by the fact that they are

> [...] expensive, hard to replace and even the smallest units represent a sizeable investment in human resources whose loss can be sudden and instantaneous and very hard for publics and governments to bear (Till 2009: 44).

This gives rise to extra requirements to protect these assets, which in turn adds to the total cost impact. This is especially true for high-end systems such as aircraft carriers, where every single ship represents such a huge portion of national wealth that the loss of just one unit is deemed politically unacceptable. They are thus marked as potentially attractive targets for asymmetric opponents, explaining the need for large carrier battle groups consisting of large surface combatants (cruisers, destroyers, and frigates), submarines, and auxiliaries, which are in effect small fleets that accompany the carrier at all times. The extent to which the American superpower status is symbolically linked to its giant aircraft carriers, at least in the perception of the public, is well reflected in a comment by journalist Robert D. Kaplan. In an article for the magazine *The Atlantic*, he writes:

> The effect of a single Chinese cruise missile's hitting a U.S. carrier, even if it did not sink the ship, would be politically and psychologically catastrophic, akin to al-Qaeda's attacks on the Twin Towers (Kaplan 2005).

Rather than being a strategic asset, such weapon systems can thus become a liability if the cost of protecting them rises, while the potential psychological damage of their loss on public morale becomes a heavy political burden.[16]

As discussed in Sect. 3.3, warships are extremely costly weapon systems, not only in their procurement, but also in their maintenance, manning, and deployment. The impact of network-centric warfare in conjunction with the rising role of defense electronics has caused cost escalation levels that far surpass normal consumer price inflation rates worldwide. The total cost impact of technological innovation for a state's budget, however, is not limited to manufacturing and operating expenditures alone. Rising R&D demands, more sophisticated component parts and sub-systems industries, as well as an efficient education system that can train the personnel needed for operating and maintaining the increasingly complex technologies, have been equally responsible for driving up the cost of sea power during the past few decades (cf. Neuman 2009: 64).

Therefore, the nominal level of defense spending a country can afford is directly indicative of the maximum possible level of naval modernization it can achieve during a given time span. Smaller budgets, even if representing a relatively large

[16] A former naval commander from Europe who had partaken in embargo operations during the Yugoslavia war described a scenario he witnessed when a Serbian submarine was reported submerged without trace in the Mediterranean Sea for several hours. In the words of the source, "the single U.S. carrier in the vicinity sought to gain distance so fast it almost crept up the Italian shore." He insisted that the actual threat to carriers even from relatively outdated diesel submarines like the one in question here is in fact even graver than publicly reported, and that two torpedo hits would probably suffice to sink a fleet carrier. This is confirmed by Scott (2011) who writes that "even the possibility of a submarine lurking undetected in the vicinity will weigh heavy in the thinking of a maritime component commander transiting a chokepoint or attempting to access the littoral."

share of a smaller country's GDP, will only buy so much equipment. But it is not only the direct cost of armaments, military installations, and personnel that deplete a country's wealth. It must be noted that militarization has other detrimental effects on the national economy as well:

> Armed forces consume wealth and are thus a burden upon the civilian economy. Soldiers, sailors and airmen produce nothing that can be eaten or used to warm, clothe or shelter anybody. Rather, such essentials have to be provided for them along with their increasingly expensive equipment [. . .]. Evidently, the nation is worse off than if it were possible for its service personnel—and their civilian support—to be employed in productive industry or agriculture (Pugh 1986: 8).

Moreover, while some governments may succeed in "mortgaging the future" to pay for militarization, it is imperative to note that, from a historical perspective, the real cost of defense cannot be circumvented in the long run, not even in the case of autocratic regimes which may succeed in temporarily hiding these costs from public view (Pugh 1986: 8–9).

For the above-mentioned reasons, analysts look to military spending trends for clues about a country's prospects for successful military modernization. Are there notable changes in spending levels over time? How large is the share of spending in relation to other state spending priorities? What funds can be dedicated to the procurement portion of the military budget in the short term and long term? These are some of the indicators that make comparisons across a wide variety of countries meaningful.

3.5.1 Interpreting Military Spending Levels

When discussing militarization levels in terms of defense spending, a distinction must be made between "normal" peace-time military spending levels and military spending levels during war years. While the so-called "defense burden"—defense spending as a share of the GNP, as defined by Pugh—reached extraordinary values of more than 50 % of GNP during the two World Wars, at least in Britain (Pugh 1986: 9–11; 16), there seems to be a fairly consistent historically "normal" range of peacetime defense burden values at roughly between 1 and 6 % of the GNP. Historical data analyzed by Pugh shows that in Great Britain, the defense burden during peace time remained within 2 and 5.5 % for centuries from the late 17th century through the 1980s. The data Pugh analyzed and compared for at least 29 different major countries for the year 1982 showed, moreover, that this is an international phenomenon. All these countries' defense burdens were found to fall between 2 and 6 % that year, with Austria and Japan showing even lower levels due to constitutional limits imposed on them after World War II (Pugh 1986: 22–25). It follows that from a historical-comparative perspective a defense burden of 2–6 % must be considered within the limits of empirical normalcy, and indeed most of the world's countries today fall within these parameters, often at the lower end or even below it. NATO actually recommends a spending level of 2 % of GNP to its member countries, and NATO officials regularly criticize the fact that most member states fall behind this mark (McTague 2014).

Military expenditure is often cited as an indicator of national power, and is often included in the various formulas used for computing national power indices. Nonetheless, this is a somewhat problematic approach. As the authors of a SIPRI paper on military expenditure cautioned in 2006:

> Military expenditure is not a measure of military capability. Nor is spending on personnel, equipment, R&D or operations and maintenance. As military expenditure data is a measure of the input of resources, it is best used for measuring the military's burden on the national economy, to assess the priorities of the government in the trade off between military and non-military activities, or as a measure of the cost of investing in the military. Very little extra information is actually revealed about a state's military strength or capability by making public its total military expenditure and its distribution on personnel, operations and maintenance, equipment and R&D (Stålenheim and Surry 2006: 2–3).

A similar point was made by a RAND study in 2000, when the relation between budgets and military capability was discussed in the following terms:

> Put simply, the question is, 'What resources does the military get, and how successfully can they be transformed into effective military power?' [. . .] A country may provide its military with generous budgets and large cadres of manpower, but if the military's doctrine is misguided, the training ineffective, the leadership unschooled, or the organization inappropriate, military capability will suffer (Tellis et al. 2000: 134).

These observations notwithstanding, it is still useful to gather comparative data on military spending, particularly if not just the GNP or GDP percentage are included, but also nominal figures and spending levels in relation to total military manpower. This is because even if budgets alone do not reveal the resulting military capability, they still constitute the economic basis for achieving military capability enhancement. The level of a country's economic power is critical for military development because "economic output [. . .] represents a mass of resources that can be utilized for various purposes by a nation's political authority" (Tellis et al. 2000: 84). The question then is how well the transformation from economic resources to military capability can be performed. The approach of comparing the basic budget-related indicators chosen in the sections below will serve to give a rough idea of typical military expenditure levels across a number of countries.[17]

China's high annual defense budget growth rates over the past two decades are often cited as evidence of a massive build-up.[18] This rising tendency contrasts starkly with stagnating defense budgets in many Western countries after the end of

[17] Nominal amounts are a better indicator of a country's actual "buying power" for major military procurement than the GDP percentage, because a large percentage of a small GDP can still be much less money in absolute terms than a small percentage of a large GDP, as indicated in Fig. 3.12. The ratio of defense spending per total military manpower, on the other hand, indicates the relative sophistication level of the force.

[18] Cf. Cole (2010: xviii) who gives a figure of 15 % as the average yearly growth rate of the PLA budget during 1993–2009.

the Cold War, a result of the so-called "peace dividend" due to lessened tensions. As a recent study has shown, however, when distortions caused by inflation and exchange rate fluctuations are taken account of, the increase in China's defense budget must still be judged to be fairly in line with government spending growth in general over the same time period, if calculated in terms of constant prices:

> [...] throughout most of the post-1978 'reform and opening up' period, rampant inflation has mitigated the real-world impact of nominally large budget increases [...]. Calculations of China's defence budget at constant prices [...] show that in many years the effective growth rate of China's defence spending is much lower than the widely cited current price ('nominal') figure suggest. [...] In other words, when calculated in real terms the average annual increases in the budget exceeded 10 per cent during only one of the ten-year periods [...]: 2000–2009 (Liff and Erickson 2013: 4).

Compared with other state expenditures, Liff and Erickson conclude, the military does not seem to dominate China's state spending priorities. They point out that Western analysts almost universally stress the menacing aspects of China's budget increases, while overlooking mitigating facts. In particular, higher inflation rates have a huge impact on the actual "real-world effects" of nominal budget increases. Furthermore, when assessing China's defense expenditure as a percentage of total government expenditure over the last years and decades, it becomes clear that—despite high GDP growth rates throughout this period—military spending as a share of total state financial expenditures has not risen, but instead declined in comparison with other budgetary priorities:

> China's investment in its military development, while increasing at a rapid clip, has been outpaced by the government's overall spending, and does not come close to dominating national priorities—in stark contrast to that of, say, the former Soviet Union (Liff and Erickson 2013: 5).

The methodological point made by Liff and Erickson is not a new one. In his pioneering effort to analyze cost aspects of naval power, Pugh already in 1986 discussed the complex ramifications of inflation, exchange rates, and other cost-related effects for the analysis of defense economics in minute detail (Pugh 1986: 72–101). Taking the studies by Stålenheim and Surry (2006) and Liff and Erickson (2013) as a starting point, more intensive research on this topic with regard to China is clearly needed. Meanwhile, the following section will try to use the currently available data to make some broad comparisons with other regional powers with the aim of establishing a measure of the empirically "normal." The methodological issues mentioned above should be borne in mind when evaluating the findings.

3.5.2 China's Level of Spending in Comparison

Even though real spending levels in China are probably much higher than indicated by the official figures, the same can be said for several other countries, e.g. Russia and Japan (cf. Stålenheim and Surry 2006: 9). Regarding the common charge of

Chinese underreporting of defense expenditures, Liff and Erickson point out that this is in fact a phenomenon common to countries worldwide and that there are currently no universally accepted norms on defense reporting:

> Even the definitions used by the US Department of Defense, NATO and SIPRI differ significantly. Consequently, much of the disparity among these organizations' estimates stems from the inclusion of very different categories of spending. One important implication of these differing definitions is that China's exclusion of certain items from its defence budget is at least to some extent an artefact of different metrics (Liff and Erickson 2013: 16).[19]

It seems that reporting standards are slowly beginning to emerge only among the most advanced industrialized countries, and that about the only thing that "different countries' defence budgets have in common is the non-uniformity in what governments define as 'defence spending'" (Liff and Erickson 2013: 12). It must be concluded, therefore, that intransparency and underreporting of defense spending is the international norm, not the exception.

Here, the question of regional yardsticks once again becomes important. Analysts often compare China's current standards of military transparency unfavorably to those of the world's most advanced industrialized nations, rather than making comparisons to China's regional neighbors or other rising powers of analogous military and economic development such as the other BRIC nations. Still, the latter might be preferable from a strictly analytic viewpoint, since

> China's military transparency is roughly comparable to that found in most ASEAN countries (including several US security partners) and India (world's most populous democracy) (Liff and Erickson 2013: 17–18).[20]

It should also be pointed out that even in the U.S. defense budget, items such as Veteran Affairs, retirement costs, veteran re-employment and training costs, portions of homeland security and U.S. National Guard activities and even defense-related nuclear energy costs are not included, even if they may be reconstructed to some extent from other publicly available information. China, on the other hand, is routinely criticized for excluding officer pensions, nuclear and strategic missile program costs, its military police budget and other items such as locally funded security activities from its official budget (Liff and Erickson 2013: 19). A double standard may therefore be at work in at least some of the voiced criticisms.

In order to avoid distortions by comparing figures based on different reporting standards, SIPRI estimations for 2012 were used in the following analysis.

---

[19] According to Liff and Erickson, "The difference between SIPRI's [2009] estimate and the upper bound of the Department of Defense's estimate was US$65.1 billion, a difference larger than China's entire official defence budget that year" (Liff and Erickson 2013: 15).

[20] See, however, the diverging analysis by Kiselycznyk and Saunders (2010: 35), who conclude that if compared e.g. to Japan, South Korea, India, and Australia, "China is significantly less transparent than its peers, especially in the areas of defense budget, force structure, and future modernization plans."

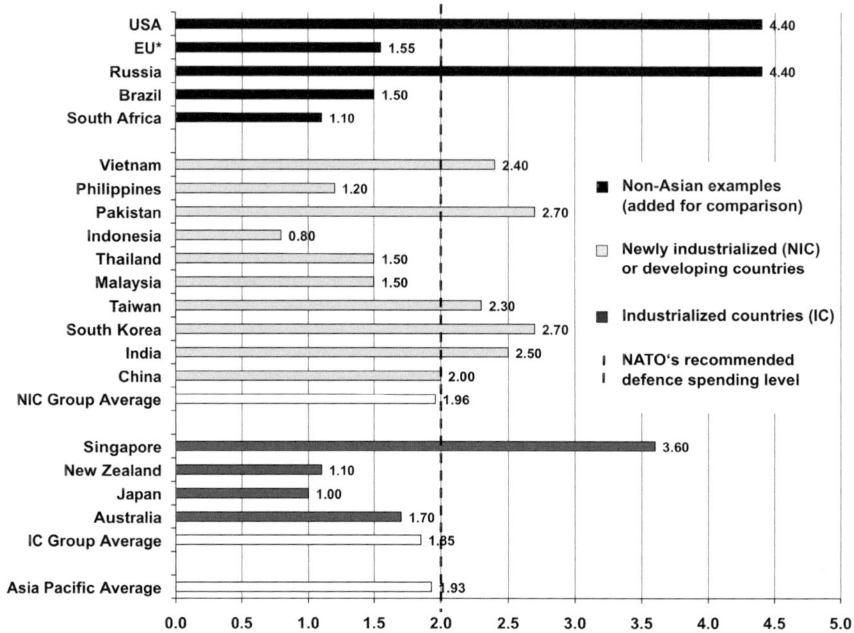

*EU data from 2011 (Denmark not included)
Source: Graph produced using data from the SIPRI Military Expenditure Database (2014) and EDA (2013)

Fig. 3.11 GDP share of defense spending, 2012 (%)

As Fig. 3.11 shows, China does not stand out in relation to its regional neighbors regarding its GDP share of defense spending. China's percentage in 2012 was surpassed by six countries within the Asia Pacific region (Vietnam, Pakistan, Taiwan, South Korea, India, and Singapore) as well as by the U.S. and Russia. China's percentage share was roughly on par with both the average GDP share of the group of Asia's newly industrialized countries as well as the Asia Pacific regional average. It was also at exactly the 2 % level NATO recommends to its member countries (which is for the most part not met).

Looking at nominal spending levels, however, yields quite a different picture (cf. Fig. 3.12). China's estimated nominal spending of US$166.1 billion in 2012 looks very large from a regional perspective: It is more than five times the amount spent on average by Asia's newly industrialized countries, almost double the amount spent by Russia, and more than double the amount spent by Japan, with India and Brazil far behind. In fact, the aggregate of fellow BRIC states Russia, India and Brazil—some US$170,017 billion in 2012—is almost equivalent to China's spending alone.

On the other hand, it is important to remember when comparing China to Japan that Japan is a special case because of its self-imposed upper limit of defense spending at 1 % of GDP. Were Japan spending at the level recommended by

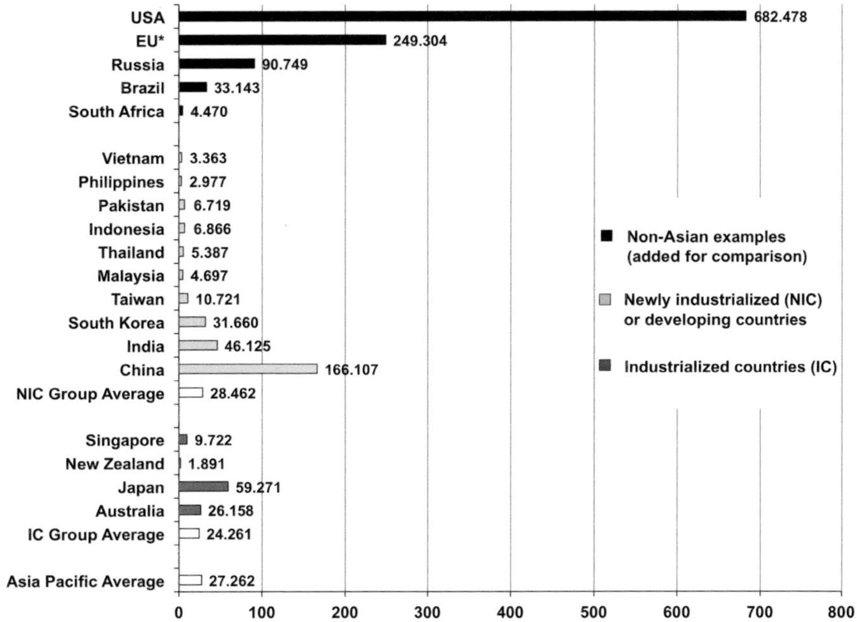

*EU aggregate figures excluding Denmark (2011)

Source: Graph drawn using data from the SIPRI Military Expenditure Database (2014) and EDA (2013)

Fig. 3.12 Nominal defense spending, 2012 (US$ bn)

NATO, i.e., 2 % of GDP, its nominal defense budget would be double the present amount and thus much closer to China's in nominal terms.

That said, no matter how China's budget may compare to the rest of the region or to the other BRIC countries, its nominal spending is still dwarfed by the amount spent by the United States. Even though the gap has been getting narrower through the past few years with rising spending levels in China and simultaneous defense cuts in the U.S., U.S. nominal spending was still estimated in 2012 to be about four times the amount of China's.

Due to China's enormous population size, national economy, and territory it is still a useful exercise to compare China's spending to the aggregate figures for the entire EU area. When viewed in terms of this yardstick, China's current level of defense spending does not seem so disproportionate. According to the most recent EDA data available to this study, the combined defense spending of the EU member states (excluding Denmark) stood at US$249.304 billion (€192.535 billion) in 2011. This is still a much larger amount than China's, despite a smaller overall GDP share of 1.55 % (cf. Guzelyte 2013: 2; EDA 2013).

The data in Fig. 3.12 also serves to illustrate how nominal defense budgets ultimately determine the maximum amount of hardware a country can afford to procure and maintain. Since many naval weapon systems are extremely costly—a single U.S.N. nuclear-powered aircraft carrier easily costs US$5 billion for the initial procurement cost, excluding the air wing, while the cost of the new U.S. *Zumwalt* class destroyer skyrocketed to at least US$3.2 billion for the first ship alone—it is easy to see why large-scale naval modernization programs are within reach only for navies with at least double-digit US$ billion size budgets.

As a rule of thumb, the total navy budget of any one country can at most be roughly one third of the total defense budget given the fact that army and air force also need to be maintained and equipped. As explained in Chap. 3, moreover, this budget cannot of course be spent entirely on procurement, which means that only a fraction is actually available each year for hardware modernization including procurement, maintenance, repairs, and refits. In the case of China, not three, but four services compete for funds—Ground Forces, Navy, Air Force, and the Second Artillery Corps (SAC) in control of China's nuclear missiles—which, in turn, diminishes the navy's share of the total defense budget to roughly 25 %.

Although detailed budget breakdowns are not generally available for many militaries, judging from the aggregate data breakdown provided by the European Defense Agency (EDA) for EU states excluding Denmark during 2006–2011, only 16 % of the budget was on average spent on defense procurement each year, with fairly constant levels over time. The largest portion of the budget—over 50 % on average—went toward personnel expenditures (see Fig. 3.13).

For China however, *Kanwa* editor Pinkov estimates that as much as 30 % of the budget might be spent on procurement (Ping 2010: 21). Given the fact that no EU country is aiming for a military build-up at present, a discrepancy of sorts is to be expected because of China's tendency towards troop reduction and military hardware upgrades over the past years compared with Europe's tendency towards maintaining the same capability level.

Fig. 3.13 Defense expenditure breakdown: average shares for EDA participant countries, 2006–2011

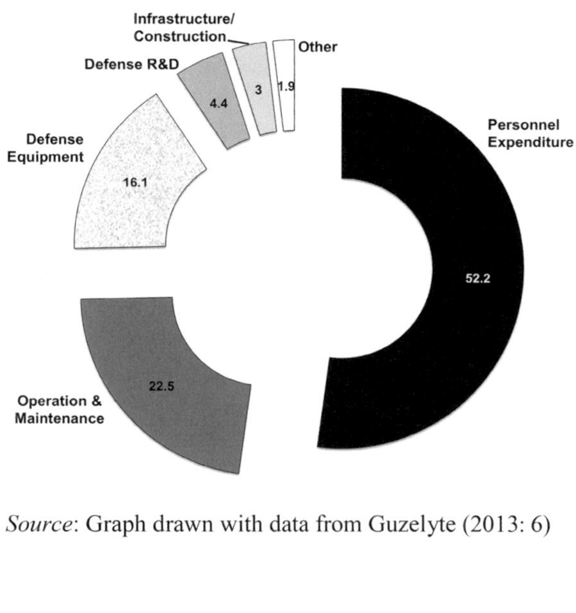

Source: Graph drawn with data from Guzelyte (2013: 6)

No.	Platform designation	Est. unit price (US$ bn)	Total (US$ bn)
	Surface fleet		
2	*Type 052C "Lüyang II"* class DDG	0.9	1.8
2	*Type 051C "Lüzhou"* class DDG	0.8	1.6
2	*Project 956 EM "Sovremenny"* class DDG	0.7	1.4
2	*Type 054 "Jiangkai I"* class FFG	0.4	0.8
2	*Type 053H3 "Jiangwei II"* class FFG	0.2	0.4
	Subtotal for surface fleet		6.0
	Submarine fleet		
2	*Type 094 "Jin"* class SSBN	~1 5_2[a]	~3_4[a]
2	*Type 093 "Shang"* class SSN	>0 8[b]	>1 6[b]
8	*Project 636 "Kilo"* class SSK	0.2	1.6
2	*Type 041 "Yuan"* class SSK	0.18	0.36
9	*Type 039A "Yuan"* class SSK	(0.15) 0.37	Orig target cost 1 35 actual cost 3.31
	Subtotal submarine fleet		> 9.87
	Minimal surface and submarine procurement total		>15.9

Source: Rough estimations published by Andrei Pinkov (cf. Ping 2010: 22–24)

[a]Rough estimate, incl. *JL-2* ballistic missile [b] Excl. cost of design, test and trials. Total cost of program over a 10 year period estimated to be at least US$4 billion

Fig. 3.14 Cost impact of Chinese naval weapon system purchases, period of 2000–2005

To illustrate the approximate magnitude of investment required for the Chinese naval build-up from the naval hardware procurement perspective alone, Fig. 3.14 gives rough estimations for the cost impact of Chinese naval weapon systems purchased during 2005–2010 as published by Pinkov, excluding the cost for finishing the aircraft carrier *Liaoning* or any other warship refits and upgrades.

Regarding the cost of the Chinese aircraft carrier program, Li and Weuve have published an estimation of the hardware procurement cost for one carrier battle group including the carrier, its escort ships and air wing amounting to an approximate total of US$10 billion. Annual maintenance, repair, training, and fuel cost would likely amount to ca. US$200 million:

> Taking into consideration the lower labor and material costs in China, the cost to build a medium-sized, conventional-powered 60,000-ton carrier similar to the Russian *Kuznetsov* class is likely to be above $2 billion. But the cost to acquire a carrier itself is just the start because a carrier needs aircraft and escorts. A Russian Su-33 carrier-based combat aircraft costs $50 million, so a notional carrier air wing of about fifty Su-33s, several airborne early warning (AEW) aircraft, and several antisubmarine warfare (ASW) and search-and-rescue

> helicopters may cost more than $3 billion. A Russian *Sovremenny*-class guided missile destroyer costs about $600 million, so an escort force consisting of a number of guided missile destroyers, frigates, and supply ships may cost more than $4 billion. [. . .] the total cost of one carrier group [is] likely to be about $10 billion [. . .]. The annual cost for regular training, maintenance, repairs, and fueling [...] is estimated to be about 10 % of the construction cost, or $200 million for each (Li and Weuve 2011: 213).

When evaluating a country's naval modernization efforts, nominal levels of defense spending are thus significant, because high-end naval weapon platforms are not only expensive in initial procurement, but entail a huge life-cycle cost over their whole service lives that is essentially out of reach for smaller navies' budgets (cf. Sect. 3.3). If the aim is not replacement procurement, but a build-up towards a higher level of capability, regular budget increases are a necessity. The fact that annual cost escalation rates for major warship types have ranged between 7 and 11 % during the past 50 years, reaching nearly double the inflation rate, ultimately threatens to limit the procurement abilities of most small and medium sized navies, as navy budgets do not tend to grow accordingly (Arena et al. 2006: xiv, xix, 2).

China's already huge and steadily rising military budget is thus a necessary precondition for a naval build-up involving the procurement of expensive new platforms. If China were able to spend ca. 16 % of its budget on procurement, as the EDA member countries do on average, this would mean that China could spend more than US$26 billion per year on new weapons systems for all four armed services. Assuming a roughly equal share for all four services, as Pinkov does, this would put the navy's procurement share at ca. US$6.5 billion (cf. Ping 2010: 21). If however the much higher procurement share estimate by Pinkov cited above is correct, the figure could be almost double this amount. On the other hand, due to China's huge military manpower, it is quite possible that the personnel expenditure share of China's budget could be even higher than the European level of 50+ percent, leaving a little less leeway for procurement. Unfortunately, no truly reliable figures are publicly available at this time of writing.

Be that as it may, it is important to realize that as shown in Sect. 3.3, initial procurement cost for a naval weapon system represents only ca. 23 % of its total life-cycle cost impact over a span of ca. 30 years (cf. Fig. 3.8). Taking the estimates listed in Fig. 3.14 as an example, this would mean that the amount of ca. US$16 billion spent during 2000–2005 on naval vessel procurement will result in a total cost impact of ca. US$69.6 billion over the systems' total service lives to be paid for while other, newer systems are being procured simultaneously.

The personnel cost aspect introduces another factor often used for assessing military prowess: total military manpower. With almost 2.3 million soldiers as of 2013, China still has the world's largest armed forces, surpassing even the U.S. (ca. 1.5 million) and India (ca. 1.3 million). By comparison, the EU countries taken together have more than 1.6 million soldiers. It is again important to point out here that a single factor like manpower is insufficient for characterizing a country's military capability. As Tellis et al. noted: "It is obvious [. . .] that the largest armies

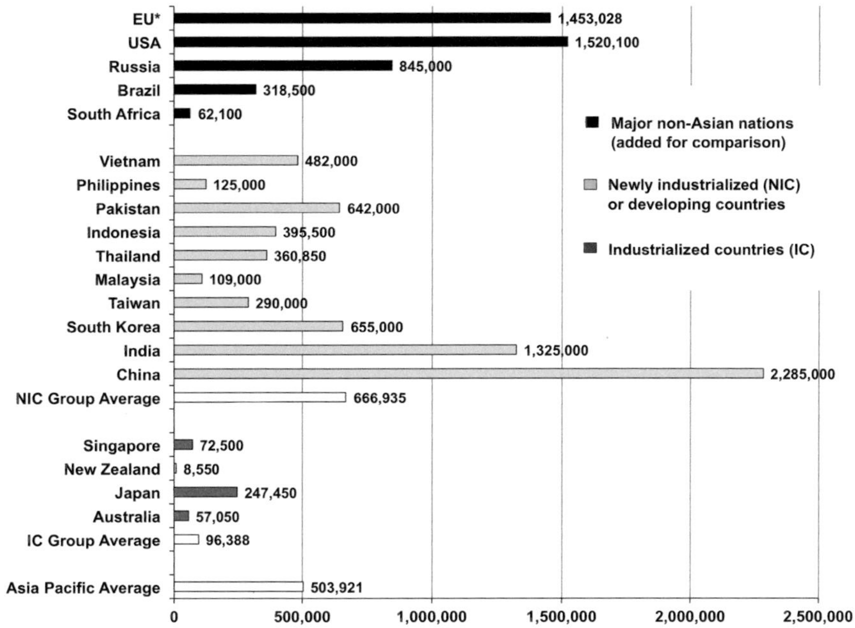

*EU data excluding Denmark, converted from EDA (2013) data
Source: Graph drawn with data from IISS (2013)

Fig. 3.15 Total regular forces personnel, 2013

may not necessarily be the most effective. In the 1960s and 1970s, Israel's small forces defeated larger opponents" (Tellis et al. 2000: 135).

Not only is a huge army not indicative in itself of high military capability, but it can also impose tremendous financial burdens. Since personnel costs eat up a significant portion of the defense budget, most industrialized countries have constantly reduced troops and chosen to focus on better training and equipment, thus putting quality over quantity. This is also the long-term goal of the PLA. Figure 3.15 illustrates the tendency of industrialized Asia Pacific nations to maintain comparatively small forces of less than 75,000 soldiers, with only Japan going against the trend with ca. 270,000 soldiers. By the same token, all the region's newly industrialized or developing countries still have much larger forces of more than 100,000 and on average ca. 670,000 soldiers. China is therefore still much more typical of the NIC group of countries than the IC group, together with North Korea (not shown in the chart), which has about 1.19 million soldiers. Again, the U.S. is an exception to the general rule and unlikely to be a suitable yardstick for China due to its global military network of bases that need to be manned and maintained, justifying to some degree a much higher personnel demand than is the case in "normal" militaries. However, it must be noted that the figure for the EU total (excluding Denmark) is also in the same order of magnitude.

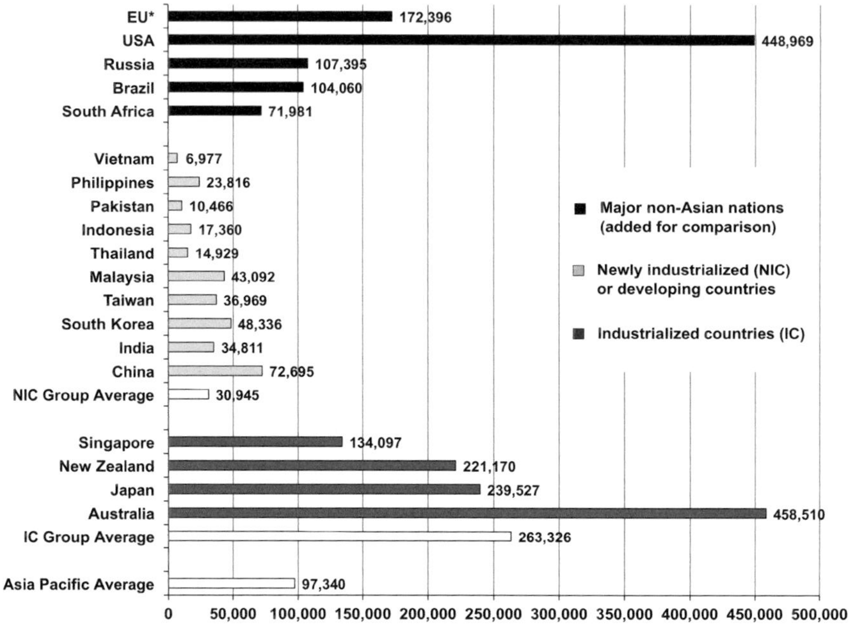

*EU figure excluding Denmark, converted from EDA (2013)
Data Source: Computed using data from the SIPRI Military Expenditure Database (2014) and IISS (2013)

Fig. 3.16 Defense budget per soldier, 2012 (US$)

Since large troop size automatically means large budgetary needs for logistics, equipment, training, and pensions, it is useful to look at the ratio of the nominal budget spent per soldier as a rough indicator of the standard of force modernization that has been reached (Fig. 3.16). The corresponding figure for the EU states excluding Denmark, according to EDA data, in 2012 was €130,457 (US $172,396). Here, it becomes clear that China is still quite typical for the group of newly industrialized countries. Singapore, Australia, New Zealand and Japan, for example, have on average 3.6 times the amount of resources per soldier at their disposal as China. Among the BRICs, China is third, after Russia and Brazil. If China continues to further reduce troop size, while concurrently raising the defense budget, this could lead to a future situation that more closely resembles the ratio achieved by the industrialized countries.

3.6 Assessing Overall Militarization

A rise in defense spending can usually be observed in situations where the external threat level is perceived to be increasing:

No.	Category	Line of reasoning	Validity
1.	"Catch-up arguments"	Very poor foundation and decades of prior neglect make catch-up a necessity; full mechanization and beginning informationization of the PLA still not accomplished	Very high
2.	"Inflation-adjusted arguments"	Inflation rate has caused a need to steadily raise salaries which account for a huge portion of military spending	High
3.	"Economic-growth-as-priority argument"	China's military spending rises in proportion with GDP growth; its share in total government spending has even been shrinking recently	High
4.	"Palliative comparison approach"	China's spending level is normal or low in relation to others if calculated per capita/as a percentage of GDP or of government expenditures, as spending per manpower etc.	High
5.	"Strategic insecurity arguments"	Unique geostrategic challenges of China make greater military spending necessary	Quite high
6.	"New historic mission arguments"	New mission types (disaster relief, anti-piracy, international joint ops) require more resources	High

Source: Compiled from Liff and Erickson (2013: 12–15).

Fig. 3.17 Categories of Chinese justifications for increasing military spending

> External pressures are important because they affect a state's *incentives* to develop the political capacity to extract and mobilize assets from its society to support its external policies. It is no surprise, therefore, that research generally indicates that the greater the external pressures, the more highly mobilized a state is likely to be. The more highly a state mobilizes, the greater its ability to penetrate society and to extract wealth from the country at large (Tellis et al. 2000: 102, emphasis in the original).

Regional threat perceptions are thus a large factor for explaining defense budget rises across countries that can even result in regional arms races. Providing a justification for their rising spending levels is therefore a political necessity for the governments of all rising powers, and the Chinese government has often tried to explain its military build-up. The above-mentioned study of China's defense spending priorities by Liff and Erickson cites six major arguments, other than rising threat levels, which are regularly made by Chinese analysts and functionaries to justify China's current defense spending levels. These are listed in Fig. 3.17 and include a variety of points that would seem to have at least some validity to a neutral observer.

It appears that none of the points cited in Fig. 3.17 can be completely refuted on a factual basis in light of the currently available information and in view of the analyses conducted in Chaps. 2–5 of this study. In fact, Western governments sometimes make similar arguments, for instance, regarding points 5 and 6. Liff and Erickson therefore conclude that

> [o]fficial data are [...] consistent with official government statements that defence modernization (i.e. spending) is subordinate to, but coordinated with, China's principal national objective: economic development (Liff and Erickson 2013: 5).

Measuring militarization levels, however, cannot be limited to comparing defense spending. The analytical approach chosen by the authors of the *Global Militarization Index (GMI)* published annually by the Bonn International Center for

Conversion (BICC) goes one step further by systematically including another dimension and comparing military spending with other state spending priorities.[21] As this analytic perspective yields the possibility of analyzing the relative weight of military development vis-à-vis other state priorities, the results of this approach will also be discussed here.

"Militarization," as defined by the GMI, is "the relative importance of the military apparatus of a state in relation to the overall society." The index tries to establish the degree of militarization by evaluating "the behavior of a state in distributing its resources to the military sector in comparison to other sectors in society" (Grebe 2011: 7). GMI values are available for 150 countries and for all the years from 1990. The weighted index assigns values between 0 (= non-existent militarization) and 1,000 (= most extreme militarization).

According to the latest edition available to this study, Israel was the most "militarized" country included in the survey with a score of 854 out of 1,000 points, with Singapore, Syria, Russia, Jordan, Cyprus, Kuwait, Azerbaijan, Bahrain, and Saudi-Arabia among the top ten. The most highly militarized "regional security complex" remained the Middle East region, with six Middle Eastern countries among the top 10 and *all* Middle Eastern countries except Qatar among the top 40. South Korea, which was still part of the top ten group in 2010, was no longer included here in 2011, while North Korea—a country not surveyed by the GMI due to lack of hard data—was suspected to be the likely true world leader in militarization, ahead even of Israel (cf. Grebe 2012: 1–4).[22]

When looking at 2012 GMI results for the Asia Pacific "regional security complex" (or rather, "super complex"), Singapore, with a rank of 2 in the world, also scores high in a regional comparison. South Korea is one of two more highly militarized Asia Pacific countries at position 18 in the world, albeit with downward tendency due to changes in its military reserves (Grebe 2012: 3–4). Index values reached by other regional powers are surprisingly average:

> China (position 82) and India (position 71) are characterized by an average degree of militarization. Australia (position 75) has been showing a moderate degree of militarization for years. Japan's degree of militarization (position 112), however, is comparatively low due to the fact that its constitution decrees that its military spending must be limited to one percent of its Gross Domestic Product. Nevertheless, Japan has modern and well-equipped armed forces (Grebe 2012: 3).

[21] The variables included in the BICC's GMI include a country's defense expenditures as a percentage of the GDP, and in relation to health spending; the number of military and para-military personnel including reserves in relation to total population size, and in relation to the number of physicians; and the number of "heavy weapons," defined as "armored vehicles (armored personnel carriers, light tanks, main battle tanks), artillery (multiple rocket launchers, self-propelled artillery, towed artillery) above 100 mm caliber, combat aircraft (attack helicopters, fixed-wing fighter aircraft), and major fighting ships (submarines, major surface combatants above corvette size)" in relation to total population; cf. Grebe (2011: 16–17).

[22] According to the GMI editors, "There is no reliable data on military spending, number of armed forces, and number of heavy weapons systems on North Korea. This makes a calculation of its degree of militarization impossible" (Grebe 2012: 1, footnote 4).

Thus, from the analytic perspective adopted by the GMI, the only Asia Pacific countries showing unusually high levels of militarization (apart from likely world leader North Korea) are Singapore and Brunei:

> There are no obvious explanations for the high level of militarization of Singapore. The country has modern weapons systems and a well trained army. Measured against the small size of the city state and the relatively small population and the overall peaceful development, one could consider the military to be oversized. A similar pattern can be observed for Brunei (2009: position 11), which for years has shown a high level of militarization [...]. Owing to its rich oil and gas reserves, Brunei has one of the highest GDPs of developing countries. Even though at the time of writing there are no direct threats for the country, government spending for the military is very high considering the small size of the kingdom (Grebe 2011: 15).

When comparing China's level of militarization to other BRIC states, the GMI editors report a clear trend in all these countries towards rising investments in the military sector in line with sustained high economic growth rates. All BRIC states have seen considerable defense expenditure growth rates over the past decade:

> Military expenditures in Brazil between 2000 and 2009 rose by 38 percent; those of Russia by 105, of India by 67, and China by 216 percent. Part of these expenditures was channeled into comprehensive modernization programs of the respective armed forces (Grebe 2011: 13).

Nonetheless, actual "militarization" growth as defined and measured by the GMI methodology has not been found to be unusual in the cases of China, India, and Brazil, notwithstanding their high budget growth rates:

> Brazil's militarization (1990: position 66; 2000: position 79; 2009: position 76) has constantly been average—with a light tendency towards a slight decrease. The same holds true for India's militarization (1990: position 79; 2000: position 87; 2009: position 79) despite a light tendency towards an increase. China's picture (1990: position 67; 2000: position 94; 2009: position 88) is similar to the one of Brazil. Russia, due to its special position in world politics, is the only BRIC state which has a constantly high level of militarization (1992: position 1; 2000: position 7; 2009: position 5). Still, even in Russia, a direct discrimination of other sectors cannot be observed at the time of writing (Grebe 2011: 13–14).

The GMI editors conclude that along with marked rises in military spending and force modernization, all the BRIC countries have also simultaneously achieved noticeable development in the non-military segments of their societies and thus experienced an overall improvement not limited to militarization. It follows that only Russia (with a rank of 4th in the world in the 2012 GMI) shows a relatively high level of militarization when compared to the United States, which ranks only 30th in the latest edition of the index despite the fact that its defense budget is by far the world's largest. China, on the other hand, is not exceptional from this analytic perspective when compared to either Brazil or India (Grebe 2012: 2).

To sum up, all the different approaches to evaluating China's military spending levels presented and discussed in this chapter show that, as of late 2013, China's military spending is not in fact outside of the realm of the "normal." This is true from both a regional or international perspective when accounting for factors such as inflation effects, troop size, budgeting priorities of the state, as well as "militarization" as defined by the GMI. Moreover, achieving a qualitative leap in terms of informationization, training, equipment, and logistics, while simultaneously fighting corruption within the Chinese military (which has been an endemic problem for decades), will probably require an even larger level of investment than exists today. This is also the position taken by Chinese proponents of enhanced military spending such as Hu Angang and Men Honghua, who recommended in 2004

> to raise sharply the percentage of defense spending in GDP to enhance the defense capabilities. The state should ensure the armed forces are fed by the government and it should be strictly forbidden to allow the armed forces to engage in business and profit-taking activities, standardize defense spending and statistics and increase transparency. The defense spending in GDP should be raised to 2.0 % in the near terms [sic] and up to 2.5 % in the middle term (Hu and Men 2004: 33).

As the cited figures show, China's current spending levels are consistent with those recommended by these two influential analysts some 10 years ago.

3.7 Economic Constraints Influencing China's Naval Build-Up

While the prospect of having a sustainable naval build-up and achieving the financial means to maintain a powerful navy depend on China's further economic development, the latter rests on rather uncertain foundations. China's strong growth record over the past few decades notwithstanding, its development model faces dire challenges and much will depend on the success of the Chinese leadership's present reform agenda. At the moment, China remains too heavily dependent on overseas export markets for its manufactured goods and faces toughening competition from countries such as Vietnam that can offer even lower labor costs to the world market. Moreover, China's financial system is far from stable (cf. Wilson 2014; Li and Hsu 2013). Future financial or regional security crises could therefore severely impact on China's export-oriented development model. Taiwanese investment on the Chinese mainland has been one of the key factors behind China's economic miracle since the 1980s. Deteriorating cross-strait relations would therefore certainly threaten both economies. Just the same, as long as China's overseas investments and foreign trade interests continue to thrive, and the present trend towards reducing cross-strait tensions persists, China will likely further enlarge its navy's capability and reach, supporting it with the necessary funds.

Nonetheless, one certain future financial pitfall that China cannot avoid confronting is the demographic cost of its one-child policy. Decreased population growth during the past three decades has come with a high price, as the structurally aging society will soon become a heavy burden on the Chinese state budget and likely have a constraining influence on the country's military build-up:

> A rapidly aging society will inevitably generate higher welfare and social service expectations, which may exacerbate extant domestic instability. *Even if* defence spending continues to grow, internal PLA factors (such as rapidly rising equipment and personnel costs), not to mention corruption, all-but-guarantee diminishing returns and will limit improvements to overall force structure and capabilities. In short, it seems clear that the future pace and scope of China's military development will depend on the health and wealth of the nation (Liff and Erickson 2013: 22, emphasis in the original).

Indeed, in the light of many other pressing state priorities China has to tackle—among them costly long-term problems such as internal security, pollution control,[23] renewable and sustainable energy generation, and social welfare—it is conceivable that a situation will arise in which domestic reform projects force China to make a choice between "guns or walking sticks."[24]

References

Arena, Mark V. et al. 2006. Why Has the Cost of Navy Ships Risen? A Macroscopic Examination of the Trends in U.S. Naval Ship Costs Over the Past Several Decades. Santa Monica: RAND.

Birkler, John et al. 2005. Differences Between Military and Commercial Shipbuilding: Implications for the United Kingdom's Ministry of Defense. Santa Monica, CA et al: RAND. http://www.rand.org/content/dam/rand/pubs/monographs/2005/RAND_MG236.pdf. Accessed February 15, 2011.

Bussert, James C. 2003. Foreign Navies Combat System Dis-Integration: Problem may be more widespread than officials acknowledge. *Signal* online. March 2003. http://www.afcea.org/content/?q=node/275. Accessed June 14, 2014.

Canadian Forces. n.y. *C.F. 'O' Class Submarines - Submarine Principles Training Notebook.* Electronic version available at Historic Naval Ships Association Website. http://www.hnsa.org/doc/oberon/principles/index.htm. Accessed January 22, 2014.

Chen, Yujie 陈宇杰. 2003. Eluosi haijun wuqi fazhan fenxi (俄罗斯海军武器发展分析 | An analysis of Russian naval weapons development). *Xiandai junshi* (现代军事) 2003/01: 19-21.

'China to invest heavily in air pollution treatment'. 2013. *Xinhua* online. December 19, 2013. http://news.xinhuanet.com/english/china/2013-12/19/c_125887562.htm. Accessed March 7, 2014.

'China's Military Rise. The dragon's new teeth: A rare look inside the world's biggest military expansion'. 2012. *The Economist* (el. ed.), April 07, 2012. http://www.economist.com/node/21552193. Accessed April 18, 2014.

Cole, Bernard D. 2010. *The Great Wall at Sea. China's Navy in the Twenty-First Century.* 2nd ed. Annapolis: Naval Institute Press.

Defense News Top 100 database. 2012. Defense News. http://special.defensenews.com/top-100/charts/rank_2011.php?c=FEA&s=T1C. Accessed Sept 17, 2013.

[23] According to a World Bank report, the physical and economic cost impact of air and water pollution in China already amounted to a staggering 5.8 % of the GDP in 2007. Xinhua reported on Dec. 19, 2013 that China would have to invest ca. 1.75 trillion yuan (ca. 290 billion U.S. dollars) for its air pollution treatment plan from 2013 to 2017 alone (cf. World Bank 2007 and 'China to invest heavily in air pollution treatment' 2013).

[24] Cf. the insightful discussion in 'China's Military Rise' (2012).

Dokkum, Klaas van. 2011. *Ship Knowledge: Ship Design, Construction and Operation.* 7th ed. Enkhuizen: DOKMAR.

EDA. 2013. Defence Data of EDA in 2011. EDA Defence Data Portal. http://www.eda.europa.eu/info-hub/defence-data-portal/EDA/year/2011. Accessed December 1, 2013.

Friedman, Norman. 2006. *The Naval Institute Guide to World Naval Weapon Systems.* 5th ed. Annapolis: Naval Institute Press.

Friedman, Norman. 2009. *Network-Centric Warfare: How Navies Learned to Fight Smarter Through Three World Wars.* Annapolis: Naval Institute Press.

Grebe, Jan. 2011. The Global Militarization Index (GMI) – A tool for evaluating development orientation of states as well as regional developments. BICC Occasional Paper, February 2011. Bonn: BICC, http://www.bicc.de/old-site/uploads/pdf/publications/papers/occ_paper_07/occasional_paper_VII_02_11_eng.pdf. Accessed September 7, 2013.

Grebe, Jan. 2012. UPDATE: BICC's Global Militarization Index (GMI) 2012: Factsheet. Bonn International Center for Conversion (BICC). http://www.bicc.de/old-site/uploads/gmi/pdf/Update_GMI_2012%20Fact%20Sheet_e.pdf. Accessed September 7, 2013.

Grevatt, Jon. 2011. Philippines Eyes Submarine Acquisition by 2020. *Jane's Defence Weekly* (el. ed.), May 17, 2011. http://www.janes.com.

Grevatt, Jon. 2013. Briefing: Closing the Gaps. *Jane's Defence Weekly* (el. ed.), July 5, 2013. http://www.janes.com.

Guzelyte, Silvija. 2013. Additional Defence Data 2011. European Defence Agency (EDA), Brussels, July 2013, http://www.eda.europa.eu/docs/default-source/documents/additional-defence-data-2011.pdf. Accessed December 1, 2013.

Hardy, James and Alex Pape. 2013. Bangladesh Announces Plans to Acquire Submarines. *Jane's Navy International* (el. ed.), January 25, 2013. http://www.janes.com.

Hu, Angang 胡鞍钢 and Honghua Men 门洪华. 2004. The Rising of Modern China: Comprehensive National Power and Grand Strategy. Paper presented at the International Conference of the Korea Institute of International Economic Policy (KIEP), 19–20 March 2004, in Seoul, Korea. http://irchina.org/en/xueren/china/pdf/mhh3.pdf. Accessed March 2, 2011.

IISS. 2013. *The Military Balance 2013.* London: Routledge for the International Institute for Strategic Studies.

Jane's Fighting Ships (el. ed.). 2014. http://www.janes.com.

Jane's World Navies (el. ed.). 2013. http://www.janes.com.

Kaplan, Robert D. 2005. How We Would Fight China. *The Atlantic,* June 2005. http://www.theatlantic.com/magazine/archive/2005/06/how-we-would-fight-china/303959/. Accessed December 6, 2013.

Kiselycznyk, Michael and Phillip C. Saunders. 2010. Assessing Chinese Military Transparency. *China Strategic Perspectives* No. 1. Institute for National Strategic Studies, Washington, D.C.: National Defense University Press. http://www.ndu.edu/press/lib/pdf/china-perspectives/ChinaPerspectives-1.pdf. Accessed March 7, 2014.

Lague, David and Charlie Zhu. 2012. China Builds Its Own Military-Industrial Complex. Reuters UK, September 17, 2012. http://uk.reuters.com/article/2012/09/17/uk-china-defence-idUKBRE88F0G720120917. Accessed October 9, 2013.

Lewis, J. A. C. 2008. Interview: Jean-Marie Poimboeuf, Chief Executive Officer, DCNS. *Jane's Defence Weekly* (el. ed.), February 06, 2008. http://www.janes.com.

Li, Jianjun and Sara Hsu. 2013. Shadow Banking in China: Institutional Risks. PERI Working Paper No. 334, University of Massachusetts. Amherst: Political Economy Research Institute (PERI). http://www.peri.umass.edu/fileadmin/pdf/working_papers/working_papers_301-350/WP334.pdf. Accessed May 20, 2014.

Li, Nan and Christopher Weuve. 2011. Chinese Aircraft Carrier Development: The Next Phase. In *Chinese Aerospace Power. Evolving Maritime Roles,* eds. Andrew S. Erickson and Lyle J. Goldstein, 209-224. Annapolis: Naval Institute Press.

Liff, Adam P. and Andrew S. Erickson. 2013. Demystifying China's Defence Spending: Less Mysterious in the Aggregate. *The China Quarterly*. doi:http://dx.doi.org/10.1017/S0305741013000295.

Lo, Ping-cheung 羅秉祥. 2012. The Art of War Corpus and Chinese Just War Ethics Past and Present. *Journal of Religious Ethics* Vol. 40.3: 404-446.

Lü, Jianwei 吕建伟, Zhonghua Liu 刘中华 and Xiaojun Zhong 钟小军. 2008. Haijun jianchuan yanzhi de jishu fengxian laiyuan jiqi houguo fenxi (海军舰船研制的技术风险来源及其后果分析 | Analysis of Technical Risk Origin & Consequence in Development of Naval Vessel [sic]). *Zhongguo jianchuan yanjiu* (Chinese Journal of Ship Research) Vol. 3 No. 2, April 2008: 31-34.

Mahan, Alfred Thayer. 1987. *The Influence of Sea Power Upon History 1660-1783*. Reprint of the 5th ed. (1894). New York: Dover Publications.

McTague, Tom. 2014. Time to pay your way: France and Germany told to 'urgently' increase military spending as the price of Nato membership. *Daily Mail* online, September 4, 2014. http://www.dailymail.co.uk/news/article-2743710/Time-pay-way-France-Germany-told-urgently-increase-military-spending-price-Nato-membership.html. Accessed January 26, 2015.

Miller, David. 2002. *The Illustrated Directory of Submarines of the World*. London: Salamander Books.

Mommsen, Klaus. 2005. Die Marinen der Welt – eine Betrachtung zur Lage der Flotten (The world's navies: observations on their fleet situations). In *Weyers Flottentaschenbuch 2005-2007* (Warships of the World Fleet Handbook), ed. Werner Globke, xix-xxxi. Bonn: Bernard & Graefe.

Neuman, Stephanie G. 2009. Power, Influence, and Hierarchy: Defense Industries in a Unipolar World. In *The Modern Defense Industry. Political, Economic, and Technological Issues*, ed. Richard A. Bitzinger, 60-94. Santa Barbara, Cal. et al.: Praeger Security International.

Phipps, Gavin and James Hardy. 2014. Taiwan says local submarine programme back on drawing board. *Jane's Defence Weekly* (el. ed.), October 2, 2014. http://www.janes.com.

Ping, Kefu 平可夫 (a.k.a. Pinkov, Andrei, a.k.a. Chang, Andrei). 2010. *Zhongguo zhizao hangkong mujian* (China constructs aircraft carriers | 中國製造航空母艦). Hong Kong: Kanwa Press.

Pugh, Philip. 1986. *The Cost of Seapower: The Influence of Money on Naval Affairs from 1815 to the Present Day*. London: Conway Maritime Press.

Sakhuja, Vijay. 2011. *Asian Maritime Power in the 21st Century: Strategic Transactions – China, India and Southeast Asia*. Singapore: Pentagon Press.

Scott, Richard. 2011. Conventional wisdom. *Jane's Defence Weekly* (el. ed.), April 7, 2011. http://www.janes.com.

SIPRI Military Expenditure Database. 2014. Stockholm International Peace Research Institute (SIPRI). http://www.sipri.org/research/armaments/milex/milex_database. Accessed April 22, 2014.

Stålenheim, Petter and Eamon Surry. 2006. Transparency in Military Expenditure and Arms Production. Paper presented at the Xiangshan Forum 22-24 Oct 2006. http://www.sipri.org/research/armaments/production/publications/unpubl_aprod/xiangshan. Accessed February 13, 2011.

Surry, Eamon. 2007. An Estimate of the Value of Chinese Arms Production. Research Note presented at the 11th Annual Conference on Economics and Security, University of the West of England and Economists for Peace and Security, 5-7 July, 2007. http://www.carecon.org.uk/Conferences/Conf2007/Papers/Surry.pdf. Accessed October 9, 2013.

Tellis, Ashley J. et al. 2000. *Measuring National Power in the Postindustrial Age*. Santa Monica: RAND.

'The Cost of Weapons. Defence Spending in a Time of Austerity'. *The Economist*, August 20, 2010: 18-20.

Till, Geoffrey. 2009. *Seapower: A Guide for the Twenty-First Century.* 2nd ed. London and New York: Routledge.

Todd, Daniel. 1985. *The World Shipbuilding Industry.* London and Sydney: Croom Helm.

Todd, Daniel and Michael Lindberg. 1996. *Navies and Shipbuilding Industries: The Strained Symbiosis.* Westport, Conn. et al.: Praeger.

Tsouras, Peter. 1986. Soviet Naval Tradition. In *The Soviet Navy. Strengths and Liabilities,* eds. Bruce W. Watson and Susan M. Watson, 3-25. Boulder, Col. and London: Westview Press and Arms and Armour Press.

UNCTAD. 2014. Liner shipping connectivity index. http://data.worldbank.org/indicator/IS.SHP.GCNW.XQ. Accessed February 23, 2015.

Wertheim, Eric. 2007. *The Naval Institute Guide to Combat Fleets of the World.* 15th ed. Annapolis: Naval Institute Press.

Wertheim, Eric. 2013. *The Naval Institute Guide to Combat Fleets of the World.* 16th ed. Annapolis: Naval Institute Press.

Wilson, Harry. 2014. Currency crisis at Chinese banks 'could trigger global meltdown'", in: *The Telegraph* online, 1 February 2014, http://www.telegraph.co.uk/finance/newsbysector/banksandfinance/10612451/Currency-crisis-at-Chinese-banks-could-trigger-global-meltdown.html (last checked 15 April 2014).

World Bank. 2007. *The Cost of Pollution in China: Economic Estimates of Physical Damages.* Conference Edition. Washington, D.C.: The World Bank. http://go.worldbank.org/FFCJVBTP40. Accessed March 7, 2014.

World Bank. 2014. Container port traffic. http://data.worldbank.org/indicator/IS.SHP.GOOD.TU/countries. Accessed February 23, 2015.

Chapter 4
Producing Naval Weapon Systems: The Industrial Basis of Naval Power

> 我们也必须清醒地认识到,世界上有些最先进的技术是买不来的。
> *"We must understand clearly that some of the world's most advanced technology is not for sale."*
> —Jiang Zemin, Chinas then State President, in 1995.
> My transl. Cited from Jiang (1995).

In order to analyze the economic basis of rising naval militarization in more detail, the following chapter will discuss the influence of industrial processes and structures on warship production. Toward this end, the development of the industrial production base, as well as the influence of transnational industrial and technological partnerships will be addressed. In this context, it is further necessary to evaluate the effects that the U.S./EU arms embargo has had so far on Chinese naval shipbuilding. In the light of ongoing discussions within the EU about lifting the embargo, it is particularly relevant to assess its impact on slowing down China's current naval build-up, and to analyze how China has attempted to compensate for these externally imposed limitations by means of both indigenous innovation and strategic industry alliances with export partners such as Russia, Ukraine, and Pakistan.

The importance of the industrial aspects of naval power, which ultimately encompasses effective logistics, can hardly be over-emphasized. Logistics is not only one decisive factor for winning wars; it may be *the* single most decisive factor, all things being equal. As British military historian John Keegan found in his study of the history of warfare, industrial output and powerful logistics were the major reasons behind the Union's victory in the American Civil War and the U.S. victory over Germany and Japan in World War II:

> It was America's industry that overwhelmed its German and Japanese enemies, though only because American shipyards also supplied the transportation to move it. [...] on average the United States, at the height of the Liberty-building programme, was launching three ships a day. Germany's U-boats, even before they were defeated by the introduction of

S. Kirchberger, *Assessing China's Naval Power*, Global Power Shift,
DOI 10.1007/978-3-662-47127-2_4

> long-range aircraft and escort aircraft-carriers, both American-built, to the Battle of the Atlantic, could not sink such an output at the rate losses were replaced (Keegan 2004: 313–314).

After showing how air cover effectively inhibited submarine assaults on transport ships in the regions covered, Keegan concludes that, in general, industrial capacity must indeed be considered the decisive factor in conventional wars.

In spite of the financial burden placed by militarization on the national economy described in Sect. 3.5, it can be observed that in the field of arms production, governments often accept economically detrimental effects for politico-strategic reasons. In a world increasingly dominated by economic thinking, it is sometimes necessary to remind oneself that there are indeed, as Hoyt writes, "significant differences between the production of military equipment, intended for defense of national territory, sovereignty, and interests, and the production of transistor radios, basketball shoes, and other civilian goods" (Hoyt 2007: 13). The international arms trade is therefore often highly political in nature, while pure economic reasoning rarely lies at the root of arms trade decisions, at least in the case of large-scale defense projects, which are usually negotiated directly between governments in so-called government-to-government ("G2G") or business-to-government ("B2G") talks. Often the relevant economic actors are state-owned or partly state-owned companies and therefore not subject to the same level of market pressures as typical producers of consumer goods. Many governments opt to retain a high degree of control over the development and trade of their indigenous arms technologies and choose their trading counterparts very carefully, with a definite view to the respective strategic implications of the deal, while very often limiting the actual amount of technical data being shared even with close strategic partners (Mayer 2012: 289–290). This is true for all major suppliers of advanced technologies, even though involuntary knowledge transfer cannot ever be entirely controlled or inhibited:

> States often attempt to maintain control over production and dissemination of critical defense technologies—the Soviets produced special low-technology versions of their weapons for export, and the US insists on end-user restrictions. Nevertheless, defense technology diffuses, through licensed production, espionage, reverse-engineering, or adaptation of dual-use goods based on related civilian technologies (Hoyt 2007: 13).

Sometimes used weapons are given away by leading arms producers to developing countries for little or no payment because this is seen as politically advantageous by the donor country (Krishnan 2008: 31). It seems that, overall, economic considerations are rarely the main driver behind strategic decisions in the global arms trade system, while traditional and long-term security partnerships between arms sellers and recipients, geostrategic competitor relationships between neighboring countries, and other security-driven incentives seem to carry more relative weight. The following discussion of China's naval progress will thus adopt a grassroots perspective of warship production and its inherent technological challenges—a perspective pertaining to what Cheung calls "the political economy of security"

(2009: ix). This term is quite fitting because of the prevalence of power relations and security concerns in the economic field of arms production.

4.1 Analyzing Production Capabilities

Generally, when assessing the quality of a country's military-technological base in any field, it is necessary to consider the country's skills in five areas:

- "Whether a county has indigenous production capabilities in the technology area;
- whether a country has transplanted production capabilities deriving from its status as a host for foreign-owned facilities;
- whether a country has trade access to foreign capabilities in a given technology area;
- whether a country engages in basic and applied research and developmental work, even if not in commercial production;
- whether a country undertakes theoretical research in the technology area in question" (Tellis et al. 2000: 64).

Since these kinds of capabilities are distributed quite unevenly across countries within the current world order, a system of "tiers" on a so-called "ladder of production" can be constructed to facilitate comparative analyses among a variety of countries. As shown in Fig. 4.1, Krause (1992) has defined five descending tiers for countries according to the measure of their innovation capability in arms production.

This system, among others, was used by Hoyt (2007: 9) in his comparative study of defense industrial policies in India, Iraq, and Israel. It is useful because it allows for a separate weighing of different technology fields, and gives relatively clear-cut criteria for countries reaching a certain level.

According to this system, a second-tier producer can "produce at (or near) the technological frontier" and is capable of *adapting* existing technology, while a first-tier producer is able to *innovate* at the state-of-the-art level. As I will discuss in this

First-tier suppliers...	"innovate at the technological frontier"
Second-tier suppliers...	"produce (via the transfer of capacities) weapons at the technological frontier and adapt them to specific market needs"
Third-tier producers...	"copy and reproduce existing technologies (via transfer of design), but do not capture the underlying process of innovation or adaptation"
Fourth-tier ("strong") customers...	"obtain (via material transfers) and use modern weapons"
Fifth-tier ("weak") customers...	"either obtain modern weapons and cannot use them, or do not even obtain them"

Source: Adapted from Krause (1992: 31–32)

Fig. 4.1 The hierarchy of arms production

and the following chapters, China must still be considered here a third-tier weapons producing country in many naval technology fields. In fact, according to Bitzinger, "[...] the technology gap between China's defence industry and the leading Western arms producers remains significant in several critical areas" (Bitzinger 2011: 15).[1] It seems that while China can definitely 'copy and reproduce (*apply*) existing technologies,' it does not in a general sense 'capture the underlying process of innovation or adaptation' in most critical naval technology fields. Otherwise, there would be no need for China to import advanced military technologies as complete weapon systems such as the ex-Soviet *Sovremenny* class destroyers, or to produce gas turbines, naval diesel engines, or jet engines under license agreements with Ukraine and other countries, as it has been doing in recent years (cf. Johnson 2013c). Nonetheless, in a few select technology fields, such as missile systems, China is probably able to produce at the technological frontier, which is a second-tier capability. In some niche competencies, China may even be a first-tier producer, e.g. in ballistic anti-ship missiles (ASBM). The true quality of Chinese naval electronics or its most advanced missile systems is hard to determine, however, given the lack of reliable performance data. Be that as it may, the available data is sufficient to conclude that China still has some obstacles to overcome to generally be included in the second-tier group of arms producers in most of the strategically important critical naval technology areas (cf. Sect. 4.5).

4.2 Autarchy and Partnerships in the Naval Weapons Industry

One of the main challenges for building and sustaining a high level of sea power is the development of matching indigenous naval shipbuilding, marine equipment, and naval defense electronics industries.[2] It is a striking commonality between all countries with blue-water type navies that they are able to maintain and develop their naval forces through an indigenous naval shipbuilding capability (cf. Todd and Lindberg 1996: 86). In some cases, these industries are also active in the naval weapon systems export market catering to countries without this capability. In general, naval export relations are mostly established between two countries within the wider context of a pre-existing military-strategic partnership or other security cooperation, and often they are accompanied by a range of further cooperation projects in other defense-related areas and/or strategic realms such as critical

[1] According to Bitzinger, "despite more than [...] at least a decade of aggressive acquisitions on the part of the PLA, the bulk of the Chinese military remains relatively backward. [...] Even many of the PLA's most recently acquired systems, such as the J-10 fighter jet, the *Yuan*-class submarine, and the *Luyang*-II destroyer, [...] are basically 1980s-era weapons systems" (Bitzinger 2011: 14, italics in the original).

[2] For an in-depth analysis of the general nature of the "strained symbiosis" between navies and domestic shipbuilders, see Todd and Lindberg (1996).

infrastructure development, (nuclear) energy cooperation, the exchange of political support in international organizations, and so on.[3]

One major incentive for rising naval powers to rely on indigenous industries rather than the export market, apart from autarchy considerations, can be the fact that the most advanced offensive power-projection weapons such as nuclear-powered submarines and aircraft carriers are generally unavailable on the export market. China's former state president Jiang Zemin alluded to precisely this circumstance in the quotation cited at the beginning of the present chapter. As a rule, such systems must be indigenously developed and built by the respective countries themselves, even if (limited) technical assistance can sometimes be obtained from a more advanced country. China has been no exception to this rule. After the Sino-Soviet split, which terminated all Soviet assistance in weapons development from 1960, China slowly developed the capacity to design and build nuclear-powered submarines and conventionally powered surface vessels up to destroyer size, and also successfully developed relevant subsystem supplier industries:

> For building ships, these industries included diesel engines, naval electronics, computer hardware and software, and radios. For the aircraft that supported seagoing vessels, these industries included aircraft engines, avionics, and more modern wing designs. Over time China's indigenous capabilities have increased to the point where it can supply the PLAN with almost all of its basic equipment needs, with advanced technology still coming from abroad, as needed (Bussert and Elleman 2011: 9).

Among the most ambitious current Chinese shipbuilding projects is a conventionally (steam) powered aircraft carrier program. As a starting point for the prototype, China was able to purchase an empty hull from Ukraine. China has been finishing and outfitting this first ship largely on its own in a lengthy process that included technical assistance from Ukraine and Russia.[4]

Compared with fellow BRIC countries Brazil and India, China has been much more restricted in its choice of export partners due to the Western arms embargo. As a result, it does not receive the same level of technical assistance from the most advanced shipbuilding nations. This is especially apparent with regard to naval propulsion: Brazil is assisted by France in the development of its nuclear submarine, and India has received Italian assistance for the propulsion system of its indigenous carrier. India, moreover, is allowed to import U.S. gas turbines, and has even recently leased a complete nuclear attack submarine (SSN) from Russia. It has thus benefited from the best of all possible worlds technologically. Reports from 2013 indicate that India plans to lease a second SSN from Russia, and has succeeded in activating the reactor of its first indigenously built SSBN, which received Russian assistance as well (Bedi and Grevatt 2013). As China is the

[3] These observations are based on systematic studies on naval export strategies of various supplier countries conducted by the author while working in the shipbuilding industry from 2007 to 2010.

[4] For an interesting discussion of technical issues during the completion of this ex-Soviet carrier hull, see Saw (2008: 33–35).

most strongly limited recipient of foreign technological assistance, India could thus be seen as its counterpoint within the BRIC countries. South Africa, which is included in the "BRICS" group, has also received extensive naval upgrades by importing state-of-the-art frigates and submarines from Germany.

Many of China's regional neighbors rely heavily on transfers of U.S. naval technology in their own shipbuilding efforts. Japan, which still keeps to its self-imposed constitutional constraint of limiting military modernization, has nevertheless adopted very large U.S. vessel designs for its indigenously built destroyers, which are practically cruisers considering their displacement of ca. 10,000 t. Quite early on, Japan also developed large LHD platforms, officially classified as "helicopter-carrying destroyers," which resemble aircraft carriers and are perhaps better called "helicopter carriers" (cf. Todd and Lindberg 1996: 18). Japan generally uses U.S. vessel designs centered around the *AEGIS* combat direction system for its large surface combatants, and a similar U.S. technological influence also exists in South Korea, especially with regard to the *KDX-III* class *AEGIS*-equipped destroyer program, and to some degree also in Taiwan, which has additionally received modern French-built *La Fayette* class frigates as an export order. South Korea, on the other hand, has been building a massive conventional submarine capability from scratch by importing advanced German diesel submarine technology in several batches. Even the small city-state of Singapore, which was already described as an unusually highly militarized country within its regional context (cf. Sect. 3.6), has imported French *Formidable* class frigates based on the *La Fayette* design, and developed a submarine capability that is currently being enhanced with new additions of modern German submarines. Other countries in Asia using Western submarine technology from France and/or Germany include Pakistan, India, and Malaysia.

China, on the other hand, has exclusively relied on, and is effectively limited to, older Soviet era designs as the basis of its indigenous naval development effort, both in its surface and submarine fleets. The only area where technological cooperation with Western and Japanese or South Korean partners remains uninhibited is commercial shipbuilding. The development and characteristics of this sector, and possible spillover effects to naval shipbuilding, will be discussed in Sect. 4.4.

4.3 The Role of Combat System Suppliers in Naval Shipbuilding

It is usually assumed that technological innovation and global technological leadership can best be achieved by privately owned companies that are active in a highly competitive market environment and can thus make efficient use of resources and opportunities (cf. Schumpeter 2006 [1912]: 162–164). State planning, on the other hand, is often seen as detrimental to successful innovation because it is supply-driven, not demand-driven. As the economist Peter Drucker once stated,

> 'Planning' as the term is commonly understood is actually incompatible with an entrepreneurial society and economy. Innovation does indeed need to be purposeful and entrepreneurship has to be managed. But innovation, almost by definition, has to be decentralized, *ad hoc*, autonomous, specific, and micro-economic (Drucker 1993: 255, italics in the original).

In his empirically based analysis of the interconnection between entrepreneurship, market forces, and innovation, Drucker emphasizes the impossibility of producing high-tech innovation by command. Rather, an overall innovation-friendly environment encompassing the respective social and economic institutions is deemed necessary:

> [...] to have 'high-tech' entrepreneurship alone without its being embedded in a broad entrepreneurial economy of 'no-tech,' 'low-tech,' and 'middle-tech,' is like having a mountaintop without the mountain (Drucker 1993: 256).

Another factor often assumed to be relevant for successful technological innovation is company size—not only in terms of the available capital for research and development, but also because of the enhanced capability for inter-industry networking among the "big players." As Todd observed regarding the role of industry leaders in his 1985 study *The World Shipbuilding Industry*:

> They dominate an industry by virtue of their large size and a high degree of interindustry linkages. By the latter is meant the web of supply and demand ties which connect the lead firms to all the principal suppliers of inputs into the industry as well as the major consumers of its products (Todd 1985: 146).

What is meant by this "domination"? Usually, such leading companies tend to have higher growth rates, larger productivity, and expanding market shares (Todd 1985: 146). Company size is also a relevant factor for innovation because the big players are less at risk from unsuccessful experiments than smaller companies, and technological innovation is inherently very risky:

> Large organisations have the wherewithal to cultivate R&D and to implement innovations on a trial basis until they overcome market resistance as well as surmount design and production flaws (Todd 1985: 153).

State-run innovation conducted by de-facto monopolists, by contrast, leads to the problem of over-bureaucratization, a lack of incentive for excellence, operational inefficiency, and a lack of pressure to be competitive in hiring the best personnel. It is noteworthy that these are exactly the problems that plagued the Chinese defense R&D system during the 1980s and 1990s (cf. Cheung 2009: 89–92). In addition, Drucker would probably emphasize the human workforce related factors. The lack of creative thinking and opportunity-driven autonomous decision-making, combined with the relative slowness that is typical of the employees of large bureaucratic state-owned enterprises to react to market changes, does not further innovation as he understands it.[5] As a recent study of China as an emerging "knowledge power"

[5] Note that Drucker's understanding of "innovation" is not limited to the development of specific products or technological solutions, but encompasses social innovations, procedural innovations, and so on (Drucker 1993: 31).

pointed out, "governments cannot simply 'produce' Schumpeterian innovations. [...] the most important role is played by individual inventors and entrepreneurs who develop bright visions and accept to take huge risks" (Mayer 2012: 294).

This assumption is relevant when analyzing China's chances for technological innovation, for even though market mechanisms were gradually introduced into the Chinese economy beginning from 1980, a large portion of the strategically relevant segments of the defense economy is still dominated by giant state-owned enterprises under the direct control of China's central government. Furthermore, China's social and political institutions have not seen as much change as some sectors of the Chinese economy, with the Leninist party-state's command structures still firmly in place behind closed doors.[6]

From an industrial viewpoint, the potential for successful military modernization seems to be greatest when the three major arms industry related "constituencies" identified by Cheung, which constantly vie for influence, authority, and dominance, are in alignment. These are: the end user of the weapons (i.e. the military); the creator/producer (i.e. the defense scientific and industrial community); and the regulator (meaning the civilian authorities responsible for procurement). In China, conflicts of interest between these groups long inhibited efficient military industrial modernization. New challenges during the mid-1990s, however, finally brought all three more and more into accord. This close alignment has been in place since the late 1990s and facilitated remarkable progress compared with the earlier periods (Cheung 2009: 17).

If the capacity to innovate at the high-tech frontier is indeed greater at large, privately owned companies operating within a competitive market environment, then an analysis of the geographic distribution pattern of the leading global defense industry players, especially those considered to be leaders in the field of defense electronics, should prove especially instructive. This study has found that there is in fact an extremely skewed global dominance pattern of very few very large U.S./EU-based defense companies (see Fig. 4.2). This pattern can be assumed to have a major impact on the technological innovation capabilities of both the members of these defense industry networks and those outside them. The U.S. share is clearly dominant, especially within the strategically relevant field of defense electronics. This is an important factor because integration into the leading echelon of defense industries can be conceptualized as a form of power. Neuman (2009: 66) calls this phenomenon a "unipolar global defense-industrial sector."

For China, the consequences of being excluded from trading with virtually all the major defense industry players except those from Russia due to the U.S./EU arms embargo are quite severe, not only in terms of the procurement of resources

[6] For a thorough discussion of this multi-faceted yet pervasive phenomenon, see McGregor (2010). More detailed descriptions on the concrete workings of the Party command system within various large organizations were given, among others, by emigré cadres Wu Guoguang (1994, 1995) and Yan Huai (1995).

Rank by aggregate revenue	Country	No. of top 50 companies	No. of top 100 companies	Aggregate of defense-related revenue of all companies (in million US$)	% share of total defense-related revenue of top 100 companies (100 % = 414,304 million)	NATO and/or EU country shares (%)
1	USA	29	45	261,115.00	63.0	63.0
2	UK	4	10	46,479.20	11.2	11.2
3	France	3	5	18,830.40	4.5	4.5
4	Netherlands	1	1	16,092.90	3.9	3.9
5	Italy	1	2	15,670.50	3.8	3.8
6	Russia	3	7	12,626.00	3.0	–
7	Japan	2	6	10,405.20	2.5	–
8	Israel	3	4	7,649.50	1.8	–
9	Germany	1	3	5,235.40	1.3	1.3
10	India	1	2	3,712.70	0.9	–
11	Sweden	1	1	3,258.90	0.8	0.8
12	South Korea	0	3	2,813.70	0.7	–
13	Norway	0	2	2,010.00	0.5	0.5
14	Singapore	1	1	1,955.20	0.5	–
15	Turkey	0	2	1,586.20	0.4	0.4
16	Canada	0	1	904.5	0.2	0.2
17	Brazil	0	1	866.8	0.2	–
18	Switzerland	0	1	844.7	0.2	–
19	Finland	0	1	779.5	0.2	0.2
20	Ireland	0	1	758	0.2	0.2
21	Spain	0	1	709.7	0.2	0.2
Total	21	50	100	414,304.00	100.0	90.1

Source: Calculated using data from the Defense News Top 100 database (2012). Rounded figures were used

Fig. 4.2 NATO/EU dominance in global arms production as reflected in the geographic distribution pattern of the world's top 100 defense companies, 2012

and access to high technology, but also in terms of military-to-military diplomacy and alliance building. Here, arms sales can be a valuable asset:

> [T]he global defense-industrial sector is a remarkably accurate indicator of the stratification of power in the post-cold war international system. As in the overall system, the United States plays the dominant role in the defense sector as well. It is a comparative advantage used by U.S. policymakers to influence the foreign policy behavior of other states. Curiously, as a policy tool, the defense-industrial sector has hitherto received relatively little scrutiny, even though it not only reflects the international order but often provides the United States with considerable leverage in it (Neuman 2009: 60–61).

Neuman goes on to conclude that "[t]he defense-industrial sector is [. . .] a powerful, if less recognized, diplomatic tool in the United States' political arsenal" (Neuman 2009: 61). Personal observations as well as various interviews with executives in the Western European defense industries cause me to concur. The same could be said about the countless U.S. military-to-military relationships formed with many countries around the world, partners and strategic rivals alike.[7]

[7] Cf. 'Military-to-Military Relationships: The Ties that Bind' (2011).

4.3.1 Regional Distribution Patterns of Leading Arms Producing Industries

What kind of consequences does this capacity for arms-sale diplomacy have for the U.S. and its military allies? First of all, U.S. producers have steadily improved their accumulated market share in the global arms trade. During 1999–2006, U.S.-based companies held a share of more than 40 % of the total value of global arms sales, with Russia and the United Kingdom in second and third places, respectively, at 15 % and 14 % (Neuman 2009: 66).

This share has since risen markedly:

> In 2006, the United States alone exported over half ($14 billion) of the world's military equipment and services [. . .]. Seeking to rearm Iraq and Afghanistan, contain North Korea and Iran, cement relations with former Soviet allies, and sustain a war on terrorism, the U.S. reportedly agreed in 2008 to sell about $32 billion in weapons and military equipment to governments in most parts of the world: the Middle East, North Africa, Asia, Latin America, Europe, and Canada. In addition, the number of countries relying on the U.S. as a primary source of major weapons systems grew with Argentina, Azerbaijan, Brazil, Georgia, India, Iraq, Morocco and Pakistan among the recent additions. If one can speak of a more globalized arms trade, it is U.S. arms that have become most globalized (Neuman 2009: 66).

Using SIPRI data, Neuman found that "in 2006, 41 U.S. companies accounted for 63 percent of the total arms sales of the world's 100 largest defense industries," while in the same year, "U.S. military R&D spending ($72 billion) was five times that of all Europe" (Neuman 2009: 66, 69). In 2014, the situation does not seem to have changed in any fundamental way. Looking at the data from 2012, a similar international distribution of revenue shares from defense sales among the top 100 largest industries could be observed by the present author.

As the table in Fig. 4.2 illustrates, the annual list provided by the *Defense News* of the top 100 defense companies according to their defense-related revenue shares as of 2012 shows a clear dominance of large U.S.- and EU-based companies. If non-EU NATO member countries and non-European U.S. strategic allies such as Japan, South Korea, Singapore and Israel are added to this picture, the situation becomes even more lopsided. All the top 50 companies are headquartered in only 12 different countries, which, with the sole exception of Russia, are either NATO and/or EU members, or at least close military allies of the U.S. Within the top 100, 9 further countries and thus 21 in all are represented. The regional distribution pattern is extremely uneven, however, because a total of 45 among the top 100 companies are U.S.-based. The relation is even more skewed when the aggregate revenues of these companies are considered in relation to this ratio. According to their combined share of total defense-related revenues of the top 100, the 45 U.S.-based companies account for 63 % of total defense-related revenues, a staggering sum of US$261 billion. No other country's aggregate share comes even close to this level, with the UK (10 % of companies and 11.2 % of revenues) coming in a distant second.

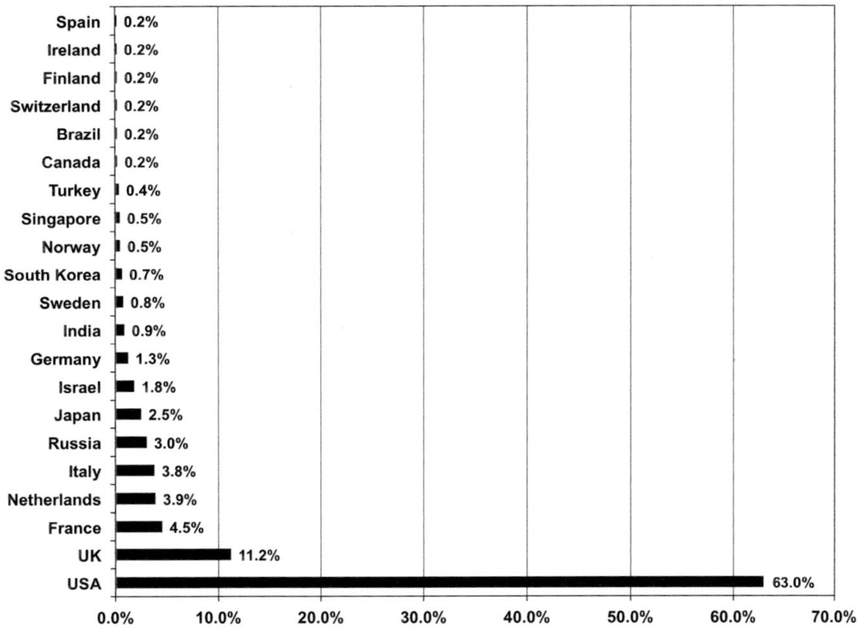

*100 % = US$414.3 billion

Source: Calculations made using data from the Defense News Top 100 database (2012)

Fig. 4.3 Percent share of aggregate defense revenues of "Top 100" defense companies by country, 2012

The graph in Fig. 4.3 depicts the unevenness of this geographic distribution pattern.

As the table in Fig. 4.2 shows, a group of only 13 NATO and/or EU member states accounted for 75 % of the top 100 companies and 90.1 % of aggregate total defense-related revenue in 2012. If U.S. strategic allies Japan, South Korea, Singapore and Israel are added, these companies represent 89 % of the top 100, and their aggregate revenue becomes an astounding 95.6 %. By contrast, the combined share of BRIC countries Russia, India and Brazil was only 10 % of the top 100 companies and a mere 4.2 % of total aggregate revenue.

It must be noted that China is absent from the list due to the intransparency of its state-controlled military-industrial complex, which does not make defense revenue data publicly available. Estimating the respective figures for China's defense industries is thus fraught with analytical difficulties. In 2007, SIPRI analyst Eamon Surry tried to estimate sales figures based on data of the official top 500 enterprises in China, which had been compiled by the Chinese CEC/CEDA since 2002 and includes the 11 major defense-related industry holdings. He could deduct from other official data that the average share of defense-related sales of all these companies must have accounted for ca. 20–35 % of the total. Using his own cross-checked estimations to make up for missing data, Surry arrived at an estimate of ca. RMB 63 to 110.2 billion (US$7.6–13.3 billion) in aggregate defense-related sales for 10 of the 11 major Chinese weapons producing companies for the year 2003 (Surry 2007: 5–9, 12–13). However, the one enterprise excluded from Surry's estimation, China Electronics Technology Group Corp. (CETC), is perhaps the most likely to eventually become a counterpart to the major Western "system houses" such as defense electronics heavyweights Lockheed Martin and Raytheon (according to its product portfolio) and could thus become a large contributor to China's total defense-related sales in the future. As invaluable as the pioneering effort made by Surry undoubtedly is, a portion of the phenomenon under study here unfortunately remains in the dark in his estimation. Nonetheless, it is clear that China's annual revenue growth rates have been high since this study was conducted. In 2013, officially published figures claimed total revenues amounting to US$230 billion for the 11 large state-owned defense corporations during 2011 (Grevatt 2013c). It is unclear, however, whether this also includes non-defense related sales of these companies, a portion of sales that is not included in the figures cited in Fig. 4.2. Regarding assets, a 2012 Reuters report cited a figure of US$315 billion for the combined assets of the ten major state-owned defense groups of China, 25 % of which were held by their more than 70 listed subsidiaries. Over 40 of these were engaged in defense-related business (Lague and Zhu 2012).

By a very rough rule of thumb, China's aggregate defense-related revenues could be estimated to be comparable in size to those of e.g. Russia or France. In any case, if reliable data were available, it can be safely assumed that China's top arms producers would most certainly be included within the top 100 (cf. Neuman 2009: 69).

4.3.2 The Powerful Role of the "System Houses"

Defense electronics is a field in naval development where technological leadership is the most difficult to achieve, and the capability for innovation is consequently the most concentrated. Only a handful of companies—all U.S. or EU-based—can presently be considered state-of-the-art or "first tier" suppliers in the field of naval combat direction systems, so that they effectively hold an oligopolistic market position vis-a-vis the rest of the world (see Fig. 3.10). These companies often work together with other industry players in specific projects or technological fields. Collaborative arms production has almost become the norm in Europe, with many of the leading arms producers being jointly owned by parent companies from various nations.[8] Moreover, in the increasingly globalized arms trade system, these

[8] Examples include the Franco-German-Dutch EADS (as of 2014, renamed Airbus Defence and Space), the Anglo-French-Italian missile producer MBDA, or the multinational satellite builder Astrium, to name just a few (cf. Krishnan 2008: 33). Moreover, Europe has seen several large-

industry players often function as the global "parent" companies for an ever larger number of local firms that become their subsidiaries:

> First-tier suppliers have already begun to acquire large numbers of second-tier companies, which might suggest that second-tier companies can only survive as subsidiaries or junior partners of global defence firms like Lockheed Martin or BAE SYSTEMS (Krishnan 2008: 34).

From the point of view of the first-tier arms producing companies, having long-term local partnerships in their primary export markets is often a decisive factor for winning large export orders, because local partners can help to fulfill so-called offset obligations, provide essential market knowledge, and have connections to local decision-makers and other critical stakeholders. It can therefore seem useful, if not indispensable, to the big players to acquire local subsidiaries when long-term activities are envisaged in the country. This can also help to keep competitors in check insofar as it denies them access to crucial local knowledge and political connections.

Because the capability of a naval vessel as a weapon system is to a large extent defined by its combat suite (as explained in Sect. 3.4), the procurement decisions of governments and navies regarding new vessel acquisitions tend to increasingly focus upon the strategic choice of combat system, with other technologies, and even the choice of building shipyard, often having second or third priority. In many large-scale naval shipbuilding projects, the combat system supplier therefore acts as prime contractor and can thus exercise a high degree of control over the selection of subsystem suppliers. This gives these so-called "system houses" considerable leverage in relation to other industry players. They therefore hold a dominant market position within the wider range of the myriad defense industries, and effectively act as "lead firms" in the sense intended by Todd. This means they can define the rules of the game to some extent, both in their interactions with government procurement agencies and other industry players, giving them considerable influence on procurement programs worldwide. To illustrate this with a practical example, the fact of having been able to conclude an exclusivity agreement with Lockheed Martin for the marketing of its *AN/SPY-1* radar outside the U.S. put a hitherto relatively unsuccessful and chronically deficitary state-owned Spanish naval shipbuilder, Bazan (today called Navantia) in the magnificent position of being the sole shipyard able to offer the required technical solution in the Norwegian *Nansen* class frigate bid and the Australian *Hobart* class air warfare destroyer bid and thus win huge contracts against strong competition.[9]

---

scale joint naval procurement projects during the last years, i.e. the Franco-Italian *FREMM* and *HORIZON* classes, the Franco-Spanish diesel submarine collaboration and the Dutch-German co-operation in the joint *F-124/LCF* class frigate project.

[9] These observations are based on studies conducted in the naval shipbuilding industry using internal sales and project data, and were confirmed by various shipbuilding and military sources interviewed by me over the process of several years. Cf. also Wertheim (2013: 495, 22).

The prevailing U.S. dominance in the international defense industries, especially in the field of defense electronics, can thus be conceptualized as a powerful leverage tool that has numerous consequences for international relations, especially to those countries without viable alternatives:

> the relative autonomy, larger size, and greater strength of its defense-industrial sector [. . .] gives the United States a substantial comparative economic and technological advantage but poses thorny dependency issues for other countries (Neuman 2009: 73).

From China's perspective, this situation is further complicated by the fact that all the major and minor players in the Western arms industries are connected to their regular subsystems suppliers, and also to other system houses, shipyards, and other leading technology suppliers via a complex maze of cross-shareholdings (see Fig. 4.4). Surprisingly little has changed since the mid-1980s (cf. Todd 1985: 318). The problem of innovation is especially impacted, because apart from the generation of vast amounts of revenue available by the few dominant market leaders for investment in R&D, there is a multitude of inter-industry linkages in typical large-scale defense projects. Todd (1985: 146) thus speaks of "the web of supply and demand ties which connect the lead firms to all the principal suppliers of inputs into the industry as well as the major consumers of its products."

These linkages are constituted through the repeated interactions of particular individuals, all of whom belong to a small community of peers. Their interactions provide a thriving communication environment, which, in accordance with the above-mentioned ideas by Drucker and Todd, can be assumed to be fertile ground for furthering technological innovation.

4.3.3 A Maze of Defense Industry Networks

The actual degree of complexity of the cross-shareholding interconnectedness among Western defense industries can be glimpsed from Fig. 4.4, which shows the situation for the European defense and aerospace industries as of 2006. Although the figure is based on relatively old data, it is still sufficient to illustrate the main point, for despite ongoing flux, the level of interconnectedness overall has not become any less complex. As the figure indicates, several national governments as well as the big system houses—among them U.S. defense giants Lockheed Martin and Raytheon—seem to act as nodes within a maze of large and small defense and aerospace companies in Europe. In their totality, they form the institutional framework for regular day-to-day interactions among a loosely knit network of people who are mostly male technical experts, often engineers, and who in many cases have prior working experience as active service members of the armed forces of their respective countries. These experts are in regular communication with each other through manifold and multi-layered long-term projects, defense industry conferences, exhibitions, and other professional commitments including sales activities. Indeed, it seems useful to conceptualize this social subsystem as an

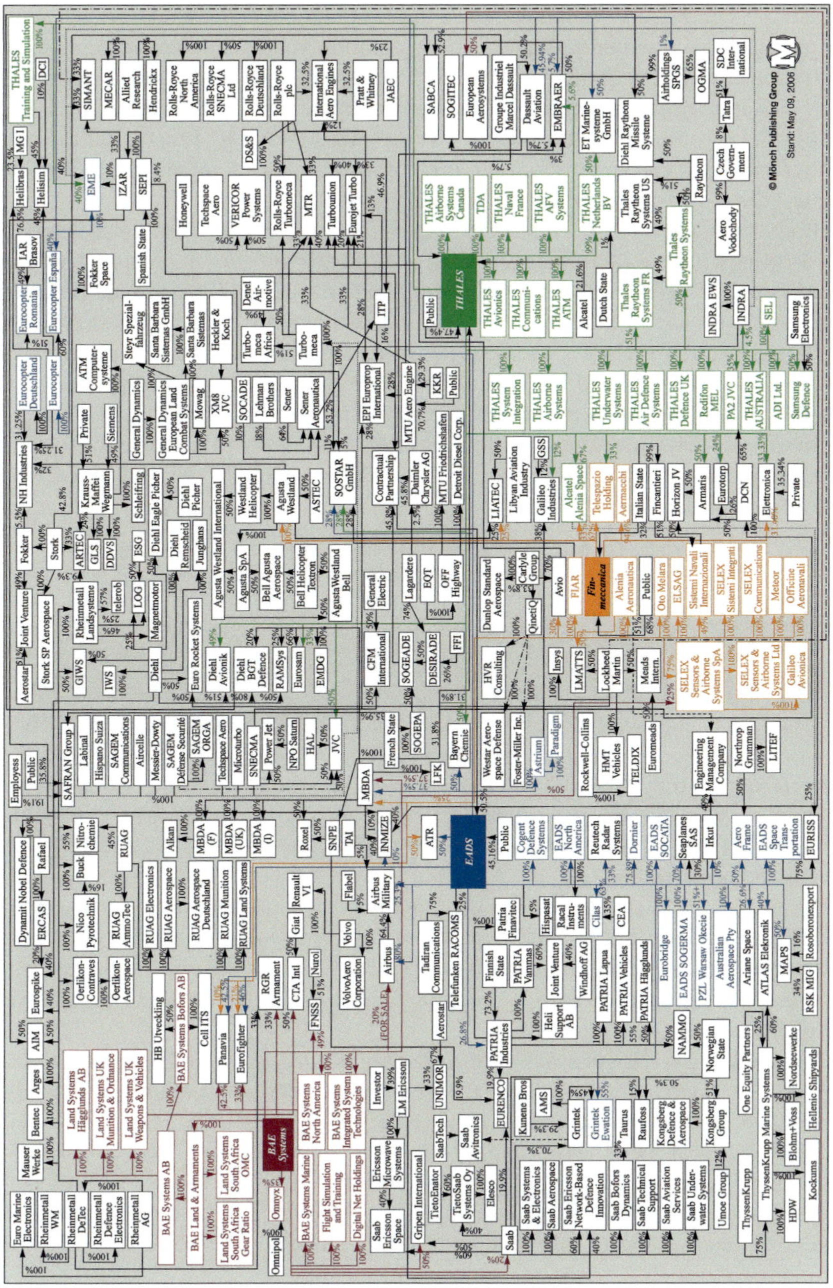

Source: © Mönch Publishing Group 2006. With permission

Fig. 4.4 European Aerospace and Defense cross-shareholdings, 2006

"epistemic community" as defined by Holzner (1972: 68–69): This group of people clearly forms a community of peers sharing similar values, interests, world views, and educational and professional backgrounds, leading to a similar frame of reference in reality construction.[10] A communicative context thus develops where innovation diffusion, i.e. "the adoption of new technology by firms other than the innovating firm" can naturally occur, often as the result of semi-conscious emulating or "bandwagoning" effects (Todd 1985: 156).

It does not seem too far-fetched to assume that closely knit and fairly homogenous groups such as aerospace & defense engineers function in a similar way as the communities of scientists described by Thomas S. Kuhn in his ground-breaking 1962 study of scientific revolutions, which first coined the term "paradigm." As Kuhn writes: "A paradigm governs [...] not a subject matter but rather a group of practitioners. Any study of [...] paradigm-shattering research must begin by locating the responsible group or groups" (Kuhn 2012: 179).

In China, such a positive innovation environment seems to have existed, at least temporarily, in the relatively isolated field of Mao-era strategic weapons programs, where

> [...] technical personnel [...] developed a close-knit professional camaraderie, which encouraged them to experiment and take risks in order to make progress. [...] In such a collegiate and motivated climate, innovation flourished and crucial technological and engineering breakthroughs were made in a wide range of areas at a rapid and sustained pace (Cheung 2009: 43–44).

Moreover, in these strategic arms programs, senior Chinese technicians, scientists and engineers were directly integrated into high-level decision-making processes and had substantial authority in project leadership, while in the conventional arms programs they tended to merely carry out administrative orders from above (Cheung 2009: 45). It seems that this factor allowed for critical breakthroughs resulting in the successful testing of the first atomic bomb in 1964, the first dropping of a thermonuclear bomb in 1967, the launching of China's first space satellite in 1970, the start of sea trials for China's first nuclear-powered attack submarine in 1971, and the launching of the first nuclear-powered missile submarine in 1981 under relatively difficult circumstances (Cheung 2009: 44–45). It can be assumed that the current aircraft carrier program enjoys a similar degree of high-level political backing (cf. Fisher 2013).

The rather fuzzy, unquantifiable factor of human connections and interactions as a factor for innovation is nevertheless extremely relevant, because, ultimately, technological development is the result of efforts by individual people working towards a definite goal. The communicative environment they work in must be considered a major enabling factor for accumulating knowledge and floating creative ideas, and also for providing frames of reference for the identification and conceptualization of complex problems (cf. Drucker 1993: 257). Studies of

[10] These observations are in part the result of the present author's experience as a participant observer in this particular industrial subsystem from 2007 to 2010.

China's defense innovation system have stressed that, apart from the more tangible factors that can be easily quantified such as R&D spending, number of patents, number of technical graduates, number of relevant publications etc., a crucial influence on the future innovation potential is in fact exercised by far less tangible, institutional or cultural factors centered around the human actors. These encompass the

> [...] activities, structures, and processes of the system, such as knowledge diffusion, learning processes, and the interactions between organizations and institutions that have been ignored in more conventional treatments of the defense industrial sector (Cheung 2009: 14).

As a practical example, it is normal for Western industry executives from various technological fields to regularly meet at the various annual or bi-annual defense fairs around the globe such as DSEi in London, DIMDEX in Doha, IDEX in Abu Dhabi, EURONAVAL in Paris, or DSA in Kuala Lumpur, to name just a few, where they can at least to some degree exchange news, facts, rumors, and ideas. China's leaders, however, have publicly complained that its representatives are not very welcome at European events such as DSEi. In China, the leadership has taken steps to establish regional and international exhibitions such as the annual high-tech fair in Chongqing and the Beijing International Technology Exhibition Fair. Still, these are comparably "modest efforts," and indicate that "the enhancement of dissemination mechanisms remains a low priority" (Cheung 2009: 146). China's marketing of weapon systems such as the *HQ-9* air-defense missile to Eastern European countries and Turkey in recent years must therefore not only be seen as normal export projects for economic gain, but also as an effort to gain a foothold within the wider European defense community. The strong U.S./NATO reaction that immediately followed Turkey's announcement of its intention to buy Chinese missiles illustrated the kind of political pressure regularly exercised within the hierarchical supplier-buyer relationships of arms technologies worldwide, and had the predictable effect of Turkey reconsidering its choice (cf. Grevatt 2013b).

In sum, the above-mentioned massive concentration of innovation capability at the top of the global arms-production system ultimately results in a hierarchical power structure with considerable consequences, not only for suppliers and buyers of weapons within this system, but for defense industry participants worldwide. To be excluded like China due to an arms embargo from the technical, managerial, commercial, and procedural knowledge circulating within this semi-closed circle can be assumed to carry disadvantages in terms of reduced innovation capacity that probably far exceeds the simple price premium that is paid for independent indigenous development. This exclusion, in other words, must be considered a critical challenge.

4.4 China's Rising Military-Industrial Complex

4.4.1 *The Quest for State-Led Innovation*

As Cheung (2009) has described in fascinating detail, China has long struggled to build a modern defense industry capable of supplying its armed forces with all the necessary weaponry.[11] The present study uses Cheung's definition of the defense technology and industrial base (shortened to "defense economy") as "the combination of people, organizations, institutional rules, technological expertise, and production capacity that provide defense-related and dual-use goods and services" (Cheung 2009: ix–x). Innovation plays a crucial role in its development. Cheung defines the Chinese "defense innovation system" as

> a key subcomponent [...] which consists of laboratories, research institutes, and academic institutions that are involved in the development, diffusion, and use of innovations for military and national security applications (Cheung 2009: x).

When analyzing China's potential to increase its innovativeness in the defense sector, one factor to consider is the fact that despite numerous reforms, restructurings, and even partial privatizations since the beginning of the Reform Era, China's state-led quasi-monopolistic industries are still largely isolated from market forces. The general negative effects of large bureaucratic organisations such as managerial problems, inflexible decision-making, over-bureaucratization and the like can therefore be assumed to be present and hamper innovativeness to some degree (cf. Todd 1985). In addition to these general factors, Cheung also mentions some further institutional obstacles to innovation specific to China:

> Fundamental aspects of the Chinese system appear to be major obstacles to the nurturing of a sustainable innovation environment. They include the absence of a robust and independent legal system, highly controlled flows of information and knowledge within society, and the lack of encouragement of pluralism that would allow for greater autonomy and self-governance within the S&T community (Cheung 2009: x).

Like all major Chinese industries deemed strategically important, the "entire arms industry in China [...] was—and remains—effectively state owned," notwithstanding some private companies that are trying to compete in the subcomponents industry. This system is a legacy of the Cold War era, when China's "centralized Soviet-inspired model of arms production" was notorious for bureaucratic inefficiencies and a lack of innovation capacity. For political reasons, it was also highly intransparent: Secrecy around arms production was often practiced because China "lacked confidence in its military strength and feared invasion" (Surry 2007: 4). This insistence on secrecy, which was exercised not only towards outsiders, but also within the system itself, inhibited the spread of knowledge across functional subsystems, and has also hindered our ability to understand the current

[11] The historical development of the arms industry during the Mao era is extensively covered in Cheung (2009: 22–51) and need not be recounted here.

state of the Chinese innovation process, especially in weapons development (Wang 1991: 27). As already indicated in the discussion of warship production processes in Sect. 3.3, this has had rather problematic consequences for foreign analysts trying to evaluate China's arms industry development:

> There has been limited examination of how the design, development, and production cycles for weapon systems and key technologies take place. The analysis of the output of the defense economy has concentrated on the development and manufacturing of complete weapons systems rather than subsystems and components [...]. Moreover, little is known about how the R&D and procurement processes work. Even more surprisingly, the examination of the roles of innovation, knowledge learning, diffusion, and absorption within the Chinese defense economy has been largely overlooked (Cheung 2009: 15).

Unfortunately, there is no immediate remedy available to the foreign analyst, who can only pay attention to any news items that surface regarding the development of critical components, and try to make plausible estimations by broad comparisons with other, better known industrial processes in other countries. For example, it can normally be safely assumed that significant delays in completing a project indicate technical difficulties; that prototypes are usually prone to a number of unforeseen problems in system integration that are only resolved in subsequent units; that the demands on project leadership are greater depending on the size and heterogeneity of the circle of companies and organizations participating in a single project; and, finally, that the likelihood of initial failures is greater given a higher quota of newly developed or experimental technologies integrated into a project. Regarding the challenge of assessing the state of China's naval electronics and system integration development, Bussert and Elleman have suggested just such an indirect approach. Specifically, they propose bridging the gap of uncertainty by deducing technical difficulties from empirical observations:

> Since China does not talk about its problems integrating diverse systems or openly admit CSI [combat system integration] problems when assembling a warship, one must study the history of delayed commissionings and equipment anomalies. Typical scenarios include having difficulty achieving multiship interoperability, problems operating a system that resulted from different nations teaming up on a design, and installing particular combat systems on only one class of ship (Bussert and Elleman 2011: 152–153).

Systematically following this approach yields interesting results, e.g. with regard to the Chinese surface combatant fleet (see Chap. 5). As a result of personal observations in various industry settings, the present author largely concurs with Bussert and Elleman that, in all likelihood, "CSI problems may be far more common and widespread than many China analysts were aware" (Bussert and Elleman 2011: 153).

In the case of China, it seems useful to distinguish between two different segments of the arms industries: one that is focused on "conventional" capabilities, producing "ordnance, aviation, naval, and military electronics equipment"; and the other that is centered on supplying strategic weapon systems such as nuclear and hydrogen weapons, nuclear submarines and strategic missiles (Cheung 2009: 20, 40). The political backing for the latter kind of strategic program has always been much greater. In contrast to almost all other fields of technical development, this has allowed for steady progress even during times of political upheaval, such as the more tumultuous phases of the Mao era when it became dangerous to be more "expert" than "red." The breakthroughs in nuclear weapons technology during the Mao era are therefore not indicative of a generally positive climate at this time within the national innovation system.

After the beginning of the Reform Era, China's overall defense economy received surprisingly little political backing during much of the 1980s for an efficiency-oriented restructuring. State leaders seem to have had other, more pressing development priorities. It was only in the 1990s that outside pressures—including the U.S. invasion of Iraq, which demonstrated the country's superior military capabilities and signaled a "revolution in military affairs" (RMA), and Taiwan's apparent move towards independence, which resulted in the 1995/1996 "Taiwan missile crisis"—led to a political re-orientation that changed the make-up of China's weapons industries (Cheung 2009: 102–105). After a major re-organization of all the state-owned industries in 1998/1999, which also brought the vast PLA-controlled commercial business empire under state control, ten large state-owned holding companies were created in the defense sector that were directly subordinate to the central government. An 11th holding company was created for defense electronics in 2002, China Electronics Technology Group Corp. (CETC, cf. Cheung 2009: 118–122). Part of the reason for this reform was to enhance intra-industry competition. As Surry writes:

> These groups [. . .] have under their control many major companies that produce both civil and military goods for the domestic and international markets. The idea behind the reorganization was to introduce a level of commercialization by encouraging competition between the companies and increasing their accountability for financial losses (Surry 2007: 3).

Nevertheless, the development of a modern defense R&D system was far from smooth, and the legacy of central planning has long dominated efforts to transform the Chinese defense industry. Cheung lists the results of internal Chinese assessments, which cited the "excessive size [. . .], production overcapacity, overstaffing, the scattered location of enterprises, lack of co-operation, separation of military and civilian production, construction of duplicate projects, and an incoherent enterprise system" as the major obstacles to be overcome. Other problems mentioned by insiders included a "lack of access to advanced foreign technology and knowledge, a highly conservative, risk-averse institutional culture, and inadequate investment" (Cheung 2009: 111). These problems seem to continue to plague China's defense economy, despite all efforts to create better inter-industry co-operation and enhance the development of a dual-use economy (Grevatt 2013c).

4.4.2 Combating Bureaucratic Inertia Through Partial Privatization

China's defense industries have long struggled to build a profitable business base. According to official Chinese sources, the nadir was reached in 1998 when "61 percent of all defense industrial enterprises bled red ink. Total net losses (after subtracting profits) of the entire defense industrial complex totaled Rmb 2.5 billion [ca. US$0.3 billion]" (Cheung 2009: 118, fn. 56). The result was a worsening debt crisis. As with other deficitary state-owned enterprises, Chinese defense companies were encouraged to raise capital on the domestic stock markets in order to reduce their dependence on state funding:

> By the beginning of 2008, sixty-two defense industrial enterprises had or were in the process of being listed on the country's two local stock markets as well as on the Hong Kong Stock Exchange. One of the biggest initial public offerings was for the civilian operations of AVIC 2, which was listed in Hong Kong in 2003 under the name of AviChina and raised U.S. $248 million. A senior COSTIND official forecast that the Chinese defense industry would raise between Rmb 50 and 60 billion from the public listing of defense enterprises by the end of this decade (Cheung 2009: 125).

There have been indications in recent years that Beijing plans to compel the subsidiaries of the major defense groups already listed on the stock markets in Shanghai and Shenzhen to buy assets from those holding groups. If followed through, they would no longer be completely state-owned, representing a definitive break with a tradition of purely state-led industry development in the higher echelons of the defense sector. The term "privatization," as it is understood in the West, would still be an exaggeration for this phenomenon, though, given the nature of China's state-led economic and financial systems and its peculiar financial market organization (cf. Heilmann 2005). In any case, this trend towards raising capital in the stock markets

> [...] would double the value of military related assets injected into these listed companies since 2007 with more in the pipeline, as Beijing presses ahead with an ambitious program to privatize most of a vast arms industry (Lague and Zhu 2012).

Besides raising capital, the listing of defense companies (or rather their subsidiaries) on the stock market has another effect. By changing a company's status from purely state-owned to (at least ostensibly) privatized, the publicly listed Chinese firms are in effect able to make successful bids for Western aerospace and defense companies. Political resistance in Western countries to take-overs by fully state-owned Chinese companies in such a sensitive technology field will probably be hard to overcome, while partial privatization may give Western decision-makers, lawyers, and public commentators the (rather faulty) impression of a lesser degree of Chinese state control over these companies' handling of sensitive technological knowledge (cf. Minnick 2007). The acquisition of existing Western companies with firmly established business practices, market knowledge, production facilities, and connections to government agencies and other related companies permits China to open up further channels for infusing advanced technological and managerial knowledge into the Chinese defense economy, and therefore has advantages that far surpass any purely technical considerations (Anderson 2011b). In fact, the 2008 financial crisis as well as strained defense budgets in many Western countries have contributed to a situation where Chinese companies are able to make attractive bids for struggling Western counterparts and thereby "penetrate international defense markets through acquisitions and military sales" (Anderson et al. 2013). According to a recent survey published in *Jane's Defence Weekly,* five purchases of U.S./European-based defense and aerospace companies were completed by Chinese defense companies in 2013, up from two in the previous year, with a further unsuccessful bid (for U.S. aircraft manufacturer Hawker Beechcraft). The successful bids in 2013 included Chongqing Helicopter's acquisition of U.S.-based helicopter manufacturer Enstrom and AVIC's acquisition of German aircraft engine manufacturer Thielert (cf. Anderson et al. 2013). Based on previous surveys, the same authors conclude that:

> Given the financial firepower of China's defence industry and its high motivation to expand and develop through access to Western technologies, its penetration in international markets through acquisitions and exports is certain to become an increasing feature in the global defence market, and in 2014 further deals can be expected (Anderson et al. 2013).

China's defense industries already accumulated profits at a level of approximately RMB43 billion (ca. US$5.9 billion) in 2007. After a decade of annual growth rates of ca. 20 %, their combined revenues were estimated to have reached around US$230 billion by 2011 (Cheung 2009: 126–127; Grevatt 2013c). These revenues in combination with China's rapid rise in state-sponsored R&D—financed by a consistently immense trade surplus during the past decades and made available to Chinese companies in the form of "multibillion-dollar soft loans as well as tax relief and other subsidies"—has created a situation where "an avalanche of cash" has had a marked impact on performance and output, regardless of the still-endemic inefficiencies of China's large bureaucratic state-owned enterprises (Grevatt 2013c; Lague and Zhu 2012). State lending by China's state-owned banks to strategically important state-owned enterprises, including those in the defense sector, has been growing fast. The Chinese government announced its intention to infuse a further US$600 billion into strategic sectors in its 12th Five-Year Plan, covering the period from 2011 to 2016. Among the defense groups that have received huge loans during the past few years were both the major shipbuilding holdings, CSSC and CSIC, which received more than US$30 billion in loans from, among others, the Export-Import Bank of China in 2009. Likewise, the aviation holding group AVIC received at least US$60 billion in 2009 alone (Grevatt 2011b). These loans do not represent a one-way-street, however, for China's defense holdings are increasingly interlinked with the state banking system through stake holding:

> [M]ost of China's state-owned military producers also have stakes in China's banks. In 2010 land systems specialist Norinco and CSSC became strategic investors in the China Everbright Bank, whose IPO took place on 18 August and raised nearly USD 3 billion on the Shanghai Stock Exchange. Additionally, CSSC and three other state-owned defence groups—AVIC, the China National Nuclear Corporation and China Aerospace Science and Technology Corporation—were revealed in late July 2010 to be key investors in the Agricultural Bank of China. In its IPO, this bank raised nearly USD 21 billion in shares on the Hong Kong and Shanghai stock exchanges [...] (Grevatt 2011b).

It appears, then, that a military-industrial-financial complex of grand proportions is in the making. This development, moreover, is no historical accident, but rather the result of a conscious political creation process engineered and controlled by China's Leninist leaders, who are effectively able to command these actors (banks and defense holdings alike) through the Chinese system of Party control of all leadership cadre positions (cf. Yan 1995). As a result, the distribution of funds to defense R&D has continued unabated. By and large, Chinese R&D expenditures have been growing at the same pace as overall government spending over at least the past 15 years. The figure for the year 2006 was already estimated to be ca. €14 billion, or about a quarter of U.S. R&D spending for the same period (Stumbaum 2009: 15). The result has been a sharp rise in the number of defense-related patent registrations over the past few years (cf. Grevatt 2011c).

Nonetheless, the exact same factor which made this organizational model possible—i.e. an unprecedented abundance of funds—gave rise to another structural problem in China's defense economy that also likely has a negative impact on innovation: widespread corruption. This is endemic in China's large state-owned industries, but especially within the military sector. Corruption in the military is perceived by the Chinese leadership as a serious concern, and has been targeted by a special campaign since the new leadership under Xi Jinping took control in late 2012 (Mulvenon 2013). The most spectacular cases investigated so far were the court martial in early 2014 of General Gu Junshan, a former deputy chief of the PLA's general logistic department, and the investigation of an even higher ranking former official—retired vice chairman of the Central Military Commission and former Politburo member, General Xu Caihou who was put under house arrest and has been expelled from the Party in late June 2014 as a prerequisite for being charged with corruption, to which he confessed in October 2014. Thanks to his control over military procurement processes, General Gu Junshan's family and associates reportedly managed to embezzle at least the equivalent of US$15.6 billion (cf. Wen 2014; Hu 2014; Bender 2015).

A political culture condoning the misappropriation of funds by way of "informal privatization" can easily become entrenched if left unchecked by an independent judiciary and a free press. If this trend becomes excessive, it can also give rise to waste and inefficiencies in the arms procurement system. Corrupt behavior among leaders and officials can also have the effect of eroding the fighting spirit of the armed forces. Although no hard data on the true extent of corruption in the defense system is currently available, comparisons with the situation in China's civilian industries suggest that corruption is both "deep-seated and widespread," and has been "a major problem in the defense industry and weapons procurement system" during the entire Reform period (Cheung 2009: 131). An index published by Transparency International in 2013 that is based on a 2-year survey, the *Government Defence Anti-Corruption Index* (GI), supports this view. The GI measures "the degree of corruption risk and vulnerability in government defence establishments" across a sample of 82 countries worldwide. Its results suggest that the institutional corruption risk in China's defense system is relatively high, with China scoring in band D− (with A the best possible score), or roughly on par with Russia, Belarus, and Pakistan but behind India, which scored slightly better at D+ (cf. Cover et al. 2013: 4, 18).

The above discussion notwithstanding, it would be a mistake to suppose that the dominant role of state-owned enterprises in China fosters a monolithic structure that is completely devoid of stimuli for creative development. As Wang (1991: 8) writes, "The impression of control can be greater than the reality." China watchers have also long noted the "fragmentation" of the authoritarian bureaucratic leadership system in China's political economy (cf. Lieberthal 1992). Even though government planning remains a dominant factor, three decades ago there were already multiple voices within China's military-industrial complex competing for influence and funds within the system through a system of bargaining:

> China's development experience in the 1980s was [...] more than the single issue of scientific planning and technological transformation; it also involved a complex process of mutual influence between political, economic, and S&T [Science & Technology] systems. There were emergent and diversified groups, bureaucracies, and institutions competing for preferences and resources surrounding science and technology policy (Wang 1991: 8).

If anything, China's defense economy is even more multi-layered and complex today. In 2012, China's defense industry sector encompassed more than 1,000 state-owned enterprises organized under the above-mentioned 11 large fully state-owned holding companies, and it employed over a million workers (Lague and Zhu 2012). If the entire defense industrial base is taken into account, including privately owned subsidiaries and other private companies, estimates are as high as ca. 10,000 firms and ca. 2.5 million employees (Grevatt 2013c). It seems that the privatization strategy described earlier is ultimately aimed at profoundly changing China's military-industrial complex:

> The long term goal is to transform some of the leading contractors, such as China State Shipbuilding Corporation (CSSC), Aviation Industry Corporation of China (AVIC) and China Aerospace Science and Industry Corporation into homegrown versions of American giants Lockheed Martin and Northrop Grumman or Britain's BAE Systems (Lague and Zhu 2012).

Thus, emulating the Western market leaders seems to be part of the Chinese defense industrial strategy. This is in itself revealing, for it points to a perception among China's leadership that a purely state-led innovation as currently practiced in China has its disadvantages. Governments can try to make up for this, as China does, by actively supporting R&D. They can also infuse some measure of competition by pitting several domestic players against each other in procurement projects—which is sometimes done in China, even though most projects probably proceed non-competitively. As cross-cultural studies have shown, however, it is by no means certain that government-sponsored R&D activities actually have the desired effect on the country's innovation performance (Todd 1985: 155). Moreover, to be as successful as the Western defense heavyweights, it would be necessary for China's industries to also attract highly qualified employees. This is no easy task given China's dynamic growth climate, which currently seems to offer much better earning opportunities to technical personnel in the civilian economy:

> Attracting and retaining the brightest, best, and most experienced defense scientists and engineers has been an enormously difficult task, [...] especially when the economic

> benefits offered by the civilian sector far outstrip those in the defense economy. It was estimated in 2002 that backbone S&T research staff in defense R&D institutes earned between one-third and one-half as much as their civilian colleagues. The gap was even wider when earnings were compared with those of scientists and engineers employed in foreign-owned enterprises (Cheung 2009: 156–158).

This discrepancy in compensation, Cheung further observes, means that "the brain drain from the private and foreign-owned technology sectors is likely to remain a long-term problem" (Cheung 2009: 158).

4.4.3 China's Progress in Critical Naval Technologies

A pronounced dip in Chinese defense imports between 2007 and 2011 of ca. 58 % according to SIPRI, and a simultaneous increase in Chinese arms exports of 95 % between 2001 and 2011 not only indicates that much progress has been made towards China's ultimate goal of self-reliance in defense production, but also a rise in the "ladder of production" discussed at the beginning of the chapter from a "third-tier producer" overall to a "second-tier supplier" at least in some fields (cf. Lague and Zhu 2012).

Nevertheless, this has been mostly the case in the field of land systems, while several critical technology areas in air and maritime defense still rely heavily on foreign imports, despite decades of intense domestic R&D activity. These glaring weaknesses include a lack of indigenous production capability for high performance jet engines, which are imported from Russia; for gas turbines, which are produced under license from Ukraine; and for naval diesel engines for surface combatants and submarines, which are imported mainly from Germany (MTU) and France (Pielstick) or produced under license. Propulsion, both in jets and warships, is of course a critical component. China's persistent dependence on outside supply sources in strategically relevant technologies like propulsion is therefore in itself rather telling. Another such area where China is still dependent on outside assistance is big military transport aircraft and helicopters, which also have strategic value in naval development. Many electronic systems aboard Chinese naval vessels, moreover, are either imported or based on earlier imports from Russia or Western countries such as France. A recent assessment of China's status thus concluded that

> [t] existing capability gap between China and the industrialised states of the West and Russia is apparent in the Chinese industry's generally unsuccessful attempts to develop and produce major systems such as aero-engines and a wide range of fourth- and fifth-generation technologies that will enable China to reach its 'informationisation' goal (Grevatt 2013c).

Official reports in China's state media in 2012 and 2013 concerning a new massive arms deal between Russia and China over the sale of 24 *Su-35* fighter airplanes and 4 *Lada* class submarines, which apparently has not yet been signed but is expected to be settled by 2015, would support this view because it indicates a still-existing gap between Chinese and Russian defense technologies. It further points to a persistent deficiency in China's innovation capacity at the state-of-the-art level in critical naval and aerospace technologies.[12] There would otherwise be no need for China to import complete weapon systems or ask for technology transfers and joint production. China's capacities in these areas thus still lag behind Russian export systems, which, in turn, are less advanced than Russian state-of-the-art technology. This view is also supported by Cheung. He concludes from his extensive analysis of China's defense industry modernization effort over the past few decades that the majority of Chinese defense products have now reached the technological standard of the late third generation (comparable to arms produced in the West during the late 1960s to early 1970s), while early fourth generation (Western 1990s production) levels have been reached in a few select areas, e.g. with some missiles, the *F-10* fighter aircraft, and the *T-98* main battle tank. According to Cheung, "[w]hile this progress is significant, the Chinese defense economy still lags as much as two generations behind the latest global standards in most areas" (Cheung 2009: 172). Figure 4.5 shows Cheung's evaluation of the major naval weapon systems and missiles according to this system of technical "generations."

Weapon system	R&D/production status	Generational category[a]
Type 051 "Lüzhou" class destroyer	Dev. since late 1990s, operational since ca. 2006	Late 3rd
Type 054 "Jiangkai I" (Ma'anshan) class frigate	Dev. since late 1990s, operational since ca. 2003	Late 3rd
Type 053H2G "Jiangwei" class frigate	Dev. since mid-1990s, operational since 1996	Late 3rd
Type 039 "Song" class diesel submarine	Dev. since early 1990s, operational since 1998	Late 3rd
Type 093 "Shang" class nuclear attack submarine	Dev. since late 1980s, operational since 2006	Early 3rd
Type 094 "Jin" class ballistic nuclear sub	Dev. since late 1980s, operational since ca. 2008	Late 3rd
DF-21 intermediate range ballistic missile	Dev. since 1980, in serial prod. since late 1980s	Late 3rd
DF-15 (M-9) theater ballistic missile	Dev. since 1985, in serial prod. since mid-1990s	Late 3rd

[a]Generational definition follows U.S. and Russian convention. Chinese generational reckoning lags one generation behind because generation one was skipped in China altogether
Source: Adapted from Cheung (2009: 173)

Fig. 4.5 Generational levels of major Chinese naval weapon systems

[12] For a Chinese state news report, see 'Zhong E qianding junshou da dan' (2013). For comments on the deal, see Johnson (2013b) and Foster (2012).

In general, notwithstanding China's ongoing privatization and restructuring efforts, its ability to counterbalance some of the negative effects of mainly or purely state-led innovation and the availability of huge funds dedicated to defense R&D, the present author suspects that China's de facto exclusion from many communication venues of the global naval industry heavyweights as a result of the arms embargo must still be considered a key disadvantage for its technological innovation effort. Were the arms embargo lifted, however, this problem could perhaps be quickly remedied.

4.5 China's Indigenous Shipbuilding and Component Industry Development

Given its limited access to advanced knowledge of the world's leading producers of naval technology, China has to make up for this through indigenous development. One possible source of naval innovation could be the country's flourishing commercial shipbuilding sector. Commercial shipbuilding has been developing at such a rapid pace over the past decade—at least in quantitative terms—that some analysts have concluded a similar rise in warship building capability will inevitably follow suit. As will be shown, however, this is not necessarily a valid assumption, especially since it is not ship hull production, but rather defense electronics and system integration which must be considered the most difficult and strategically relevant fields in naval shipbuilding innovation. Will China, then, still be able to count on spillover effects from its civilian shipbuilding sector to its advanced naval programs?

Judging from past and present experience regarding the development of "second-tier" industrial arms production capabilities in other countries, this expectation is perhaps too simplistic and optimistic. "Technical innovation does not fall like manna from heaven" as a by-product of commercial success, and countries attempting to develop autarchy in arms production generally face a steep learning curve, in which it does not get any easier to move up, even as more capabilities are successively being developed (Bitzinger 2003: 35–37).

Nevertheless, a highly developed infrastructure of civilian shipyards is at the very least a boon for securing a maintenance and repair infrastructure for the fleet, while in a war scenario, the ability to quickly produce large numbers of new vessels can be a decisive factor. In order to determine how sustainable China's naval modernization can be in the long run, it is therefore instructive to look at the state of China's overall shipbuilding industrial development and assess its strengths and weaknesses.

4.5.1 China's Commercial Shipbuilding Industry

Since the early 1980s, China's shipbuilding industry as a whole has expanded steadily. By 2005, it already encompassed more than 1,200 shipyards responsible for a wide variety of activities. Most of these are commercial, with only a handful of the larger yards active in naval shipbuilding (Medeiros et al. 2005: 115). China's two giant state-owned shipbuilding groups, China State Shipbuilding Corp (CSSC) and China Shipbuilding Industry Corp (CSIC), are complementary in their focus rather than in true competition with each other. Together, they operate more than 120 subsidiary companies, many of which are involved in both naval and commercial construction. CSIC companies concentrate more on submarines, whereas CSSC companies are more strongly involved in naval surface vessels (Grevatt 2013a). China's major naval shipyards are shown in Fig. 4.6.

In terms of output size, China became the world's second largest shipbuilding nation after South Korea after 2007. It became the world leader in 2010, a goal originally envisaged by the government for the year 2015. Growth in China's shipbuilding sector was so rapid that government officials started to fear that there would already be overcapacities from 2010 onwards (cf. Luttmer 2008: 58–59; Murray 2014). Just the same, competing with the highly advanced South Korean and Japanese civilian shipbuilding industries in qualitative rather than quantitative terms will likely prove to be an entirely different matter for China's shipyards.

Judging from its low starting point, China's technological progress in shipbuilding since 1980 has indeed been impressive. According to a 2005 RAND study, however, its quantitative progress in commercial shipbuilding had not been accompanied by a simultaneous rise in build quality and design capability at a level that was comparable to the most advanced yards in South Korea, Japan, and Europe (Medeiros et al. 2005: 130). Among the most commonly listed "chronic" problems of Chinese commercial ships at that time were poor construction quality and delays[13]:

> There is poor cost control; production still uses outdated and inefficient equipment and technologies, and there is poor management of large shipbuilding projects (Medeiros et al. 2005: 133–134).

The persistence of at least some of these deficiencies is evident from the fact that Chinese shipyards have so far only been able to produce the simpler types of commercial ships without foreign assistance. As of 2013, only a small number of the more complex commercial ship types (i.e. LNG tankers) have been built in China, with South Korea and Japan still clearly dominating this high-end market

[13] According to the above cited RAND study, it was still customary for foreign commercial ship buyers in 2005 to dispatch teams of 8–10 advisors to China "to supervise construction of their vessels to ensure that they meet the buyers' standards and are completed in a timely manner", a practice, if still continued, that would be indicating severe quality issues (Medeiros et al. 2005: 135).

	Shipyard	Location (Province)	Naval production 1996-2006 (displ. t).	Primary products built so far
Controlled by State Council	**China State Shipbuilding Corp. (CSSC)**			
	Jiangnan Shipbuilding Co.			
	▪ Jiangnan Shipyard	Shanghai	46,500	SSK (Type 039), DDG (Type 052B)
	▪ Qiuxin Shipyard	Shanghai	8,400	FAC (Proj. 022, Type 037-I)
	Hudong-Zhonghua Shipbuilding Co.			
	▪ Hudong Shipyard	Shanghai	56,000	FFG (Type 054/054A), LPD (Type 071)
	▪ Zhonghua Shipyard	Shanghai	33,600	LST (Type 072-II), Landing Craft (Type 073-III)
	Guangzhou Shipbuilding Co.			
	▪ Guangzhou Shipyard International	Guangdong	23,000	var. Auxiliaries
	▪ Guangzhou Huangpu Shipyard	Guangdong	14,500	FFG (Type 053H3, Type 054/ 054A), FAC (Type 037-II)
	Other:			
	▪ Wuhu Shipyard	Anhui	9,300	Landing Craft (Type 073-III, Type 074)
	China Shipbuilding Industry Corp. (CSIC)			
	Dalian Shipbuilding Industry Co.			
	▪ No. 1 Yard (Ex-Dalian Shipyard)	Liaoning	34,400	CV (*Liaoning*, new Type 089); DDG (Type 051B/051C), LST (Type 072-II)
	Other:			
	▪ Bohai (Huludao) Shipbuilding Heavy Industries	Liaoning	44,000	SSBN (Type 094), SSN (Type 093)
	▪ Wuchang Shipyard	Hubei	53,600	SSK; LST (Type 072-II)
Controlled by CMC	**PLA Shipyards**			
	▪ Dinghai Naval Dockyard	Zhejiang	3,000	Landing Craft (Type 074A)
	▪ Lüshun Naval Shipyard	Liaoning	5,600	Landing Craft (Type 073-III)
	▪ Qingdao Naval Dockyard	Shandong	6,700	Landing Craft (Type 073-III, Type 074A)

Source: Adapted from Collins and Grubb (2008: 14)

Fig. 4.6 Major Chinese naval shipyards (location, output, and products) since 1999

segment. So far, among all the Chinese shipyards, only Hudong Zhonghua Shipyard in Shanghai has the technical expertise to build LNG tankers. The first such ship was built in 2008 and around 20 vessels in all have been built so far. By comparison,

South Korea, the global leader in building LNG tankers, has built five times as many within the same period. Since Chinese domestic demand for such ships is on the rise, however, a larger market share for China can be expected in the future, thus giving China's domestic producers a strong incentive to overcome the technical difficulties of LNG tanker production (cf. Hall 2013).

Another factor to note is that while some observers cite the (still relatively small) number of Chinese LNG projects as proof that Chinese shipbuilding is now on par technologically with more advanced nations, these projects have relied heavily on foreign assistance in system integration, indicating an actual lack of domestic capability. The French shipbuilder Chantiers de l'Atlantique, for example, provided extensive assistance for the first Hudong-Zhonghua LNG project, which required keeping a team of 50 French technicians on-site throughout.[14]

Foreign involvement seems to have had a strong impact on commercial shipyard development in China. Shipbuilding cooperation projects with foreign yards have so far included the Japanese-led Nantong-Kawasaki-COSCO shipyards, the German-led Shanghai Edward shipyard, the Singapore-led Yantai Raffles shipyard, and the Korean Samsung-Ningbo shipyard (Medeiros et al. 2005: 116). Technology transfers within such projects as well as indigenous developments have furthered modernization in a range of shipbuilding areas, such as

> [...] increased emphasis on hull-block construction, investment in major new 'greenfield' shipyards, and bolstering of Chinese firms' ability to produce marine diesels and gas turbines—all of which have military implications. Other areas of technological focus include enhancing systems integration abilities and fostering the growth of China's currently weak ship-subcomponents industry, lest China become merely a 'world-class hull builder' (Collins and Grubb 2008: 2).

Regarding hull quality, however, it must be noted that independent experts from Lloyd's List warned in 2008 that an estimated 30–40 % of the commercial vessels built at new Chinese shipyards were of substandard build quality, and that some must even be considered positively dangerous (cf. 'Sécurité Maritime' 2008).

These problems are important to consider when making assumptions about possible spillover effects from commercial into naval shipbuilding. As a classic study has noted, however, the ability to construct quality ship hulls is only the tip of the iceberg:

> [S]hipbuilding capability, by itself, is not sufficient to conceive a warship. As an assemblage of parts, the vessel is only as good as the individual parts from which she is constructed. Superb assembly skills on the ways or in the dock avail the navy little if the propulsion machinery of the completed ship is found wanting because of inadequate engine-building expertise. They may be of little use too if the ship's electronics gear and weapons are prone to malfunctioning on account of a dearth of capability in the ancillary radar, instruments, computing, and ordnance fields. [...] the navy needs to look beyond the shipyard to the chain of components producers when reckoning the capability of its industrial base (Todd and Lindberg 1996: xii).

[14] Cf. Grubb and Collins (2008: 47), Medeiros et al. (2005: 120), and Murray (2014).

By and large, Collins and Grubb note that commercial interests and export orientation so far seem to be the dominant factors in Chinese commercial shipbuilding, with naval production taking the back seat. Chinese shipyards already account for ca. one fifth of global shipbuilding, and as much as 80 % of Chinese shipbuilding is destined for export. Therefore, it seems that "China's ambitious drive for new and world-premier shipbuilding capabilities is *not* at present primarily driven by military factors but rather by commercial incentives" (Collins and Grubb 2008: 4, emphasis in the original).

4.5.2 The Prospects for Spillover Effects from Commercial to Naval Shipbuilding

There are striking systemic differences between commercial and military shipbuilding that not only preclude any automatic knowledge transfer from the one industry to the other, but also call for quite different production facilities. With the exception of some naval auxiliary vessels such as tankers, oilers, or tugs, which are often built to commercial build standards, military ships typically differ so much from commercial vessels that they effectively demand a different industrial base for their production and maintenance, thus precluding any direct knowledge transfer from civilian to military shipbuilding. This effect is most pronounced with regard to submarines. As already discussed in Sect. 3.2, commercial ships are essentially little more than "very large steel boxes with relatively small engines and small crews, sufficient only to move the ship from port to port but nothing else," while warships have to operate and survive in a hostile environment, perform a variety of highly complex mission types, and give shelter and accommodation to relatively large crews. While commercial ships typically contain large portions of relatively empty space, naval vessels are universally

> [. . .] compact and outfitted with complex sensors, weapons, and communication and power distribution systems (often in duplicate). [. . .] Military ships have complex hotel systems to serve a crew that may number in the hundreds or thousands of people, and many have structural features to enhance performance and survivability (Birkler et al. 2005: 30).

Military requirements regarding hull survivability and shock resistance give rise to a set of demanding build standards, and usually use a different choice of steel plates requiring dedicated welding techniques. Welding and other construction skills acquired by the workforce in commercial shipbuilding projects, therefore, are not always directly transferable to military shipbuilding:

> Whereas merchant ship hulls are typically made of mild-steel plates of consistent thickness for ease of construction and lower cost, stringent weight and strength concerns in naval vessels often force the use of more advanced steel types and a wide variety of plate thicknesses. These demands require sophisticated steel fabrication and welding techniques that are rarely found in commercial shipbuilding (Collins and Grubb 2008: 29).

Moreover, in contrast to the relatively short average mission durations of fighter aircraft, for example, the capabilities of naval vessels "must be sustained for several months at sea, requiring a significant amount of equipment and provision storage." This is the chief reason for making naval vessels substantially different in design and construction than other kinds of weapon systems such as planes or tanks, and is ultimately one of the major cost drivers (Arena et al. 2006: 5).

These advanced mission requirements have several other consequences for industrial production. The task of cramming a large amount of highly sensitive electronic equipment alongside extremely heavy and sizeable missile systems into the limited space of a warship is formidable. Systems integration is not just a problem of making available the necessary space; rather, the question is how to operate every system on the ship without impacting negatively on the performance of the others. As physicist Friedman explains,

> Because a ship is relatively small, and because it must carry numerous antennas in close proximity to its superstructure, its radars may often be affected by reflection off both the ship and other antennas. For example, a whip antenna vibrating near a radar antenna may impose a modulation on the radar signal through its sidelobes (Friedman 2006: xiv).

This complex, sensitive, and heavy payload, in turn, gives rise to a need for an entirely different magnitude of propulsion power necessary to reach the desired top and cruising speeds, which further adds to the total payload. The level of complexity even among supposedly "basic" naval subsystems can be illustrated by an example given by Grubb and Collins (2008: 46):

> Whereas a typical supertanker may have 200 major pieces of mechanical and electrical equipment between two dozen systems on the entire ship, a modern destroyer can have this level of complexity in its propulsion plant alone.[15]

Due to their potential to mutually influence signal transmission, modern weapons and sensor systems pose even greater integration challenges than propulsion. These challenges multiply if components from different countries of origin have to be integrated within the same hull. This is usually the case in China due to its heavy reliance on Russian as well as French and indigenously developed (or -reverse-engineered) sensor technologies.

In summary, the main complicating design factors in warship building are the relatively large and heavy propulsion systems in the aft; the arrangement and integration of extremely heavy weapon and sensor systems; and the necessary accommodation and life-support needs of the sometimes very large crew complement. Additional problems are posed by structural integrity and shock requirements, protection against nuclear, biological, and chemical threats, signature reduction demands ("stealth"), firefighting facilities, and weapon storage facilities.

[15] The challenges of naval propulsion design were witnessed by the German Navy first hand during 2008–2012, when five newly-built *K130* class corvettes could only be commissioned after massive delays due to faulty gears that proved extremely difficult to repair. This problem occurred despite the fact that Germany is considered to be among the world leaders in conventional naval propulsion technology.

Changes in just one of the principal sub-systems of a ship often require design changes affecting the vessel's entire hull and can thus prolong the build phase considerably (cf. Birkler et al. 2005: 35–36). This is because in ship design, all factors tend to have an influence on a number of other factors. As Pugh explains,

> [...] a ship's displacement together with its maximum speed and length is sufficient to set narrow limits upon the power it requires. For a given standard of technology, this gives machinery weight and rate of fuel consumption. The role of the ship will fix sea-keeping standards (strongly influencing length), range (combining with fuel consumption to give fuel and stores needed) and weapons fit (thence, weapon weight). Sea-keeping and machinery, weapon and fuel weights together with length largely determine hull weight which, when added to those other weights, gives displacement [...]. Design is full of such closed loops at all levels of detail. They constrain the designer who is far from having a free choice of those parameters which determine cost if he is to produce a balanced and efficient design (Pugh 1986: 104).

The choice of a heavier helicopter than was originally envisaged during the design phase, for instance, can ultimately necessitate a complete redesign of the hull. This is because the flight deck and hangar must be larger, and this combined with the added deck reinforcement needed for safely operating a heavier helicopter can ultimately impact on the ship's weight and statics so much that changes in all the hull's major lines may be required. It is easy to see that the art of designing warships is not called "naval architecture" without reason.

Another complicating factor influencing the production process pertains to the enormous amount of different subsystems supplied by various producers. In effect, this means that a large number of people from heterogeneous organizations and companies need to communicate effectively and coordinate their efforts in a timely manner, which in practice is often an organizational challenge.[16] For these and many other reasons that cannot be fully discussed here, it was at the most a slight exaggeration when an Australian naval officer quoted by Till (2009: 96) remarked that "[...] short of major incursions into space, [...] navies continue to represent a uniquely complex and challenging manifestation of state effort."

Collins and Grubb (2008: 23) have identified altogether six areas where China's commercial shipbuilding prowess could prove influential for enhancing China's naval power. These are shown in the table in Fig. 4.7.

As Fig. 4.7 summarizes, the current prospects for quasi-automatic transfers of knowledge and technologies between China's commercial and naval shipbuilding ventures seem somewhat limited. There are, however, areas where more transfers could be expected in the future, e.g. in modern build methods, marine propulsion, and subcomponents technologies. In the latter case, China has not yet managed to

[16] During an industry event in 2008, I saw a chart depicting the envisaged organizational structure of the British-French carrier program. The transnational complexity of overlapping hierarchies and competencies shown was such that a senior colleague with naval shipbuilding project leadership experience off-handedly remarked, "It looks like they are busted—this will never work." It remains to be seen whether his assessment was fair, but in any case it illustrates rather nicely the problems posed by project complexity as perceived by people with industry experience.

	Area	Description	Presently seen potential for enhancing naval shipbuilding
1	Advanced ship-production methods	Modular shipbuilding methods (hull-block construction method), adopted especially at the fully modernized construction facilities and China's newly built "greenfield shipyards"	+ Reduced build-times, higher output, reduced cost – Trend towards shipyard specialization on either naval or commercial building also seen in China
2	System integration	Integration of a multitude of mechanical, electrical, cargo, and habitability systems as necessary in complex commercial vessel types (LNG tankers, cruise liners, but not bulk carriers) could offer limited carry-over effects to naval vessels (excluding however naval electronics and combat systems)	– So far little experience with more complex comm. vessel types; persistent dependence on foreign assistance for system integration in these projects
3	Hull construction and metallurgy	Basic welding and construction skills could be transferable, as could ship design software; however, commercial vessels use mild steel plates of uniform thickness, naval vessels use advanced steel of varying plate thickness requiring widely different welding techniques	– Persistent construction quality issues in Chinese commercial ships (material quality and welding) hint at related problems also in naval vessels
4	Subcomponent technologies	Habitability installations, deck equipment, non-combat related mechanical systems and COTS technologies in computer processors, networks, navigation and bridge control systems would be transferable and pose cost benefits	– So far no autarchy achieved; in 2008, only 40% of subcomponents in commercial ships are of Chinese origin, even less in the complex vessel types; fragmented status of domestic components industry
5	Marine propulsion technologies	Marine diesel engines and gas turbines are the "dual-use technology most directly transferable between commercial and military shipbuilding"	– Persistent dependence on imported and/or license-produced foreign designs of both diesel engines and gas turbines
6	Human workforce factors	Shipyard work experience and growth in the commercial industry fuels human resources development; foreign investments and transfers of technology for advanced projects lead to enhanced overall skills	+ Exposure to and experience in basic and advanced comm. shipbuilding projects of a growing marine work force adds considerably to the talent pool also for naval shipbuilding

Source: Compiled from Collins and Grubb (2008)

Fig. 4.7 Areas of possible commercial-military synergies in Chinese shipbuilding

achieve a satisfactory level of development. As of 2008, only 40 % of subcomponents in Chinese-built basic commercial vessels, and considerably less than this in the more complex vessel types, were of domestic origin, compared with 85–98 % on average in the case of Korean- and Japanese-built vessels. This unsatisfactory situation is often lamented by Chinese shipbuilding executives (Collins and Grubb 2008: 29–30). At present, Chinese domestic sub-component industries seem to be more effective in producing comparatively "low tech" products such as pumps, gears, fixed pitch propellers, and the like, while more complex components such as engines, more complex electronics, and navigation equipment are still frequently imported from foreign suppliers (cf. Collins and Grubb 2008: 31).

China's most severe obstacle to overcome for gaining self-sufficiency is the above-mentioned lack of domestic capacity for building the more complex types of marine propulsion systems. China has traditionally imported diesel engines from the Soviet Union (either direct imports, engines built under license with technical assistance, or reverse-engineered copies) for both commercial and naval vessels. Other past sources of supply include East Germany, Czechoslovakia, Poland, and Yugoslavia (Bussert and Elleman 2011: 9). It seems that the results of early Chinese indigenous design efforts were less than satisfactory in terms of their performance:

> The quality of these Chinese engines was rather suspect; the engines of several Romeo-class submarines sold to Egypt reportedly required extensive refurbishment after their delivery voyages in 1983–84 (Collins and Grubb 2008: 34).

In recent years, in contrast to the mostly gas turbine- and nuclear-powered naval vessels of the U.S. Navy, Chinese naval vessels have continued to rely primarily on imported or license-produced medium-speed diesel engines. These engines have mostly been supplied by European firms as in the case of the French SEMT-Pielstick PA- and PC-series, which were built into dozens of Chinese naval and commercial vessels alike. Others were sourced from MTU Friedrichshafen GmbH, Germany, which has granted the Shaanxi diesel engine works a licence agreement for its *Series 1163* high-speed diesel engines, which are aboard the *Type 052A "Lühu"*, the *Type 051B "Lühai"*, and the modern *Type 052B* and *Type 052C "Lüyang"* class destroyers. Bussert and Elleman also note the physical proximity of the Siemens AG office to the Guangzhou shipyard, which in their reckoning is suggestive of a licence agreement in the realm of diesel engine production for future Chinese warships (Bussert and Elleman 2011: 29).

Such diesel engines offer an attractive trade-off regarding performance and weight and consume much less fuel than gas turbines. Just the same, even though some of these engines are produced in China under license agreements, no major indigenous design has been developed so far. Significantly, Chinese-designed diesel engines account for less than 1 % of low-speed diesel engines on the larger commercial vessels built in China (Collins and Grubb 2008: 33–34). This problem is not unique to the maritime field. In aerospace development, too, "China's efforts to develop a sufficiently powerful and reliable fighter aircraft power plant have taken several decades but, to date, have not been successful" (Grevatt 2013c).

Gas turbines, which allow larger surface vessels to attain higher top speeds, but are comparatively expensive in procurement, maintenance, and fuel consumption, are one of the major high-tech areas that are affected by the arms embargo. Because the U.S. stopped exporting U.S. General Electric *LM-2500* gas turbines after 1989, China had to cancel its *Type 052A "Lühu"* destroyer program after only two hulls had been built (Bussert and Elleman 2011: 9). Ukrainian gas turbines became the only viable alternative:

> An alternate source was the huge Nikolayev plant in Ukraine that provided GT-25000M main propulsion for all major Soviet warships. The combined design and production team

of Mashproekt and Zarya sold four gas turbine engines to China in July 1997. Luhai was the first PLAN warship with combined diesel and gas turbine propulsion (CODOG) [sic; CODOG = combined diesel *or* gas], and two of the Ukrainian gas turbines supplemented the Shaanxi copy of 12V 1163 TB83 diesels on Luhai. The Ukrainian manufacturer Mashproekt has a license agreement for GT-25000 gas turbines with Xi'an Aero-Engine Corporation and the Harbin marine boiler and turbine works, which is the second-largest marine generator plant in China. Because of its central location in Manchuria, and also because of the railway construction carried out by both Russia and Japan, it is conveniently sited just up the railroad line from Dalian Shipyard (Bussert and Elleman 2011: 29).

For most of its newer destroyer classes, China has thus switched to Ukrainian *DA-80* turbines and is still entirely dependent on imports and license agreements for building these foreign designed gas turbines (Collins and Grubb 2008: 36–37). Most other Chinese surface ships, however, either continue to rely on diesel (all frigate classes) or steam propulsion (*Type 051C "Lüzhou"* class and *Sovremenny* class destroyers, the aircraft carrier *Liaoning*), a reliable but outdated propulsion system according to Western technical standards. Ongoing Chinese R&D regarding jet engines could possibly offer synergies to indigenous gas turbine development, but no breakthrough has been announced so far and the relevant programs have been greatly delayed (Grevatt 2013c).

A lack of gas turbines means that much larger and heavier diesel engines need to be installed in order to reach comparable top speeds in large surface vessels, taking up more space in the aft than gas turbines require. It should be noted, however, that gas turbines do not necessarily offer a significant strategic advantage in proportion to their relatively high price, maintenance and fuel costs. Most navies rarely drive their vessels at top speed anyway, and therefore seldom actually use the gas turbines in a CODAG or CODOG propulsion arrangement, relying instead on their more economic diesel engines for cruising speed.[17] On the other hand, the development of indigenous and reliable diesel propulsion to reduce dependence on foreign suppliers might be an even more pressing concern.

At the moment, only the nuclear propulsion systems aboard Chinese submarines are completely of Chinese origin, which of course represents an impressive accomplishment. According to reports collected by Wertheim, however, there were initial problems with excessive internal radiation levels in the *Type 091 "Han"* and *Type 093 "Shang"* class nuclear submarines, endangering the safety of the crew. In the meantime, though, this issue may have been alleviated by Russian technical assistance (cf. Wertheim 2007: 110–111; Wertheim 2013: 109–110).

Apart from the obvious strategic concerns raised by China's dependence on crucial foreign spare parts and subcomponents, and on technical assistance from abroad for their maintenance and repair, the deficiencies in the country's domestic

[17] A U.S.N. annual speed/time profile provided by Alexander et al. (2010: 70) shows that U.S.N. *CG 47* cruisers used top speed for only 3 % of the yearly operating time. Fifty-four percent of the time the ships used speeds of less than 15 knots. According to shipbuilding experts I interviewed in 2011, this speed/time profile is typical of most navies.

subcomponents industry development have other detrimental effects, such as in the field of commercial shipbuilding:

> The cost of purchasing subcomponents abroad directly offsets hard-currency profits, raises the exposure of China's shipbuilders to monetary exchange-rate fluctuations, and diminishes the 'pillar industry' effects of shipbuilding—that is, promotion of secondary industries in China. Moreover, delivery delays of outsourced components often interrupt shipyard production flows, further eating into profits through decreased efficiency (Collins and Grubb 2008: 30).

To conclude this overview of the Chinese marine subcomponents industry, it seems clear that China still faces problems in both electronic system integration and domestic propulsion system development. As O'Rourke notes: "None of these obstacles can be overcome swiftly, and none can be overcome merely by throwing more money at the problem" (2008: 21). It is therefore hardly conceivable that China will be able to simply leapfrog over its present shortcomings. Rather, a lengthy process of addressing problems on a case-to-case basis is to be expected.

Returning to the question of possible spill-over effects from commercial to military shipbuilding, the major strategically relevant effect could conceivably derive from the growth of an increasingly experienced domestic marine work force, and also perhaps from the exposure of parts of this workforce to foreign knowledge as a result of commercial shipbuilding joint-ventures or complex foreign-assisted shipbuilding projects. Nevertheless, in naval shipbuilding, much more durable and intensive technological partnerships than these are normally required to successfully complete a project of more than rudimentary complexity.

One final factor that should be weighed carefully is the theoretically possible conversion of China's growing merchant vessel fleet to military use in times of conflict:

> [. . .] during wartime merchant ships might be adapted to serve as auxiliaries, amphibious assault ships, and minelayers. Accordingly, the growing Chinese-flag merchant fleet represents a degree of latent military potential. Indeed, Chinese analysts have closely examined Britain's rapid conversion of forty-nine merchant vessels to military roles during the 1982 Falklands War. It should also be noted that the Chinese government maintains a vessel tracking system known as China Ship Reporting System, or 'CHISREP,' to which all Chinese-flag merchantmen must ostensibly report, no matter their location. Knowledge of where these vessels are at all times across the globe could have military utility (Collins and Grubb 2008: 4).

In addition to this, the impressive build-up of commercial building and repair capacities could certainly form the backbone of large-scale serial wartime production of naval vessels, should this become necessary—a key factor for winning conventional wars according to military historian Keegan (2004: 314).

In order to assess the state of China's access to foreign knowledge, the next section will review China's future potential and past history of forming international industrial partnerships with strategic technological allies, and their impact on China's technological development in the naval sector.

4.6 China's Transnational Shipbuilding Partnerships

An often-overlooked aspect of military innovation involves the strategic long-term alliances that are formed between key industrial players (usually, but not always state-owned), the national governments that back them, and the governments of customer countries that aim to achieve technological progress. These alliances are in many cases embedded within a system of large-scale strategic government-to-government exchanges that can be part of a much wider framework, often also encompassing civilian technology transfers such as nuclear power plants or civilian infrastructure development. They typically demand so-called "offsets," which can pertain not only to arms-related industries, and they can also sometimes include political bargaining in the international arena such as the exchange of political support within international organizations for the respective government's strategic aims.[18] Partners in such long-term arrangements frequently experience periodic friction, e.g. because of alleged reverse engineering of advanced technologies by the recipient country, supposedly "insufficient" transfers of technology by the donor country, and mutual conflicts over cost overruns or delays. Despite occasional tensions, however, these relationships usually prove extremely durable, especially if a formal security alliance or defense co-operation agreement also exists. One major reason for this may be the fact that the more complex types of military hardware require constant access to a myriad of spare parts and the producer's maintenance services. The seller of the item (or rather, the government of the seller's home country) thus has undeniable political leverage over the recipient:

> The export of more complex equipment [. . .] usually implies that the exporting countries provide the support capacities for the equipment, which could be withdrawn at any time. This means that there is more uncertainty on both sides: the arms suppliers have less control over the degree to which technology is transferred, while the recipients cannot be sure, whether the support for imported systems can be secured in the long term (Krishnan 2008: 7).

The political ramifications associated with large-scale arms transfers should therefore not be underestimated; they strongly resemble marriages of convenience within a cultural climate where divorce is not really an option.

4.6.1 Foreign-Induced Disruptions in China's Naval Shipbuilding

Since the founding of the People's Republic in 1949, China as a recipient of naval technology has been on the receiving end of the above-described punitive mechanism on several occasions.

[18] These observations are based on extensive market and competitor analyses conducted by me during 2007–2010 in the shipbuilding industry using public domain as well as internal company data.

Similar to India, China's naval shipbuilding effort has relied to a large extent on Soviet, and later—after a brief Western interlude—Russian technical assistance. Nevertheless, this assistance has never been entirely comprehensive. First, in 1960, the Soviet Union withdrew its technical assistance because of the Sino-Soviet split, forcing China to develop its nuclear weapon programs on its own. Then, in 1989, after a brief 9 years of access to Western arms markets, the U.S./EU arms embargo brought a halt to several running naval projects, since vital subcomponents such as American gas turbines could no longer be obtained. China has officially been excluded from strategic arms trade partnerships with the leading Western system houses and naval shipyards ever since, and is limited to trading with these powers in dual-use technologies.[19] From 1989 to the present, China's major strategic arms trade partnerships have included Russia and Ukraine (as suppliers) and Pakistan (as a recipient). These relationships are therefore the most influential external factors with regard to China's military-technological progress. Their influence on China's naval modernization should not be underestimated. As Cheung notes:

> Another insight on the important linkage between innovation and access to foreign knowledge and technology supports the findings of developmental-state theorists who point out that openness to foreign knowledge is one of the most critical factors in the ability of late-industrializing states to catch up. This is especially relevant in the crucial area of absorptive capacity (Cheung 2009: 18).

At the same time, China's export activities in Pakistan and elsewhere provide China with opportunities to come into contact with more advanced technologies, as these customer countries also typically have access to Western weapon systems. One question to be addressed is therefore the following: How much of China's naval innovation is truly indigenous, and to what extent is China's technological progress in fact dependent upon these various partnerships?

As the previous section showed, China's dependence on foreign sources of supply for several critical subsystem technologies still persists. The remainder of the chapter will therefore outline China's role as a recipient of naval technology transfers, before then discussing the effects of China's emerging role as a supplier of naval technology to less advanced powers as a result of its strengthened arms exports.

4.6.2 China as a Recipient of Naval Technology Transfers, Pre-1991

The historical development of the PLAN since the founding of the PRC and during the Mao era has been covered in detail by Cole (2010: 7–18) and need not be recounted here. Figure 4.8 gives an overview of the major stages in Chinese naval

[19] For a thorough discussion of the strategic implications of dual-use technology transfers to China from a European viewpoint, see Stumbaum (2009).

Source: Drawn using data from Collins and Grubb (2009), Cole (2010) and Bussert and Elleman (2011)

Fig. 4.8 Historical stages of China's naval modernization effort

shipbuilding development with regard to major disruptions caused by the withdrawal of outside assistance.

When the turmoil of the Mao era had ended, China embarked on naval modernization in the 1980s and followed a "two-track" approach. Initially, China was almost fully dependent on Soviet transfers of technology. The relationship was uneasy, however, even before the Sino-Soviet split, because the Soviets were reluctant to share their then state-of-the-art equipment with China:

> Even during periods of so-called Sino-Soviet friendship, the USSR withheld modern and even standard FC [fire control] equipment on Chinese exports, with the goal of keeping the Chinese navy weaker than their Soviet counterparts, which increased after the 1960s split. For example, the Soviets removed Strut Curve and Drum Tilt fire control radars from the Osa-class missile attack craft sent to the PRC in the early 1960s and also withheld the 1952-vintage Owl Screech and Hawk Screech gun fire control radars (Bussert and Elleman 2011: 5).

China's arms industries were therefore forced early on to make up for these deficiencies, which prompted indigenous development based on the reverse-engineering of imported naval systems (Bussert and Elleman 2011: 6). This strategy prevailed after the Sino-Soviet split. China's existing (but technologically outdated) Soviet-based vessels were modified, while follow-up programs were developed simultaneously, using new designs that were based on the old ones. All these attempts used Soviet design principles as their starting point. There is thus still a powerful technological legacy today stemming from this early dependence upon vintage (i.e. World War II-era) Soviet vessel designs in the Chinese fleet. In fact, as of 1989, the list of naval electronics built into Chinese vessels shows that there was little change in basic equipment such as navigational radars ever since China began to import these types from the Soviet Union during the 1950s and 1960s. During the 1980s, the same radar systems were still installed on almost half of the PLAN's ca. 1,000 naval vessels (Bussert and Elleman 2011: 2). Even though the past two decades have witnessed considerable technical improvement overall, it is nonetheless striking that relatively old technologies such as the 1950-vintage Soviet "Knife Rest" early warning radar are still in use today, even aboard newly developed ship classes (Bussert and Elleman 2011: 4).

After the Mao era, a more diverse choice of technologies became available for integration into China's newly designed ships (Medeiros et al. 2005: 138). As a result of Sino-Soviet tensions and warming relations between the U.S. and China during the 1970s and 1980s, the Reagan administration finally began to relax high tech transfer regulations towards China. The relative technological backwardness of China's navy at that time, especially in the field of naval electronics, is clear when considering the spending share on electronics per sailor. As of 1985, China spent only US$560 per sailor on shipboard electronics per year, while Russia and the U.S. spent US$4,000 and US$5,700, respectively. According to Bussert and Elleman,

> This discrepancy perhaps explains why PRC warships went to sea without air search or fire control radars and why innovative domestic designs seemed to remain in prototype stages for years (Bussert and Elleman 2011: 14).

In 1984, China was granted eligibility for the U.S. system of Foreign Military Sales (FMS). The major U.S. arms export support program, it uses a financing tool named Foreign Military Finances (FMF) for arms purchases by foreign governments from U.S. companies. Several arms deals were concluded in the following years until 1989 (Zhang and Hyer 2001: 96–98). At the time, this novel U.S. arms transfer strategy, which included assistance in several naval programs, was "aimed mainly against the USSR" (Bussert and Elleman 2011: 2). From the Chinese perspective, this was very welcome, not least because it allowed for an upgrade of China's production facilities:

> [...] defense enterprises and their civilian counterparts concentrated on the import of large-scale plant equipment that was based on the requirements and orders of government ministries. This approach was reminiscent of the wholesale importation of Soviet factories during the 1950s. The primary objective of the procurement strategy was to renovate outdated manufacturing capabilities. The technological hardware acquired for the defense economy included production equipment to produce color televisions, numerically controlled lathes, and oceangoing vessels (Cheung 2009: 61–62).

Apart from such infrastructure upgrades, there was also an influx of new knowledge that became available to Chinese experts for the first time through direct interaction with Western counterparts. Consequently, for a few years only during the second half of the 1980s,

> China enjoyed good military-to-military relations with the West, [and] delegations from the defense economy regularly visited advanced Western countries to gather information and selectively acquire off-the-shelf weapons systems. A few joint-venture projects were permitted in technologically critical areas in which the defense economy was unable to meet the requirements (Cheung 2009: 91).

Such cooperation included joint projects in avionics and in the construction of fighter-aircraft power plants.

All these exchanges came to a sudden halt after June 4th, 1989. Hundreds of already concluded contracts were suspended, and at least 300 export agreements were blocked. U.S. sanctions were only slightly relaxed after China signed the Non-Proliferation Treaty in 1994. Since then, the remaining U.S. sanctions have been used as "leverage to encourage China to cooperate with the United States on a host of international security concerns such as nuclear proliferation and missile technology control" (Zhang and Hyer 2001: 100). They pertain mainly to four especially sensitive areas: missile technology, nuclear technology, intelligence-gathering technology, and anti-submarine warfare technology. All of these are at least to some degree relevant for naval development (Zhang and Hyer 2001: 104).

4.6.3 Trading with Russia and Ukraine, Post-1991

Fortunately for China, the almost simultaneous demise of the Soviet Union in 1991/1992 presented a new avenue for gaining access to foreign technology. Cash-strapped Russia proved much more willing to export advanced arms technologies to China in the 1990s than the Soviet Union had ever been during the preceding decades. The reasons for this were almost purely economic. Until the mid-1980s, the Soviet Union had been a major arms producing state on par with the U.S. The sharp economic decline suffered by Russia after 1992 led to a very difficult situation for the Russian defense industrial base. Its share in the world's arms production dropped dramatically, from one third of the world's total production prior to 1985 to a mere 4 % by 1996. According to Tsai, in 1992,

> [...] nearly 21 percent of Russian defense enterprises stood on the verge of bankruptcy, while 43 percent would soon be classified as barely viable. By 1993, the financing of arms production had been reduced by nine times, and Russian military orders were fulfilled by no more than 10-15 percent of the defense industry's production capacities (Tsai 2003: 121).

Moreover, the Soviet Union's major naval shipyards were inherited by Ukraine, not Russia, after 1992. And, several large-scale naval shipbuilding projects destined for the Soviet Navy were left unfinished, including the *Varyag* aircraft carrier hull later acquired by China and re-christened the *Liaoning*. The Russian arms industries have since recovered somewhat, but still, according to Russian news reports from 2009 cited by Neuman,

> [...] only 36 percent of Russia's strategic defense companies are solvent—and many of these survive only due to export orders—while about another 30 percent are on the verge of bankruptcy (Neuman 2009: 74).

Russia's arms industries, whose products are marketed abroad by the Russian export agency Rosoboronexport, are with very few exceptions state-owned. Their dependence on export orders is nevertheless high:

> In spite of the growth of domestic procurement, export orders have continued to account for about 60 percent of sales for the leading Russian defense companies. These sales and the revenues they earn are considered so important that some R&D and production programs are tailored specifically for foreign customers rather than the Russian military itself (Neuman 2009: 62).

The politico-strategic importance of the arms industries in Russia's domestic politics becomes obvious when the sector's size and economic impact is taken into account. Estimates for the year 1994 assume that the defense industrial sector's output contributed to nearly 60 % of Russia's entire industrial production and had about four million employees working at ca. 2,000 facilities, with 80 Russian towns—so-called "defense industry towns"—almost entirely dependent on the local defense industries for their economic viability (cf. Tsai 2003: 121).

Sales information published by Rosoboronexport in 2014 indicates that Russia exported arms technologies worth US$13.2 billion to 60 countries worldwide in 2013, and, further, that a group of only six countries accounted for ca. 75 % of these sales: Algeria, China, India, Indonesia, Venezuela and Vietnam. Seventeen percent of the Russian exports were naval technology (Dunai and Smith 2014). Assessments published by *Jane's Defence Weekly* in 2011 conclude that Chinese defense imports from Russia during 2000–2011 had an annual value of ca. US$1–3 billion. On the whole, China accounted for ca. 20 % of Russian export orders during 1997–2007 (Anderson 2011a). Russia is thus China's primary source of modern defense equipment in quantitative (as well as qualitative) terms, while the value of Ukrainian exports to China was taxed at an average of a few hundred million US $ per year (Grevatt 2011a). Meanwhile, the fact that Russia's naval industries have not yet fully recovered is illustrated by Russia's relatively new status as an importer of Western naval technology. The deal concluded with French shipbuilding group DCNS in 2011 over the sale of two *Mistral* class LHDs (or helicopter carriers) indicates a lack of Russian design capability for producing such advanced amphibious assault ships. This has been freely admitted by Russian officials, even though the LHDs' level of technical complexity is much lower than guided-missile destroyers, for instance. Russia's naval industries therefore cannot be considered technologically on par with the most advanced Western shipyards, a fact that is confirmed by Russian official reports. According to data and analyses cited by Dunai and Anderson, R&D spending in Russia remains at markedly lower levels than in all the Western countries (ca. 33 % of Germany's and 5 % of U.S. R&D spending), while productivity in the Russian shipbuilding industry was also deemed to be much lower. Build times, namely, were estimated to be between two and two and a half times longer when compared to foreign competitor yards, and required construction hours were estimated to be 3–5 times higher (cf. Dunai and Anderson 2013).

Many Soviet era designs are still in use today, however. Some new Russian ships, such as the *Steregushchy* (export version: *Tigr*) class corvette recently sold to

Algeria,[20] are functional, proven designs offering relatively great firepower considering their comparatively low price. For developing navies still in the process of informationization, such vessels can offer marked cost control and risk minimization advantages over more expensive Western high-tech designs that overall require a higher level of crew training and maintenance quality to operate safely.

China, along with its strategic rivals India and Vietnam, has therefore been one of Russia's major export customers for quite some time, which has led to a powerful legacy of Soviet/Russian defense technology in China's navy. From an institutional viewpoint, this means that the PLAN is probably best disposed to absorb advanced technologies from this and other post-Soviet sources such as Ukraine due to established maintenance and training routines.

Moreover, there are indications that China's naval shipbuilders still have a few things to learn from Russian naval technology. In 1998, China decided to import Soviet-era vessels such as the four steam-powered *Sovremenny* class destroyers as complete weapon systems during a time when several indigenous destroyer programs were simultaneously underway. Starting from 1994, China opted to purchase altogether 12 *Kilo* class diesel submarines, even though it was building its own indigenous submarine designs at the same time. This trend has continued, and indeed another similar sized large-scale arms deal is reportedly under negotiation. This deal would include the sale of 24 advanced *Su-35* fighter aircraft and 4 *Lada* class diesel submarines, which would represent a noticeable improvement over the existing *Kilos,* which nevertheless are still deemed the most capable conventional submarines in the current Chinese inventory (cf. Foster 2012; Johnson 2013b). China could thus take advantage of the major "naval and maritime redevelopment plan" officially announced by Russia in early 2013, which includes plans for heavy investments in Russia's shipbuilding industry infrastructure as well as large series of new buildings for the Russian navy. According to this ca. US$42 billion plan, Russia aims to build 50 modern surface vessels and 20 submarines by 2025 (Dunai and Anderson 2013). If implemented, it could offer synergy effects to Russia's major naval customers including China. Russia, on the other hand, also likely evaluates the arms trade relationship with China in strategic terms:

> While commenting on Russian arms transfers to China, one Russian military officer stressed that apart from the purely economic advantages, supplying weapons to China could provide a chance of influencing China's military and political strategy and maintaining ally-like relations with that country. As such, Russia might increase its political leverage in the region and in relation to the U.S. (Tsai 2003: 123).

Notwithstanding such speculation, the Russian-Chinese export relationship has never been free of friction (cf. Tsai 2003: 173–175). There have been several indications in recent years that Russia's defense industry has become more cautious in its dealings with China due to repeated reverse-engineering issues with Russian

[20] According to a recent report, two Russian *Tigr* class corvettes are currently under construction for Algeria following a contract concluded in 2011, and are due to be handed over in 2015 (Kingston 2014).

systems. Reverse engineering, as explained by Stumbaum, "defines the process of discovering the technological principles of a device, object or system through analysis of its structure, function and operation" (Stumbaum 2009: 11, fn. 17). Chinese engineers seem to employ this strategy fairly regularly, as the following excerpts from *Kanwa Asian Defence Magazine* interviews with Russian arms industry sources suggest.

In its December 2009 issue, *Kanwa* cited several Russian defense industry representatives, who implied that China regularly imitates Russian equipment. O. Azizov, the then Director of the Navy Equipment Department at Russia's official arms export agency Rosoboronexport, stated at a press conference that China obtained the *Mineral-ME* serial radar system and the *Fregat-ME* radar systems as a result of their acquisition of the *Sovremenny*-class destroyers. Shortly afterwards, "almost identical 'cloned versions' of these radar systems made an appearance on China's indigenous 054A [*'Jiangkai II'* class] FFGs" (cf. 'Frictions between Russia & China' 2009: 15). An unnamed representative from Russia's Typhoon Design Bureau elaborated:

> China purchased a total of 8 Mineral-ME radar systems along with a large amount of spare parts, sufficient to assemble several additional sets. We had some suspicion at that time. Shortly after that, the Chinese invited our technical expert group to visit and pleaded us to help assemble 'back-up radar' in China. They had sufficient parts, but our expert group refused, as the assembling work was not easy and required special equipments as well as approval from the authority. Following that, our experts received several reports from the Chinese side, claiming that the systems we provided had encountered technological problems. After our technical experts examined the systems, we noticed some of the parts were not original, and this meant that they had started imitating the component parts of the radar. Through such contacts with our experts, they intended to identify the problems in their imitation projects.

A different representative of the Russian arms industry was also cited as follows:

> [T]hey did not acknowledge that they imitated our radar. They said that they designed the radar independently and had full intellectual property rights of the equipments. We believe that anyone with common sense would know from the physical appearance of those systems that they were the same systems [...] the distrust in China within the [Russian] industry is growing, and we have decided at strategic level [sic] that the Chinese market will no longer [be] a priority ('Frictions between Russia & China' 2009: 15).

Kanwa reporters also noted that one of the 12 *Kilo* class submarines seemed to have been continuously berthed for at least 2 years. This suggests that China may have used this boat for reverse engineering and maintenance testing purposes. According to the report, the Russian Ministry of Defense has listed the Chinese *Type 041 "Yuan"* class submarine as an "imitation" of the *Kilo* class (cf. 'Maintenance of 956E DDGs' 2009: 17).

Regarding maintenance and upgrade procedures of the *Kilo* class submarines, numerous conflicts between China and Russia have also in effect lead to a closer *entente* between China and Ukraine. According to an anonymous Russian industry source, talks about maintenance procedures between Russia and China dragged on for years because China insisted that all the necessary maintenance and upgrade work be done in China with the help of Russian experts, who were expected to train Chinese personnel. The Russians, however, wanted to perform the work on specialized yards in Russia. The source further said that China then turned to the Sevastopol shipyard in Ukraine, where it acquired *Kilo* class maintenance blueprints, albeit for an older version, and started the maintenance work in China independently (cf. 'Russia and China have a widening difference' 2009: 16). The degree of success achieved through this unusual route is uncertain. The *Kilos* are rarely seen on mission, which could be a result of maintenance problems. Strained relations with Russia's naval shipbuilding industry also mean that China might be receiving less support from Russia for its carrier program, apart from the supply of some subsystems. Ukraine, on the other hand, has been instrumental in China's carrier program, and provided extensive technical assistance. This extended not only to the completion of the ship hull, but also to training.

For example, as of November 2013, Ukraine had offered Chinese fighter pilots access to its aircraft carrier training base NITKA near the city of Saky on the Crimea peninsula, which is equipped with all necessary facilities to simulate landings on a carrier such as the *Liaoning*. This base was inherited by Ukraine after the Soviet Union's demise, even though the Ukrainian Navy has no carrier capability of its own and therefore no use for NITKA. Russia, for its part, still has no comparable installation on its own territory to train its pilots. NITKA was therefore periodically leased to Russia under an agreement, which was suspended however for a time after 2008 because of Russia's invasion of Georgia. In late 2013, Ukraine reportedly made plans to offer NITKA for exclusive use to the PLA Navy Air Force (PLANAF), and to India as a back-up plan, because Russia was unwilling to pay a higher rent. Chinese officials have reportedly shown a great deal of interest in this facility and its personnel's expertise, despite the fact that China is currently using an indigenous training facility near Xi'an which was modelled on NITKA (cf. Johnson and Hardy 2013).

China's uncomfortable situation of quasi-dependence on Russian arms sales after 1991 clearly contributed to Ukraine's naval expertise becoming an ever more important bargaining lever, as it provided China with an alternative source of vital technical assistance. This situation could now be on the verge of changing, as Russia's de facto annexation of the Crimea peninsula in early 2014 included the takeover of all the naval facilities (including NITKA) in that region. Given Russia's seizure of most of Ukraine's naval assets—ships, shipyards, and training facilities, which during the Cold War constituted about 30 % of the Soviet Union's total shipbuilding industry—and that Ukraine may opt to associate more closely with the West in order to gain protection from Russia, a situation could emerge in which Ukraine's technical expertise is either under the influence of Russia or subject to the U.S./EU arms embargo and thus no longer available to China (cf. Moore 2014; Kofman 2014). In either scenario, a one-sided Chinese dependence on a single supply source—Russia—would be the probable outcome. The dramatic events of early 2014 following Russia's annexation of Crimea, which are still unfolding as of the time of this writing, are thus almost certain to have a tremendous relevance for

China's future defense industries. These developments, however, will require a more detailed evaluation in the future once the situation has been played out.

4.6.4 Coping with the Arms Embargo

Chinese leaders have often publicly complained about their "unfair" treatment by Western countries in the context of the arms embargo. They point out that a double standard seems to be at work when China is still blacklisted almost 25 years after the events on Tian'anmen square in 1989, while many other authoritarian regimes are allowed to receive massive Western arms transfers. The official Chinese position on the significance of the events of 1989 and the arms embargo is concisely summed up in the following quote from an article on the CCP's official news network website that cites excerpts from a book on the theoretical foundations of "socialism with Chinese characteristics" by the *People's Daily's* deputy editor-in-chief, Chen Junhong (陈俊宏, *1954):

> In June 1989, China pacified an internal political uprising. This was a matter of domestic Chinese politics. Nonetheless, Western countries with the U.S. as their lead distorted the facts and used this as a pretext to impose 'sanctions' on China. The leaders of the G7 and the European Council also issued declarations imposing 'sanctions' on China. This caused major setbacks in the relations between China and the U.S.-led developed Western countries (Chen 2012, my transl.).

Judging from the often somewhat emotional reactions to the embargo that the author of this study personally witnessed when interacting with Chinese functionaries while working in the naval shipbuilding industry, it seems that the embargo is not only perceived as an obstacle to China's naval progress, but also as a humiliating punitive measure that has been meted out arbitrarily and without justification by Western nations. Together with the perceived strategic "encirclement" by U.S. military bases described in Sect. 2.3, the embargo is apparently seen as an instrument of strategic containment rather than a true attempt to take a stance on human rights abuses committed in 1989.

In order to close the technology gap, China's arms industries have therefore resorted (no doubt with high-level political backing) to a variety of coping strategies, among them "academic exchanges and 'state-sponsored espionage'" in which foreign dual-use technologies are often exploited for their military potential (Anderson 2011b). Apart from naval export orders and voluntary technology transfers by willing partners such as Russia and Ukraine, "[l]ess voluntary transfers include forced technology transfer in Joint Ventures, certification practices, and espionage activities" (Stumbaum 2009: 11, fn. 17). A proverbial complaint often made by foreign technology providers thus concerns the alleged Chinese propensity for reverse engineering foreign systems, Western and Russian alike. To this end, China seems to have repeatedly obtained access to "captured" military technology, e.g. unexploded and exploded shells of U.S. cruise missiles collected from Iraq, Serbia, Afghanistan, and Sudan; and wreckages of aircraft such as the *F-117* stealth

Source	No.	System	Year(s) of delivery	Vessels
France	2	100 mm Naval Gun	1989	1 Jianghu II FFG
	14	Castor-2 Fire Control Radar	1994–2002	2× Lühu, 1× Lühai, 3× Lüda I DDG, 8× Jiangwei II FFG
	6	DRBV-15 Sea Tiger Radars	1987–1999	2× Lühu, 2× Lühai, 2× Lüda I DDG
	5	DUBV-23 Sonars	1991–1999	2× Lüda, 1× Lühai, 2× Lühu DDG
	2	DUBV-43 Sonars	1994–1996	2× Lühu DDG
	336	R-440 Crotale SAM	1990–2002	2× Lühu, 1× Lühai, 3× Lüda DDG
	28	AS-365/AS-565 Dauphin Helos	1987–1991	Various
Italy	17	RTN-20S Fire Control Radars	1991–2001	2× Lühu, 1× Lüda III, 1× Lühai DDG, 6–7 Houjian PTG
Russia	10	Ka-27PL (Helix-A) Helos	1997–2000	Various
	4	Fregat/Top Plate Air Surv. Radars	2004	2× Lüzhou, 2× Lüyang I DDG
	8	MR-90/Front Dome FC Radars	2004	2× Lüzhou, 2× Lüyang I DDG
	144	48Ng/SA-10 Grumble SAM	2002	2× Lüzhou DDG
	264	9M317/SA-17 Grizzly SAM	2005	2× Lüyang I DDG
	Kilo-class SSK, Sovremenny-class DDG (as complete weapon systems)			

Source: SIPRI data cited from Grubb and Collins (2008: 48)

Fig. 4.9 Foreign weapon systems on PLA Navy vessels

attack aircraft shot down in Serbia in 1999. In 2011, U.S. defense officials voiced concerns that Pakistan could allow China access to the wreckage of an unidentified stealth helicopter that was left behind during the Bin Laden assassination on Pakistani soil on May 1st, 2011 (Parsons 2011).

It is noteworthy that some European countries, above all France, tend to interpret the EU arms embargo rather loosely, as can be seen from the table in Fig. 4.9, which lists foreign weapon systems aboard Chinese vessels. A more complete list of non-Russian military transfers to China since 1989 can be found in Archick et al. (2005: 37–41). France has delivered many naval components to China and is pressing hard for lifting the EU embargo, along with Spain, which reportedly went as far as offering a basic carrier design to China in 1995.[21] According to Bell:

> France and Spain remain the two strongest European backers of lifting the embargo, although Spain appears unwilling to move unilaterally and risk damage to its standing within the bloc. Both member states appear strongly motivated by the prospect of boosting sales for their domestic defence companies. Officially the UK is against lifting the embargo, although there seems to be room for manoeuvre should the issue arise again (Bell 2011).

DCNS, France's state-owned shipbuilding giant, is also active in the business field of civil nuclear power plants, and in 2009 concluded deals with China Nuclear Power Engineering Company (CNPEC) for deliveries of heat exchangers (cf. Rouillé 2009). This could be seen as a market-opening strategy to establish a timely foothold in anticipation of the lifting of the arms embargo. Such an approach would be consistent with the general thrust of the French arms export strategy observable in developing and newly industrialized countries all over the world,

[21] Cf. Medeiros et al. (2005: 149), citing reports in *Jane's Defence Weekly* from Feb. 18, 1995, p. 18.

e.g. in the Maghreb or in Latin America, which often combines large-scale nuclear co-operation projects with naval and other arms exports, and is backed by extensive political support from France's government, armed forces and secret service. All these institutions work in close co-operation, while at the same time attractive financing conditions are offered by French state-owned banks.[22] Along similar lines, France was also the first NATO country in 2011 to conclude an export deal for naval vessels with Russia—two *Mistral* class amphibious assault ships (LHDs, ca. 21,000 t at full load)—in the face of massive opposition from other NATO powers, and it has also concluded partnerships with Russian naval design bureaus (cf. Anderson 2007). This fairly aggressive export strategy seems to be part of the French national industrial policy aimed at securing the welfare of France's rather oversized state-owned shipbuilding and defense industry work force (cf. Lewis 2007).

According to SIPRI data cited by Stohl (2006), despite the embargo, France was responsible for 73.2 % of all EU weapon deliveries to China between 1989 and 2004. From 2009 to 2011, France's defense sales to China doubled, reaching an amount of ca. 10 % of total Chinese arms imports. Even the UK, which is still officially against lifting the embargo, reportedly had defense-related sales amounting to 5 % of the Chinese total in 2009 (Bell 2011).

Given EU estimates that the Chinese arms market for European systems could amount to US$15 billion[23] if the ban were lifted, the above-described patterns of behavior may seem understandable from a national industrial policy perspective. Nonetheless, the question remains whether such considerations should be decisive in a strategically important question such as the one at hand, in view of the potentially grave repercussions that lifting the embargo could have for the entire region. Whatever the original reasons for imposing the embargo may have been and whether or not the embargo is indeed "fair," its existence until now has had an undeniable dampening impact on China's military build-up that the Asia Pacific region has by now become accustomed to. Lifting it would almost certainly result in a surge of French and other European arms exports to China, thus accelerating the pace of China's military modernization to a degree that could endanger the fragile balance of power in Asia, both in actual fact and in terms of regional threat perceptions. Given that China's government is still unwilling to change its official stance on the Tian'anmen protests, that it has not yet concluded a formal peace treaty with Taiwan, and that it seems unwilling to compromise with other nations in the various islands issues in the East China and South China Seas, it would seem

[22] On the official government-lead streamlining of French arms export-related actors under President Sarkozy, see Lewis (2007). For an example of a typical comprehensive naval arms export package including offsets and financing support as offered by France to Bulgaria during 2007, which included an exchange of French political support for freeing Bulgarian hostages in Libya, see Koinova (2007). For the extent of the Franco-Brazil strategic partnership, see Rezende (2008).

[23] This figure is mentioned in Smith (2004).

ill-advised from a Western perspective to consider lifting the embargo even partially, especially if only in exchange for limited and short-term economic gains.

In contrast to a number of EU states, the U.S. has consistently pushed for maintaining the embargo. Moreover, the U.S. embargo is legally binding, whereas the EU embargo only has the character of "trade restrictions [that] take the form of a non-binding statement of common position with legislation enacted at a national level" (Bell 2011). Initiatives within the EU for a relaxation or lifting of the embargo have met with strong criticism from the U.S., which cites both the human-rights situation and overriding security concerns as reasons to keep it in place. In spite of such U.S. concerns, an Australian company—AMD Marine Consulting—has chosen to supply the Chinese navy with design support for developing a catamaran platform capable of carrying antiship missiles: the *Type 022 "Houbei"* class missile catamaran (250 t at full load) featuring water jet propulsion, shallow draft, and stealth features, with each vessel able to carry eight *C-803* missiles. As of 2008, 30 such vessels apparently based on the civilian *AMD 350* catamaran design had already been built for the PLAN, with 68 such ships listed by 2013 (Wertheim 2013: 124–125). A large force of such fast, stealthy, missile-carrying craft could play a significant role in a potential Taiwan crisis scenario with U.S. and allied forces (likely including Australian vessels), making this particular naval export decision by an Australian company rather peculiar (cf. 'AMD Building Catamarans' 2008). In any case, this example illustrates the less than perfect control exercised by Western governments over individual dual-use technology exporting companies.

In contrast to China, other rising powers such as India, Brazil, and even Russia have been official recipients in the past few years of advanced Western naval technologies. While Brazil has secured extensive French assistance in its nuclear submarine program, India is in the especially fortunate position of enjoying the best of both technological worlds, namely Western and Russian. While India remains one of Russia's principal customers for advanced naval technology, which includes an unprecedented leasing agreement for a nuclear-powered submarine, the U.S. significantly relaxed its export restrictions in 2010 and has since concluded many arms and dual-use technology deals with India, one involving "transport aircraft, maritime reconnaissance platforms, *M777* 155 mm/39 caliber lightweight howitzers and missile systems" as well as gas turbines ('Annual Defence Report' 2010). India's other sources of advanced technologies have been France (diesel submarines), Germany (diesel submarines), and Italy (carrier propulsion technology).

4.6.5 China as a Supplier of Naval Technology

Partly to make up for its comparatively limited access to some kinds of high technology, China's government has decided to promote Chinese arms exports to a number of less developed countries. Several reasons may be identified for this decision. China, namely, can use arms export relations as a power tool to ensure political support by recipient countries, e.g. in the Taiwan question; as a means to gain a greater exposure to international defense markets and secure stronger access to advanced dual-use technologies; as well as a source of additional revenue for its domestic R&D (Grevatt 2010). The political aspects are probably of paramount importance, however. After all, as Pugh notes:

> Few people, or nations, seek to antagonize those from whom they buy, to whom they sell or to whom they owe money, especially if the suppliers, customers and creditors have the means to enforce their concepts of fair trading. A poor nation probably needs its trade with a rich one more than the rich nation needs the reciprocal trade just as it lacks military muscle should 'push come to shove'. Thus, it is inclined to appease and acquiesce rather than to stand on principle and object (Pugh 1986: 24).

Given China's need to ensure the compliance of a majority of nations with its "One China" policy, among other things, its arms trade relationships with lesser-developed countries can be seen as a useful tool for enhancing China's political status in the world.

As a result of the successful boosting of its defense industries, China has increasingly become an exporter of military and, first and foremost, naval equipment, especially over the past 10 years. This development has high-level political backing: China's 12th Five-Year Plan (2011–2016) specifies that China's defense industries should gain a stronger presence in foreign defense markets (cf. Grevatt 2010). According to the SIPRI estimates cited by Johnson, China's total arms exports increased 162 % between 2008 and 2012. Since 2013, China has supplanted the UK as the world's fifth largest arms exporter, behind the U.S., Russia, Germany, and France (cf. Bokhari 2013). Unnamed Russian and Ukrainian arms industry sources interviewed by *Jane's Defence Weekly* recently voiced apprehension about China's arms industries becoming a global force that was poised to "overwhelm" competitors (Johnson 2013a). China's naval exports are marketed by two large industrial trading houses, each of which is associated with one of the two major shipbuilding holding groups. China Shipbuilding & Offshore International Company (CSOC) manages the CSIC's products and services, and China Shipbuilding Trading Company (CSTC) is responsible for the CSSC's export activities (Pape and Scott 2013).

China's shipbuilding exports, however, have so far been confined to a relatively small number of countries, with sales to Pakistan alone accounting for ca. 55 % of China's recent defence export sales (Bokhari 2013). In most cases, naval exports have been restricted to a market segment interested mostly in inexpensive and low-tech naval weapon systems—what Pape and Scott call the "less sophisticated and more value-driven end of the market." Another group of customers consists of countries that have trouble gaining access to other suppliers due to political restrictions. China has thus mostly been able to offer "affordable choices to cash-strapped nations that cannot afford or are unable to access materiel from other sources (Iran and Myanmar have historically fallen into the latter category)" (Pape and Scott 2013).

As of 2014, China's ability to integrate Western systems into its export offers is severely limited due to the arms embargo. The weapons and electronics on offer by China's two naval export trading houses, CSOC and CSTC, still mostly consist of relatively outdated systems that were cloned or adapted from Russian or Western models from the pre-embargo era. As a consequence, China's exports are mainly restricted to countries that are not interested in more advanced systems for financial reasons or due to training limitations; to countries willing to accept Chinese turn-key solutions, which often consist of technologies about a generation behind current PLA navy standards; or to countries such as Thailand, which are willing to perform "post-delivery integration of more complex Western combat systems on Chinese-built hulls" and thus are effectively interested mainly in the hull (Pape and Scott 2013).

In the past, Chinese naval export sales have often consisted of transfers of used PLAN equipment. One of the earliest sales successes for an older generation of frigates was the transfer of two *Type 053H "Jianghu I"* class frigates (ca. 1,700 t at full load) to Egypt in 1984/1985 (Wertheim 2007: 171–172). Egypt has also purchased submarines from China, four Chinese-built *Type 033 "Romeo"* class during the early 1980s, as well as six Soviet-built submarines already during the 1960s (Wertheim 2013: 169). A total of 20 of these submarines were also delivered to North Korea during the 1970s.

Bangladesh has been another early and ongoing frigate customer of China's naval shipyards. One *Type 053H1 "Jianghu II"* class frigate (ca. 1,700 t at full load) originally built for the PLA Navy was transferred to Bangladesh upon its completion in 1986 (the *Osman*, Pennant No. F18). The option to acquire a second ship of the same class was not pursued by Bangladesh in 1990 (Wertheim 2007: 40). Most of these export projects were probably price-driven. Bangladesh has nevertheless remained a loyal customer of Chinese naval technology to this day, having signed a contract for two new *Type 053 H2 "Jianghu III"* class frigates (ca. 1,960 t at full load) as recently as 2012. Other naval vessels that Bangladesh recently purchased from China include two newly built corvettes (possibly *Type 056*) and two large (ca. 650 t at full load) guided-missile patrol craft (Hardy and Pape 2013).

In 1991/1992, four *Type 053H2 "Jianghu III"* (ca. 1,900 t at full load) class frigates were also exported to Thailand. According to a naval shipbuilding industry source interviewed by me who personally saw these ships in Thailand shortly after their delivery, there must have been serious and highly visible hull construction deficiencies at that time. The expert noted extensive rusting, commenting that "these brand-new ships looked as if they'd just been brought up from the bottom of the sea." Such an assessment indicates serious steel quality and welding issues at the time of the ship's construction. Wertheim likewise remarks that they are "[s]aid to be flimsily constructed and to have very poor damage-control features" (Wertheim 2013: 726). Two additional larger frigates of a different design (ca. 2,980 t at full load) were ordered and delivered shortly afterwards. Here, too, construction deficiencies were noted by observers. According to Wertheim (2013: 725), they are "lightly built, with the firing of the 127-mm gun causing structural damage forward." At the time of delivery, Thailand essentially opted to use only the hulls. It removed the original engines while installing a CODOG propulsion plant that uses U.S. G.E. *LM-2500* gas turbines and integrating American missiles, namely the *RGM-84A Harpoon* SSM and *Mk 48 Sea Sparrow* VLS SAM (Wertheim 2013: 725). Exports of used *Type 053* class frigates have continued nonetheless. In 2012, China transferred two of its used *Type 053H1 "Jianghu II"* class frigates (pennant nos. 554 and 557) to Myanmar. Built in the mid-1980s, they will thus reach the end of their maximum service lives by the mid-2020s.

In recent years, China has exported more sophisticated systems, and Chinese naval shipbuilders have aimed at a much wider geographic reach. Over the last decade, China has actively pursued naval sales projects in Latin America. In an interesting turn of events, China won a repair contract for Venezuela's six *Lupo* class frigates, specifically for their U.S.-made *LM-2500* gas turbines, thus taking advantage of the fact that Venezuela, like China, is also negatively affected by a U.S. arms embargo. China has apparently not only learned to cope with the lost maintenance support for its own *LM-2500* gas turbines that were used until recently for some of China's destroyers, but it can now even offer such support itself (Foster 2013). In Brazil, China won a civilian contract for building an oceanographic research vessel (Pape and Scott 2013).

One completely new defense export market outside of China's traditional sphere of influence that was recently targeted by Chinese defense companies is Turkey, a NATO member. China's offer of its *FD-2000 (HQ-9)* long-range air and missile defense system initially found favor with the Turkish government, which, however, did not ultimately award the contract to China in early 2014 because of U.S. and NATO pressure (Grevatt 2013b). Moreover, China Shipbuilding and Offshore International Co. has reportedly been showing a model of an amphibious assault ship—ostensibly designed for Turkey—featuring a radar mast that closely resembles the Thales *I-Mast* integrated radar system, thus hinting at China's ability to incorporate such state-of-the-art electronics if necessary (Foster 2013). This unusual development could be connected to the Chinese export contract for three light frigates concluded with Algeria in 2012. For these Chinese-built ships, the combat system and radar will be supplied by Thales Netherlands, and the export license for the installation of these systems aboard the Chinese-built hulls has already been granted to Algeria, though some parts of the hardware will also be installed in China (Pape and Scott 2013). Traditionally oriented towards Russian military technology, Algeria has the potential to become China's most sophisticated naval customer on the African continent, with several extensive Sino-Algerian shipbuilding and nuclear power co-operation agreements already underway (cf. Zanardi 2014: 92–95).

China has otherwise had notable success in selling to the African continent over the past few years:

> Sub-Saharan Africa has become a major target for Chinese military hardware. This can be explained on two counts: first, defence sales form part of Beijing's wider programme to grow its political and economic influence in the continent; second, the key attributes often claimed of Chinese equipment—keenly-priced, and relatively simple to operate and maintain—fit well with the perceived needs of many African states (Pape and Scott 2013).

Smith (2008) further notes that "China [...] has a well-documented interest in Africa's resource wealth [...]. Developing a military export capability potentially provides China with an additional source of leverage when negotiating these critical access arrangements." So far, most sales projects in Sub-Saharan African countries have pertained to smaller and/or less heavily armed vessels such as Offshore Patrol Vessels (OPV) and patrol craft suitable for maritime policing missions, in line with the fact that most navies in that region only operate these kinds of vessels. Contracts concluded in the past few years include an order for two OPV (ca. 1,800 t at full load) from Nigeria in 2012, which contains a shipyard infrastructure upgrade agreement and partial technology transfer. The second vessel is also to be built partially in Nigeria. Other contracts have been settled with Namibia, for another, larger Chinese-built multi-purpose OPV (ca. 2,800 t at full load) delivered in 2012, and with Ghana, for an order of several smaller (ca. 250 t at full load) patrol craft (Pape and Scott 2013).

China's most important and strategically relevant naval equipment customer today, however, is undoubtedly Pakistan. The relationship far exceeds a simple supplier-donor model like the examples cited above. Since 2005, Pakistan has placed several large-scale major arms equipment orders with China. It has negotiated a technology-transfer agreement in which some vessels were constructed locally with Chinese assistance, and formed several defense-related joint ventures with China to produce, among other things, the *JF-17 Thunder* aircraft and the *Al-Khalid I* main battle tank. Pakistan has also developed into a major strategic partner of China for the joint development of the armed forces, with ongoing talks about the establishment of a joint committee on defense production, and massive Chinese investments in key infrastructure projects involving harbors as well as shipbuilding facilities. Discussions on joint submarine development have been officially confirmed at least since 2010 (cf. 'Annual Defence Report' 2010; Pape and Scott 2013).

The naval component of China's defense partnership with Pakistan has existed since 2009, when China exported four *Type 053H3 "Jiangwei II"* class (ca. 3,190 t at full load) frigates. Today, the relationship encompasses plans for the joint production and overseas marketing of the *JF-17* fighter airplane. The aim of the joint Sino-Pakistani plan is

> [...] to sell 300 JF-17s to markets across Africa, Asia and the Middle East by 2017, marketing the aircraft as a cost-effective alternative to surplus US Air Force or other second-hand F-16s that they might otherwise consider (Bokhari 2013).

If followed through, the most strategically relevant joint production project would be a new submarine deal, which, as of late 2014, seems very likely. In the past, Pakistan received extensive assistance from the French in building a submarine construction facility at Karachi, and gained a great deal of technical expertise from importing three French-built *Agosta 90B* class diesel submarines (Pakistan's *Khalid* class), among which one was equipped with the French *MESMA* air-independent propulsion system, in addition to its two older French *Agosta 70* boats (Pakistan's *Hashmat* class). The new Chinese-Pakistani submarine deal would be the largest defense contract so far between the two countries. Negotiations have been ongoing for several years, and Pakistani officials recently projected the conclusion of a contract by the end of 2014 for the purchase of up to six Chinese-designed subs, probably an export variant of the *Type 041 "Yuan"* class SSK marketed abroad as the *"S20,"* with technical negotiations reportedly almost settled and the financing details still to be resolved. China's history of offering huge loans to developing countries for financing Chinese products and services, as for example in many Chinese Sub-Saharan African infrastructure and natural resources projects, indicates that Pakistan's shortage of funds is not necessarily a hindrance for concluding such a contract. Similar to the French defense export strategy, the Chinese defense partnership with Pakistan also extends to civil nuclear power plant construction. Two plants close to Karachi are planned with support from a Chinese state-owned bank loan of US$6.5 billion (Bokhari 2014).

Pakistan's past and present access to advanced U.S. and European weapon systems makes joint naval shipbuilding projects especially attractive to China as a possible route of gaining access to those technologies. Unconfirmed but persistent industry rumors circulating at least since 2008 (which I overheard in various international venues) indicate that Pakistan has concluded a secret agreement with China about sharing its technical information on foreign-imported systems with Chinese experts. If true, this would make the Sino-Pakistani partnership all the more strategically significant for China's military modernization. Similar information is reported by Parsons (2011), who writes that Pakistan has allowed Chinese experts access to U.S. military technology (e.g. the *F-16A* fighter aircraft) at least since the early 1990s. French *MESMA* AIP technology would doubtlessly have been of especial interest for China's indigenous AIP development.

The political aspects of the partnership are also noteworthy. China can use its access to Pakistani territory—including the management of Gwadar port by a Chinese company, the possible future use of this location as a Chinese naval base, and plans to connect Gwadar via train lines directly with Chinese territory—to balance India's growing naval influence in the Indian Ocean and beyond. Pakistan, for its part, can use its relationship to China to exert pressure on its major military ally, the U.S., within the context of a complex and somewhat strained relationship, especially following Al-Qaeda leader Bin Laden's assassination by U.S. forces on Pakistani soil in 2011. The Sino-Pakistani defense industry partnership therefore offers mutual benefits of far-reaching proportions, creating a win-win-situation that is likely to persist in the foreseeable future.

Ukrainian and Russian defense industry executives have noticed the increasing ability of Chinese engineers to quickly adopt new technologies (Johnson 2013a). The increased exposure of Chinese systems and technical experts to formerly closed export markets, enhanced sales activities, and a stronger Chinese presence at various industry events around the globe, are expected to be ongoing trends in the future.

4.7 Summary: Industry-Related Factors Influencing China's Naval Prowess

To sum up this chapter, at least six crucial factors can be identified when assessing the prowess of China's industrial naval shipbuilding base from the point of view of sea power. First, China's remarkable commercial shipbuilding success during the past decades is not automatically transferable to military shipbuilding, even if it does enhance some aspects of it, notably production processes and work-force related learning effects. Second, as far as can be ascertained from accounts of direct exposure to the more complex Chinese commercial shipbuilding projects, there are still some indications of persistent hull design and construction deficiencies, which could, if not remedied, have a negative influence on the seaworthiness and survivability of China's naval vessels as well. Third, system integration in the more complex commercial vessels seems to be an area still requiring development; it may therefore be assumed that naval shipbuilding is likewise affected, which would be detrimental to the vessels' combat value (e.g. through limited AAW capability; it must be noted, however, that the true state of China's indigenous combat direction systems is very difficult to assess at present). Fourth, a still surprisingly high degree of dependence on foreign imports, e.g. in naval propulsion systems such as diesel engines and gas turbines, weapon and sensor technologies, and various other subcomponent technologies, points to vulnerabilities to disruptions in the supply chain. Fifth, the ongoing Western arms embargo effectively excludes the Chinese naval shipbuilding community from full exposure to, and participation in, various international learning opportunities, thus quite possibly slowing down innovation, while export restrictions on key naval technologies force China to gradually develop its own indigenous solutions to already-solved problems. Sixth, and finally, indications of strain in the critical relationship with Russia, China's single most important source of advanced naval technology, suggest possible risks of disruption in the spare parts and maintenance supply for a number of key projects and technologies (e.g. the *Sovremenny* class destroyers, *Kilo* class submarines, and their various subsystems). The impending loss of Ukraine as a naval technology supplier could lead to a stronger Chinese dependency on Russian support. Nonetheless, China's increasingly active role in the global naval export markets, especially in Africa, Latin America, Southeast Asia, and in Pakistan, and its attempts to also forge connections with NATO partners, such as Turkey, point to a heightened degree of contact between China's shipbuilders and various customers and subsystem suppliers around the world that could to some degree begin to counteract the arms embargo's exclusionary effects.

References

Alexander, Dwight et al. 2010. Hybrid Electric Drive Evaluation for CG 47 Class Guided Missile Cruisers. *Naval Engineers Journal* No. 2, 2010: 67-70.

'AMD Building Catamarans for China's Navy'. 2008. *The Sydney Morning Herald* online, February 1, 2008. http://news.smh.com.au/business/amd-building-catamarans-for-chinas-navy-20080201-1pgi.html Accessed April 19, 2014.

Anderson, Guy. 2007. DCNS and Rosoboronexport reach R&D and naval engineering accord. *Jane's Defence Weekly* (el. ed.), November 07, 2007. http://www.janes.com.

Anderson, Guy. 2011a. Bilateral Military Trade Makes Agenda in China-Russia Talks. *Jane's Defence Weekly* (el. ed.), June 20, 2011. http://www.janes.com.

Anderson, Guy. 2011b. Business Analysis: Pentagon Says China Still Using Means Foul and Fair to Gain Foreign Technology. *Jane's Defence Weekly* (el. ed.), August 30, 2011. http://www.janes.com.

Anderson, Guy, Jon Grevatt and Matthew Smith. 2013. Annual Defence Report 2013: Industry. *Jane's Defence Weekly* (el. ed.), December 4, 2013. http://www.janes.com.

'Annual Defence Report 2010 – Asia Pacific'. 2010. *Jane's Defence Weekly* (el. ed.), December 13, 2010. http://www.janes.com.

Archick, Kristin, Richard F. Grimmett and Shirley Kan. 2005. European Union's Arms Embargo on China: Implications and Options for U.S. Policy. (Updated May 27, 2005). CRS Report for Congress, Congressional Research Service, Washington, D.C.: The Library of Congress. http://www.fas.org/sgp/crs/row/RL32870.pdf. Accessed February 1, 2011.

Arena, Mark V. et al. 2006. Why Has the Cost of Navy Ships Risen? A Macroscopic Examination of the Trends in U.S. Naval Ship Costs Over the Past Several Decades. Santa Monica: RAND.

Bedi, Rahul and Jon Grevatt. 2013. India, Russia in Talks Over Second SSN Lease. *Jane's Defence Weekly* (el. ed.), October 22, 2013. http://www.janes.com.

Bell, Matthew. 2011. Briefing: Is It Time For the US and EU to Lift Arms Embargoes Against China? *Jane's Defence Weekly* (el. ed.), March 24, 2011. http://www.janes.com.

Bender, Jeremy. 2015. This Former General Is One Of The Biggest Targets Of Xi Jinping's Anti-Corruption Push. *Business Insider Australia* online. January 24, 2015. http://www.businessinsider.com.au/general-xu-caihou-guilty-for-corruption-2015-1. Accessed January 27, 2015.

Birkler, John et al. 2005. Differences Between Military and Commercial Shipbuilding: Implications for the United Kingdom's Ministry of Defense. Santa Monica et al: RAND. http://www.rand.org/content/dam/rand/pubs/monographs/2005/RAND_MG236.pdf. Accessed February 15, 2011.

Bitzinger, Richard A. 2003. *Towards a Brave New Arms Industry?* Oxford: Oxford University Press.

Bitzinger, Richard A. 2011. Modernising China's Military, 1997-2012. *China Perspectives* No. 2011/4: 7-15.

Bokhari, Farhan. 2013. Analysis: Pakistan Modernization Feeds China's Export Growth. *Jane's Defence Weekly* (el. ed.), April 12, 2013. http://www.janes.com.

Bokhari, Farhan. 2014. China, Pakistan Set for Submarine Deal by End of Year, Say Officials. *Jane's Defence Weekly* (el. ed.), February 3, 2014. http://www.janes.com.

Bussert, James C. and Bruce A. Elleman. 2011. *People's Liberation Army Navy Combat Systems Technology, 1949-2010*. Annapolis: Naval Institute Press.

Chen, Junhong 陈俊宏. 2012. Lengjing guancha, chenzhuo yingfu, taoguang yanghui, juebu dangtou, you suo zuowei (冷静观察、沉着应付、韬光养晦、决不当头、有所作为 | Observe coolly, cope with things calmly, conceal your strength and bide your time, do not take the lead, accomplish some things). Renminwang Lilun Pindao, October 28, 2012. http://theory.people.com.cn/n/2012/1028/c350803-19412863.html. Accessed February 19, 2013.

Cheung, Tai Ming. 2009. *Fortifying China. The Struggle to Build a Modern Defense Economy*. Ithaca and London: Cornell University Press.

Cole, Bernard D. 2010. *The Great Wall at Sea. China's Navy in the Twenty-First Century*. 2nd ed. Annapolis: Naval Institute Press.

Collins, Gabriel and Michael C. Grubb. 2008. A Comprehensive Survey of China's Dynamic Shipbuilding Industry: Commercial Development and Strategic Implications. *China Maritime Studies* No. 1, U.S. Naval War College, Newport: Naval War College Press. http://www.usnwc.edu/Research---Gaming/China-Maritime-Studies-Institute/Publications/documents/CMS1_Collins-Grubb.aspx. Accessed September 24, 2013.

Cover, Oliver et al. 2013. *GI 2013: Government Defence Anti-Corruption Index 2013*. London: Transparency International UK. http://www.transparency-se.org/GI-2013-Main-Report-Phase-04.pdf. Accessed March 24, 2014.

Defense News Top 100 database. 2012. Defense News. http://special.defensenews.com/top-100/charts/rank_2011.php?c=FEA&s=T1C. Accessed Sept 17, 2013.

Drucker, Peter F. 1993. *Innovation and Entrepreneurship: Practice and Principles*. 3rd ed. - New York et al.: Harper.

Dunai, Peter and Guy Anderson. 2013. Russia Outlines USD42 Billion Naval and Maritime Redevelopment Plan. *Jane's Navy International* (el. ed.), January 17, 2013. http://www.janes.com.

Dunai, Peter and Matthew Smith. 2014. Rosoboronexport Boss Outlines 2013 Export Numbers. *Jane's Defence Weekly* (el. ed.), January 28, 2014. http://www.janes.com.

Fisher, Richard D. 2013. Xi visit adds weight to Chinese carrier plans. *Jane's Defence Weekly* (el. ed.), September 3, 2013. http://www.janes.com.

Foster, Robert. 2012. Russia to sell, co-produce Lada-class submarines to China. *Jane's Navy International* (el. ed.), December 21, 2012. http://www.janes.com.

Foster, Robert. 2013. China's CSOC Reveals Its Latin American Shipbuilding Ambitions. *Jane's Defence Weekly* (el. ed.), May 21, 2013. http://www.janes.com.

'Frictions between Russia & China on Imitation of Russian Navy Equipments'. 2009. *Kanwa Asian Defence* No. 62, Dec. 2009: 15.

Friedman, Norman. 2006. *The Naval Institute Guide to World Naval Weapon Systems*. 5th ed. Annapolis: Naval Institute Press.

Grevatt, Jon. 2010. China's Five-Year Plan Aims to Put Defence Industry on Global Stage. *Jane's Defence Weekly* (el. ed.), December 30, 2010. http://www.janes.com.

Grevatt, Jon. 2011a. China and Russia Agree to Expand Defence Relations. *Jane's Defence Weekly* (el. ed.), August 9, 2011. http://www.janes.com.

Grevatt, Jon. 2011b. China to Double Lending to Strategic Industries. *Jane's Defence Weekly* (el. ed.), March 8, 2011. http://www.janes.com.

Grevatt, Jon. 2011c. Rapid Rise Seen in Chinese Military Patent Registrations. *Jane's Defence Weekly* (el. ed.), May 5, 2011. http://www.janes.com.

Grevatt, Jon. 2013a. China Approves Shipbuilding Reforms, Restructuring. *Jane's Defence Weekly* (el. ed.), August 13, 2013. http://www.janes.com.

Grevatt, Jon. 2013b. Turkey's Selection of Chinese SAMs viewed as 'Stepping Stone' to Europe. *Jane's Defence Weekly* (el. ed.), Sep 30, 2013. http://www.janes.com.

Grevatt, Jon. 2013c. Briefing: Closing the Gaps. *Jane's Defence Weekly* (el. ed.), July 5, 2013. http://www.janes.com.

Grubb, Michael C. and Gabriel Collins. 2008. Chinese Shipbuilding: Growing Fast, But How Good Is It? *US Naval Institute Proceedings* March 2008: 44-51.

Hall, Simon. 2013. Asian LNG Tanker Builders Vie for Market Share. *The Wall Street Journal* online. December 20, 2013. http://online.wsj.com/news/articles/SB10001424052702304866904579269290895618508. Accessed March 7, 2014.

Hardy, James and Alex Pape. 2013. Bangladesh Announces Plans to Acquire Submarines. *Jane's Navy International* (el. ed.), January 25, 2013. http://www.janes.com.

Heilmann, Sebastian. 2005. Regulatory Innovation by Leninist Means: Communist Party Supervision in China's Financial Industry. *The China Quarterly* Vol. 181, March 2005: 1-21.

Holzner, Burkart. 1972. *Reality Construction in Society*. 2nd revised ed. Cambridge, Mass.: Schenkman Publishing.

Hoyt, Timothy D. 2007. *Military Industry and Regional Defense Policy: India, Iraq, and Israel*. London and New York: Routledge.

Hu, Shuli. 2014. Xu Caihou case shows Chinese leaders must clean up the military. *South China Morning Post* online. July 9, 2014. http://www.scmp.com/comment/article/1550296/xu-caihou-case-shows-chinese-leaders-must-clean-military. Accessed August 24, 2014.

Jiang, Zemin 江泽民. 1995. Nuli shishi kejiao xingguo de zhanlüe (努力实施科教兴国的战略 | Strive hard to implement the strategy of rejuvenation through science and education). Speech delivered on May 26, 1995 at the high-level Chinese National Conference on Science & Technology. Full Chinese text available at http://news.xinhuanet.com/ziliao/2005-03/16/content_2703878.htm. Accessed March 7, 2014.

Johnson, Reuben F. 2013a. China's Domination of Global Arms Trade 'Inevitable', Say Russian, Ukrainian Industry Insiders. *Jane's Defence Weekly* (el. ed.), March 22, 2013. http://www.janes.com.

Johnson, Reuben F. 2013b. Russian Officials Deny Chinese Deal on Su-35 s, submarines. *Jane's Defence Weekly* (el. ed.), March 26, 2013. http://www.janes.com.

Johnson, Reuben F. 2013c. Kamov role in WZ-10 design suggests China still dependent on Russian innovation. *Jane's Defence Weekly* (el. ed.), March 14, 2013. http://www.janes.com.

Johnson, Reuben F. and James Hardy. 2013. Ukraine Preparing to Offer China Use of NITKA Carrier Training Facility. *Jane's Defence Weekly* (el. ed.), November 14, 2013. http://www.janes.com.

Keegan, John. 2004. *A History of Warfare*. 2nd ed. London: Pimlico.

Kingston, Tom. 2014. Algeria Prepares To Receive LPD Amid Defense Spending Boost. *Defense News* online, January 31, 2014. http://www.defensenews.com/article/20140131/DEFREG04/301310036/Algeria-Prepares-Receive-LPD-Amid-Defense-Spending-Boost. Accessed December 31, 2014.

Kofman, Michael. 2014. An Uneasy Ménage à Trois: Reliance on Russian and Ukrainian Weapons Puts China in a Tight Spot. *Foreign Affairs* online, Dec 4, 2014. http://www.foreignaffairs.com/articles/142417/michael-kofman/an-uneasy-menage-a-trois. Accessed January 2, 2015.

Koinova, Elena. 2007. Sarkozy's corvette deal with Bulgaria. *The Sofia Echo*, October 8, 2007.

Krause, Keith. 1992. *Arms and the State: Patterns of Military Production and Trade*. Cambridge: Cambridge University Press.

Krishnan, Armin. 2008. *War as Business: Technological Change and Military Service Contracting*. Aldershot: Ashgate.

Kuhn, Thomas S. 2012. *The Structure of Scientific Revolutions*. 4th ed. with an introductory essay by Ian Hacking. Chicago and London: The University of Chicago Press.

Lague, David and Charlie Zhu. 2012. China Builds Its Own Military-Industrial Complex. Reuters UK, September 17, 2012. http://uk.reuters.com/article/2012/09/17/uk-china-defence-idUKBRE88F0G720120917. Accessed October 9, 2013.

Lewis, Peter. 2007. President urges radical overhaul of French export policy. *Jane's Defence Weekly* (el. ed.), November 14, 2007. http://www.janes.com.

Lieberthal, Kenneth G. 1992. Introduction: The 'Fragmented Authoritarianism' Model and Its Limitations. In *Bureaucracy, Politics, and Decision Making in Post-Mao China*, eds. Kenneth G. Lieberthal and David Lampton, 1-30. Berkeley et al.: University of California Press.

Luttmer, Nora. 2008. Der Kampf um die Spitzenposition (Fighting for Pole Position). *Schiff & Hafen* February 2008: 58-59.

'Maintenance of 956E DDGs'. 2009. *Kanwa Asian Defence* No. 62, Dec. 2009: 17.

Mayer, Maximilian. 2012. Exploring China's Rise as Knowledge Power. In *Power in the 21st century. International Security and International Political Economy in a Changing World*. Ed. Enrico Fels, Jan-Frederik Kremer and Katharina Kronenberg, 287-311. Berlin et al.: Springer.

McGregor, Richard. 2010. *The Party. The Secret World of China's Communist Rulers.* New York: Harper Collins.

Medeiros, Evan S. et al. 2005. A New Direction for China's Defense Industry. RAND Project Air Force. http://www.rand.org/pubs/monographs/MG334.html. Accessed March 2, 2011.

'Military-to-Military Relationships: The Ties that Bind'. 2011. *The Economist,* Feb 26', 2011: 55-56.

Minnick, Wendell. 2007. China Finds Growth in Leninist Tack to Aerospace: System Lacks Transparency, Say Experts. *Defense News* online, August 20, 2007. http://minnickarticles.blogspot.de/2009/09/china-finds-growth-in-leninist-tack-to.html. Accessed April 21, 2014.

Moore, Gregory J. 2014. Ukraine: A Military-industrial Complex to Die For. *Asia Times* online, May 27, 2014. http://www.atimes.com/atimes/Central_Asia/CEN-01-270514.html. Accessed October 31, 2014.

Mulvenon, James. 2013. Comrade, Where's My Military Car? Xi Jinping's Throwback Mass-Line Campaign to Curb PLA Corruption. *China Leadership Monitor* No. 42 (07 Oct 2013): 1-5. http://www.hoover.org/publications/china-leadership-monitor/article/158546. Accessed March 23, 2014.

Murray, Geoffrey. 2014. China charts course into LNG shipbuilding. *Global Times online,* February 10, 2014. http://www.globaltimes.cn/content/841637.shtml. Accessed March 3, 2014.

Neuman, Stephanie G. 2009. Power, Influence, and Hierarchy: Defense Industries in a Unipolar World. In *The Modern Defense Industry. Political, Economic, and Technological Issues,* ed. Richard A. Bitzinger, 60-94. Santa Barbara et al.: Praeger Security International.

O'Rourke, Ronald. 2008. *China Naval Modernization: Implications for U.S. Navy Capabilities.* New York: Nova Science Publishers.

Pape, Alex and Richard Scott. 2013. Enter the Dragon: China's Naval Export Drive Picks Up Speed. *Jane's Defence Weekly* (el. ed.), November 14, 2013. http://www.janes.com.

Parsons, Ted. 2011. Analysis: Could China Profit From Bin Laden Helo Wreckage? *Jane's Defence Weekly* (el. ed.), May 6, 2011. http://www.janes.com.

Pugh, Philip. 1986. *The Cost of Seapower: The Influence of Money on Naval Affairs from 1815 to the Present Day.* London: Conway Maritime Press.

Rezende, Pedro Paulo. 2008. Brazil, France Schedule Signing of Status of Forces Agreement. *Jane's Defence Weekly* (el. ed.), February 06, 2008. http://www.janes.com.

Rouillé, Sabrina. 2009. DCNS signe un contrat avec la Chine (DCNS signs contract with China). *Ouest France*, April 23, 2009. http://www.ouest-france.fr/dcns-signe-un-contrat-avec-la-chine-398602. Accessed April 18, 2014.

Russia and China have a widening difference on battleship maintenance'. 2009. *Kanwa Asian Defence Monthly* No. 62, Dec. 2009: 16.

Saw, David. 2008. Evolution & Revolution – Chinese Naval Developments. *Asian Defence & Diplomacy* Nov./Dec. 2008: 27-40.

Schumpeter, Joseph. 2006 [1912]. *Theorie der wirtschaftlichen Entwicklung* (The Theory of Economic Development). Reprint of the 1st ed. of 1912, ed. and introduced by J. Röpke and O. Stiller. Berlin: Duncker & Humblot.

Sécurité Maritime: Des chantiers chinois qui font froid dans le dos" (Maritime Safety: Chilling Chinese Shipyards). 2008. *Mer et Marine,* June 18, 2008. http://seaandnavy.com/fr/content/securite-maritime-des-chantiers-chinois-qui-font-froid-dans-le-dos. Accessed April 19, 2014.

Smith, Charles R. 2004. French Arms Sales to China. *Newsmax*, Nov. 11, 2004. http://www.newsmax.com.

Smith, Matthew. 2008. China Strengthens Hold in African Eport Arena. *Jane's Defence Weekly* (el. ed.), July 14, 2008. http://www.janes.com.

Stohl, Rachel. 2006. Wrangling Over Arms Sales to China. *Foreign Policy in Focus online,* November 21, 2006. http://www.fpif.org/articles/wrangling_over_arms_sales_to_china. Accessed March 1, 2011.

Stumbaum, May-Britt U. 2009. Risky Business? The EU, China and Dual-Use Technology. Occasional Paper No. 80. The European Union Institute for Security Studies (EUISS). http://www.iss.europa.eu/uploads/media/op80.pdf. Accessed October 9, 2013.

Surry, Eamon. 2007. An Estimate of the Value of Chinese Arms Production. Research Note presented at the 11th Annual Conference on Economics and Security, University of the West of England and Economists for Peace and Security, 5-7 July, 2007. http://www.carecon.org.uk/Conferences/Conf2007/Papers/Surry.pdf. Accessed October 9, 2013.

Tellis, Ashley J. et al. 2000. *Measuring National Power in the Postindustrial Age*. Santa Monica: RAND.

Till, Geoffrey. 2009. *Seapower: A Guide for the Twenty-First Century*. 2nd ed. London and New York: Routledge.

Todd, Daniel. 1985. *The World Shipbuilding Industry*. London and Sydney: Croom Helm.

Todd, Daniel and Michael Lindberg. 1996. *Navies and Shipbuilding Industries: The Strained Symbiosis*. Westport, Conn. et al.: Praeger.

Tsai, Ming-yen 蔡明彥. 2003. *From Adversaries to Partners? Chinese and Russian Military Cooperation After the Cold War*. Westport and London: Praeger.

Wang, Yeu-Farn 王禹凡. 1991. *China' Science and Technology Policy: 1949-1989*. Stockholm: University of Stockholm, Dept. Of Political Science.

Wen, Philip. 2014. China corruption purge claims top general Gu Junshan. *The Sydney Morning Herald*, April 1, 2014. http://www.smh.com.au/world/china-corruption-purge-claims-top-general-gu-junshan-20140401-zqpa1.html. Accessed April 15, 2014.

Wertheim, Eric. 2007. *The Naval Institute Guide to Combat Fleets of the World*. 15th ed. Annapolis: Naval Institute Press.

Wertheim, Eric. 2013. *The Naval Institute Guide to Combat Fleets of the World*. 16th ed. Annapolis: Naval Institute Press.

Wu, Guoguang 吴国光. 1994. Command Communication: The Politics of Editorial Formulation in the *People's Daily*. *The China Quarterly* Vol. 137, March 1994: 194-211.

Wu, Guoguang 吴国光. 1995. 'Documentary Politics': Hypotheses, Process, and Case Studies. In *Decision-Making in Deng's China: Perspectives From Insiders,* eds. Carol Lee Hamrin and Suisheng Zhao, 24-38. Armonk: M.E. Sharpe.

Yan, Huai. 1995. Organizational Hierarchy and the Cadre Management System. In *Decision-Making in Deng's China: Perspectives From Insiders,* eds. Carol Lee Hamrin and Suisheng Zhao, 39-50. Armonk: M.E. Sharpe.

Zanardi, Claude. 2014. Les relations militaires sino-algériennes (Military relations between China and Algeria). In *Chine-Algérie: Une relation singulière en Afrique* (China and Algeria: A singular relationship in Africa), eds. Thierry Pairault and Fatiha Talahite, 81-99. Paris: Riveneuve Éditions.

Zhang, Qingmin and Eric Hyer. 2001. US 'Dual Track' Policy: Arms Sales and Technology Transfer to China Mainland and Taiwan. *Journal of Contemporary China* (2001), 10(26): 89-105.

'Zhong E qianding junshou da dan: Gou 24 jia Su-35 he 4 sou Lada ji qianting' (中俄签订军售大单:购24架苏35和4艘拉达级潜艇 | Big arms deal between China and Russia concluded: Purchase of 24 Su-35 and 4 Lada class submarines). 2013. Sina.com.cn, March 25, 2013. http://mil.news.sina.com.cn/2013-03-25/0936719576.html. Accessed March 24, 2014.

Chapter 5
The PLA Navy's Capability Profile

"Quantity has a quality all of its own."
—Aphorism, commonly attributed to Joseph Stalin

The present chapter describes the level of development that has been reached by the PLA Navy as of 2014 with regard to the criteria outlined in earlier chapters. Rather than making a purely numerical force-size comparison with other navies, or merely listing the equipment currently in service (as is often done for the sake of convenience[1]), I will use a set of qualitative criteria as a guideline for organizing and evaluating the available data on the state of China's naval development:

- Force structure (fleet size, vessel type and age distribution patterns)
- Technological factors influencing combat proficiency
- Primary function, main mission, and overall mission proficiency of the navy
- Fleet sustainability (logistics, maintenance, training)

These factors will be assessed separately, drawing on the heuristic tools described previously: A system of nine naval combat proficiency levels outlined in Sect. 3.2 (cf. Fig. 3.4) and the system of ten naval capability ranks shown in Sect. 3.1 (cf. Fig. 3.1).

An additional, somewhat elusive criterion that would nonetheless be quite instructive is the "reputation" that a navy appears to enjoy both within its region and abroad. As Till (2009: 117) explains, this could be potentially relevant for assessing threat perceptions. However, since this criterion is extremely difficult to

[1] Cf. for example the table provided by Holslag (2010: 44). It is much more sophisticated than most because U.S. PACOM and Russian Pacific Fleet data are included for comparison, and vessel numbers are given for three different years to show growth trends.

S. Kirchberger, *Assessing China's Naval Power*, Global Power Shift,
DOI 10.1007/978-3-662-47127-2_5

operationalize in a consistent and meaningful way for different navies, I will refrain from attempting to examine it here; in order to shed some light on this issue, future research might nevertheless consider doing fieldwork on the question of the power perceptions of regional navies within the respective defense communities and populaces of the Asia Pacific security complex. This chapter will start by giving an overview of the major milestones that were reached during the PLAN's historical development from 1949 and then conduct a step-by-step evaluation of the factors cited above.

5.1 Historical Development Trajectory of the PLAN: Major Milestones

Since its founding in 1949 following a directive from Mao Zedong, the PLA Navy has undergone several major transformations. Originally conceived as a force purely meant for inshore defensive operations within the context of the "People's War" doctrine, the PLAN has successively added ever more advanced capabilities. As a result, it has conducted offshore defensive and area-denial operations as well as strategic deterrence and also gradually expanded its operations into more distant waters, as evidenced by its heightened level of activity in the South China Sea and in waters beyond the "First Island Chain." The PLAN's expansion was especially visible due to China's continuous participation in anti-piracy missions off the Gulf of Aden beginning in late 2008 and its contribution to the successful evacuation mission of Chinese workers from Libya in 2011.

In 2009, a Xinhua News Agency article by Wang Cong celebrated the achievements of the PLA Navy since its founding 50 years earlier by listing its major technological and organizational breakthroughs. These are indicated in Fig. 5.1.

All in all, between 1985 and 2011, 33 flotillas altogether consisting of over 40 different ships have visited more than 30 countries on 5 continents (Bussert and Elleman 2011: 30–31). The milestones proudly described in the cited *Xinhua* piece provide a good illustration of how the Chinese Navy has steadily expanded in fleet size, geographic reach, and capability.

04/1949	The "birth" of the PLAN (Mao Zedong and Zhu De make declaration of intent to build up the navy). Early equipment consists of 183 units from the Kuomintang's coastal defense fleet that has been abandoned, defected, or surrendered to the PLA.
11/1949	Dalian Naval Academy established.
1949–1955	Surface ship force, coastal defense force, aviation, submarine force and Marine Corps established.
11/1949	First frigate flotilla established.
1950	First artillery battalion established.
1951	First flotilla of torpedo boats and 13 coastal artillery regiments established.
1954	First flotilla of destroyers and submarines and first division of naval air force established. 12/1954: first division of Marine Corps established (cancelled again in 1957).
1953–1957	First Five-Year Plan: China purchases a large number of vessels, aircraft, and weapon systems from the Soviet Union. By 1955, the PLAN has more than 800 vessels, 515 aircraft, and 600 anti-aircraft guns.
1955–1960	Donghai Fleet, Nanhai Fleet, and Beihai Fleet successively established.
1957	Ten academies and six preparatory schools established. By the end of 1957, 14,900 officers and 21,500 soldiers have been officially trained.
1960	Coastal anti-aircraft guns are gradually replaced with missiles.
1971	First Chinese-designed, conventionally powered submarine launched.
1974	First nuclear-powered submarine commissioned.
1979	Marine Corps, consisting of armored troops, combat divers, chemical defense corps, and amphibious troops, reestablished.
1980	First successful landing of a ship-borne helicopter on board a vessel.
1981	First ship (the refitted DDG *Jinan*) equipped with ship-borne helicopter platform and satellite positioning system.

Fig. 5.1 (continued)

1982	First successful launching of a submarine-to-surface missile from a submerged conventionally powered submarine.
1983	First nuclear-powered ballistic missile submarine commissioned.
1985	First visit abroad by a Chinese flotilla (destroyer *Hefei* and supply ship *Fengcang*) in Pakistan, Sri Lanka, and Bangladesh.
1988	China becomes the fifth country in the world to launch carrier rockets from a submerged nuclear submarine.
1989	First port call of a PLAN vessel, the training ship *Zheng He*, at Pearl Harbor.
1990s	PLAN's naval air force equipped with *"Feibao"* fighter bombers, reconnaissance airplanes, anti-submarine aircraft, early warning and patrol aircraft, and tanker airplanes.
1997	Four-month long cruise of a small flotilla (destroyers *Harbin* No. 112, *Zhuhai* No. 166 and oiler *Nancang*), which visits Hawaii and San Diego.
5/2002	Departure of destroyer *Qingdao* (No. 113) and oiler *Taicang* for a 33,000-mile, round-the-world voyage. Ten countries visited, including the USA and Russia.
2003	First joint exercise with a foreign counterpart since 1949: Chinese and Pakistani naval forces conduct joint search and rescue exercises off the coast of Shanghai.
2004	First joint military exercises between PLAN and French naval forces off Qingdao.
2005	First joint military exercises between China and Russia (code named *Peace Mission*). The 1-week maneuvers, starting in Vladivostok and later moving to East China's Shandong Peninsula, involve 10,000 troops from both countries. Joint military SAR exercises with Pakistani, Indian, and Thai naval forces in 2005; PLAN's first participation in a joint military exercise in overseas waters.
2007	Eighty-seven day long cruise of *Type 052B* destroyer No. 168 and oiler no. 887 to St. Petersburg/Russia and three NATO nations. Destroyer no. 112 and oiler no. 887 carry out joint exercises with Australia and New Zealand.
12/2008	A three-ship flotilla (destroyers *Wuhan* and *Haikou* and supply ship *Weishanhu*) sails to the Gulf of Aden on December 26 for anti-piracy missions. By April 2009, the flotilla has escorted more than 100 ships off the Somali coast.

Fig. 5.1 (continued)

2009	PLAN destroyer *Guangzhou* takes part in the 10-day "Peace 09" exercise at the Arabian Sea off the southern Pakistani port of Karachi in March. The PLAN deploys a second flotilla consisting of the destroyer *Shenzhen* and the frigate Huangshan to the Gulf of Aden on April 2, 2009. The supply ship *Weishanhu*, part of the first flotilla, remains in the Gulf.
2011	Successful evacuation operation of Chinese nationals from Libya; China sends a frigate and military transportation aircraft plus private chartered aircraft.
2012	First operational aircraft carrier, the ex-Soviet *Varyag*, commissioned into service as a training vessel and rechristened the *Liaoning*.
2014	First active participation at the biennial RIMPAC exercise with four vessels. In September, a PLAN Type 039 "Song" class submarine makes a port call in Sri Lanka.

Compiled from Wang (2009) with later additions from Bussert and Elleman (2011: 30) and various news sources

Fig. 5.1 Development milestones of the PLA Navy

5.2 Comparative Force Size in Relation to Territory

Today, the PLAN's fleet is one of the world's largest, both in terms of vessel numbers and naval manpower. The high number of vessels in China's fleet, when compared both regionally and internationally, is often commented on by foreign observers. Stalin's pseudo-Hegelian dictum that "quantity has a quality all of its own" is an apt reminder that large numbers of any kind of weapon system can automatically create an intimidating effect, regardless of their actual quality or capability.

As Fig. 5.2 shows, China's fleet includes a very large number of large surface combatants (destroyers, frigates, corvettes, and a carrier), ca. 86 in total, and of submarines, both conventionally and nuclear powered, ca. 64 in total. China's submarine fleet is one of the largest in the world, comparable in size with the U.S., Russian, and North Korean submarine fleets, and the country's number of large surface combatants is also among the highest in the world.

Although in naval shipbuilding quantity alone is not thought to be indicative of actual capability, the fact remains nevertheless that "to some extent the possession of given numbers of large surface combatants is [...] regarded as a rough indication of relative strength" (Till 2009: 115). Redundancy enhances fleet survivability and ensures that all capabilities are theoretically available at any given time. The average vessel spends only about one-third of its total time on mission, another third on repair or maintenance, and the last third on training. Operating only a single submarine or aircraft carrier is therefore not considered tantamount to having a full-fledged submarine or carrier capability, because these vessels are unavailable for duty a large portion of the time. For this reason, a certain minimal number of each vessel type is necessary for having a basic capability continuously available.

	China	Japan	South Korea	India	Taiwan	North Korea	USA
Total EEZ (km^2)	2,285,872	4,469,020	475,469	1,630,356	1,149,189	115,649	6,219,065[a]
Navy personnel[b]	255,000	41,900	97,000	58,350[a]	45,000	46,000	512,800
Aircraft carriers	1 (+2)	0	0	1 (+2)	0	0	11
Helicopter carriers	4 (+3)	2 (+2)	1 (+2)	0	0	0	16 (+6)
Cruisers	0	0	0	0	0	0	22
Destroyers	27	41	11	8 (+7)	4	0	62 (+7)
Frigates	51	6	9 (+6)	15 (+8)	22	1	22 (+12)
Corvettes	3 (+3)	0	25	6	0	1	0
Fast attack craft	254	6	83	32	76 (+11)	15	0
Total large surface combatants	86	49	46	29	26	2	133
SSBN	3 (+4)	0	0	0 (+3)	0	0	18[c]
SSN	6 (+x)	0	0	1 (+1)	0	0	54 (+24)
Diesel subs	ca. 55	19	12 (+6)	14 (+6)	4	ca. 70	0
Total submarines	ca. 64	19	12	15	4	ca. 70	72

Source: EEZ figures from the *Sea Around Us* project (www.seaaroundus.org) at the University of British Columbia, Canada. Vessels counted using Wertheim (2013) and *Jane's World Navies* (2013)
[a]Excluding island and offshore territories
[b]Including Marines, naval aviation, and conscripts
[c]Including SSGN

Fig. 5.2 Comparative data on selected Asia-Pacific naval forces as of 2013

In the case of aircraft carriers, a navy must be further prepared to deploy with each carrier a small flotilla of surface combatants, submarines, and auxiliaries for protection and logistics support during a mission, thus forming the so-called carrier battle groups (which in the U.S. Navy typically consist of several large surface combatants, two submarines, and a supply ship). These vessels, of course, are then not available for other missions. Such escort capabilities need to be added to the standard strength of the fleet. Higher-than-average numbers of both surface combatants and submarines are therefore not only to be expected, but necessary in all carrier-operating navies.

The question however remains: How much redundancy is "sufficient" for any given purpose, and how much is considered "excessive"? This is not easy to determine a priori because different countries' coastlines, maritime territories, and regional security challenges can vary greatly, both in dimension and character. A simplified approximation can nevertheless be useful for evaluating the number of vessels relative to the size of the maritime territory to be surveyed and patrolled on a regular basis. Figure 5.3 shows the comparative EEZ size for a number of countries.

Maritime borders are distinct from land borders, which can be clearly demarcated and at least partly surveyed using permanent installations such as fences, walls, spring guns, or even full-fledged "death zones" between enemy states. Maritime borders are not visible or clearly marked and need to be constantly guarded and patrolled by ships or aircraft in order to be enforced. The necessary number of maritime assets to accomplish this task is therefore directly dependent upon the size of the claimed maritime territory. The larger a country's territorial waters and claimed EEZ, and the more complicated the coastline structure (as, for

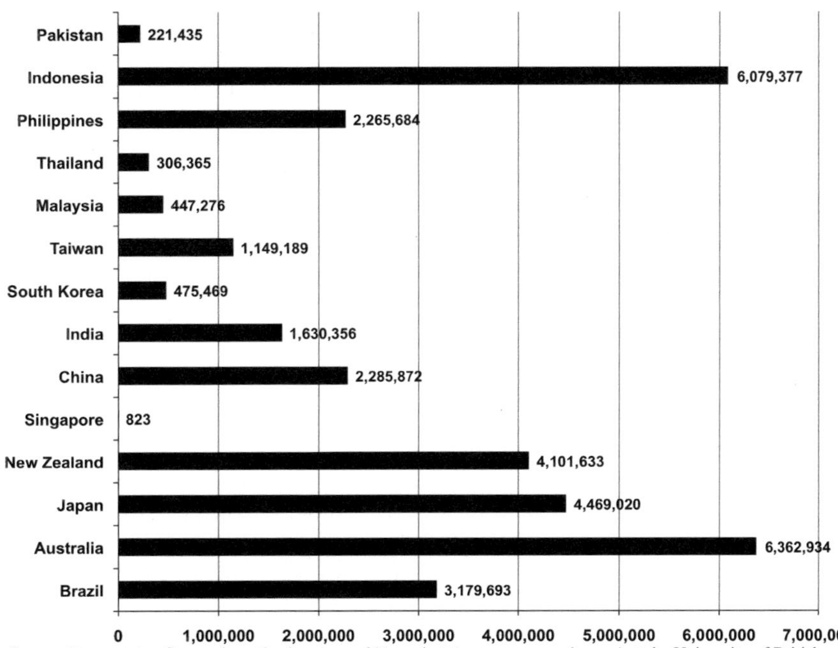

Source: Drawn using figures from the *Sea Around Us* project (www.seaaroundus.org) at the University of British Columbia, Canada

Fig. 5.3 Size of claimed Exclusive Economic Zone (km^2)

example, with large archipelagic countries such as the Philippines or Indonesia), the larger the fleet of patrol vessels a country theoretically needs to perform the most basic naval tasks of coastal defense. As Friedman (2006: ix) explains, compared with other mission types, coastal patrol generally requires a relatively high number of ships, because:

> Sight lines and their equivalents are quite short in coastal waters. Tactical situations are very complex. No single unit can conduct surveillance over a very large area, as it can in the open sea (Friedman 2006: ix).

The commitment to coastal defense therefore bears on the minimal size of an effective navy and/or coast guard, while the additional capability of performing expeditionary missions or being able to permanently assign vessels to carrier battle groups is only possible when fleet redundancy levels surpass the minimal number of necessary patrol vessels. Moreover, vessels for blue-water missions usually differ from standard coastal patrol vessels insofar as they require a higher displacement to be seaworthy enough for extended deployment. The task of patrolling the full range of a 200 nm EEZ already requires much larger vessels with better seakeeping and higher endurance than traditional coastal patrol missions of the territorial waters.

Since governments need to protect both their land and maritime borders, it might be expected that the length of a country's maritime border and the size of its claimed EEZ could be used as rough indicators of a country's inclination to finance and maintain sizeable naval forces. Empirically, however, this does not seem to be the case. Whereas in Sect. 3.1.2, it was possible to empirically establish a correlation between the size of a country's world trade share and the capability level of its naval forces (cf. Fig. 3.3), no such correlation seems to exist between claimed EEZ size and fleet size. There are many countries in the Southern hemisphere, especially in Sub-Saharan Africa, with long coastlines and vast EEZs yet very limited naval forces belonging to the lower capability ranks in the Todd/Lindberg classification system. In Asia, some of the huge archipelagic countries—notably Indonesia and the Philippines—nominally have the largest coastlines and EEZs, but nowhere near the most sizeable or the most capable navies or coast guards. There is obviously no direct correlation between fleet size and the size of a country's commitment to coastal defense.

Just the same, the size of a country's claimed EEZ, which usually refers to the 200 nm zone adjacent to the country's coastline and is therefore normally dependent on coast length, does give a rough indication of the area to be patrolled. It can thus be used as an approximate indicator of the size of a navy's coastal defense commitment, or at least the extent of its naval ambition, all other factors being equal. This value can therefore be used as a general yardstick for comparing fleet sizes. It is clear that the larger the area, the higher the number of vessels that are needed to accomplish regular patrols.

In order to gain a rough idea of what might be termed a "theoretical EEZ patrol capability" across a sample of 11 Asia Pacific countries, I have determined the approximate number of surface vessels in each country that are deemed capable of extended patrol missions, and correlated this value with the country's coastline and EEZ size. All relatively large and seaworthy surface combatants above ca. 70 m of length over all, and/or from ca. 800 t full load displacement upward, were generally deemed to be theoretically capable of extended patrols. I have included surface combatants such as destroyers, frigates, corvettes, and large patrol vessels, but excluded offensive power-projection vessels such as carriers and amphibious assault vessels, as well as unarmed oceanographic survey ships and auxiliaries. Vessels within the given parameters currently in service with either the country's navy or coast guard were counted using standard warship catalogues. This number was then used to divide up the claimed EEZ territory of the respective country, in order to then determine the approximate share of the maritime territory that is theoretically to be surveyed by every single large surface combatant. The results are shown in Fig. 5.4.

As can be seen here, while most countries within this small sample arrived at values ranging between 10,100 and 50,000 km^2 to be surveyed per major patrol vessel, a few countries clearly stood out with highly divergent values. Specifically, the Philippines, Vietnam, and Australia showed much less favorable ratios. One

Country	**EEZ (km^2)**	**No. of patrol vessels, >800 ton fl/>70 m l.o.a.**	**Size of territory per patrol vessel (share of EEZ per PV >800 ton in km^2)**
South Korea	475,469	47	10,116.36
Pakistan	221,435	14	15,816.79
Thailand	306,365	16	19,147.81
China	2,285,872	85	26,892.61
Malaysia	447,276	16	27,954.75
India	1,630,356	54	30,191.78
Taiwan	1,149,189	28	41,042.46
Japan	4,469,020	93	48,053.98
Vietnam	1,396,299	7	199,471.29
Australia	6,362,934	12	530,244.50
Philippines	2,265,684	3	755,228.00
Average	1,909,991	34.1	154,923.67

Source: EEZ data from the *Sea Around Us* project (www.seaaroundus.org) at the University of British Columbia, Canada. The no. of patrol-capable vessels (>800 ton fl/>70 m l.o.a.) was determined using the fleet lists in *Jane's Fighting Ships* (2014) and Wertheim (2013)

Fig. 5.4 Size of patrol fleet in relation to claimed EEZ, 2013

might conclude from these figures alone that especially the Philippines, given its extremely complex maritime territory, must be considered to have a vastly undersized naval force. Australia, on the other hand, would seem to benefit from a relative lack of maritime security threats on its Southern coastline, and can thus be considered a special case. If the more extreme cases are excluded from the sample, the remaining eight Asian countries have an average patrol area to vessel ratio of ca. 27,400 km^2. It is worth noting that this is almost the exact value that has been reached by China (and Malaysia). According to this yardstick, and based on a regional comparison, the Chinese surface fleet size therefore seems to be fairly average in relation to the size of China's claimed EEZ (which however includes large contested areas in the South China Sea), with South Korea, Pakistan, and Thailand having relatively large fleets. Among all countries in the sample, South Korea has the most favorable ratio and could therefore be seen as having an above-average sized, and perhaps even excessively large, surface fleet. This is easily explained, however, given South Korea's peculiar security challenges due to its unresolved conflict with North Korea, and is also consistent with South Korea's relatively high degree of overall militarization as measured by the *Global Militarization Index* (GMI) discussed in Sect. 3.6.

While the size of China's surface fleet (however imposing it may appear at first glance) can therefore be seen as more or less "normal" in relation to its vast maritime claims, the same cannot be said for China's submarine fleet. China clearly operates an unusually high number of submarines compared with most other countries in the world (cf. Fig. 5.2). Regionally, China's submarine fleet dwarfs the fleets of its closest neighbors (cf. Fig. 5.5).

Relative to the size of maritime territory, only South Korea and Pakistan have a comparably favorable submarine ratio. Compared with the regional average, China's territory share per submarine is 8.5 times smaller, and thus much more favorable. Even the U.S., with its sizeable all-nuclear submarine fleet, does not have the favorable ratio of China. With the United States' EEZ (excluding island and offshore territories) of 6.219 million km^2 and its submarine fleet standing at 72 vessels, the resulting U.S. EEZ per submarine share is ca. 86,376 km^2, or 2.4 times the size of China's. If all U.S. island and offshore territories are included in the calculation, the total EEZ amounts to ca. 11.3 million km^2 and the resulting EEZ per submarine share is then ca. 157,653 km^2, a much larger value than China's, and larger even than India's.

However, contrary to popular opinion, China's large number of submarines is not a new development connected with the ongoing military build-up. In fact, the submarine fleet has not been expanded in recent years and, historically, has actually been shrinking massively. As Tien (1992: 279) wrote in 1992, China at the time still operated about 120 submarines. As of 2014, it was 70, though the operational status of many of the vessels is unknown. Kristensen (2007) provides more detailed data

Country	EEZ (km^2)	No. of submarines in service	Size of EEZ share per submarine (km^2)
China	2,285,872	64	35,716.75
South Korea	475,469	12	39,622.41
Pakistan	221,435	5	44,287.00
India	1,630,356	15	108,690.40
Malaysia	447,276	2	223,638.00
Japan	4,469,020	19	235,211.58
Taiwan	1,149,189	4	287,297.25
Vietnam	1,396,299	2	698,149.50
Australia	6,362,934	6	1,060,489.00
Thailand	306,365	0	n.a.
Philippines	2,265,684	0	n.a.
Average	1,909,991	11.7	303,677.99

Source: EEZ data from the *Sea Around Us* project (www.seaaroundus.org) at the University of British Columbia, Canada. The no. of submarines was determined using the fleet lists in *Jane's Fighting Ships* (2014) and Wertheim (2013)

Fig. 5.5 Relative size of submarine fleets in the Asia Pacific region, 2013

over a time span of 30 years. According to the assessments he cites, there were still 120 boats in operation in 1985. This was followed by a steady decline to ca. 70 boats in 1995 and then to ca. 60 around 2001. All in all, he concludes that the Chinese submarine fleet has

> [. . .] declined by approximately 50 percent since the mid-1980s, mainly due to retirement of old and obsolete classes. Construction of new classes is underway but is not anticipated to lead to an increase (Kristensen 2007).

The relatively high ratio of submarines to surface ships in the Chinese fleet is thus not historically unusual, and is something that China shares with other Leninist countries, e.g. North Korea. As Howarth puts it: "Since its foundation, the PLA Navy has always had a numerically strong submarine arm; it is therefore organizationally disposed to maintaining this capability" (Howarth 2006: 13). A fuller analysis of these issues will be conducted in the following sections, which will start with a more detailed discussion of the surface fleet, before then discussing submarine development issues.

5.3 Fleet Age Structure and Principal Naval Weapon Systems

The following section analyzes the PLAN's current force structure according to vessel types and classes, age structure, and technological standards.

The analysis will thus try to evaluate China's progress in naval shipbuilding, starting with the "work horses" of the navy, i.e. the major surface combatants such as destroyers and frigates, which are the primary vessels in any navy responsible for extended patrols and anti-air-warfare. Afterwards, strategic weapon systems—i.e. offensive power-projection vessels such as carriers and nuclear submarines—will be discussed. This will then be followed by a closer examination of the development of China's large conventional submarine fleet. Combat system and naval electronics issues pertaining to these vessel types will be described where appropriate.

5.3.1 The PLAN's Major Surface Combatants

Figure 5.6 provides a visual overview of the type and age structure of China's present-day destroyer fleet. Starting with the oldest vessels still in service shown at the bottom, it can be seen that Chinese destroyers have grown in length and displacement. This indicates a development towards the level of seaworthiness needed for operations outside coastal waters, as well as a tendency to incorporate larger, more capable radar systems required for anti-air warfare (AAW).

As already explained in detail in Sect. 3.3, each arrow in the figure symbolizes a single vessel, whereby the arrow's length indicates its approximate maximal service life. Decommissioned or not-yet commissioned vessels are depicted in light gray. To the left, the vessels are further identified by a pennant number. Since it is much more difficult to ascertain technical and operational information about China's naval vessels compared to most other navies, various sources give conflicting information on the same vessels regarding even basic facts such as the date of commissioning, main machinery, hull dimensions, and displacement. These conflicts, moreover, cannot be satisfactorily resolved in all cases. Fig. 5.6, as well as similar graphical depictions, should thus be taken as helpful approximations that are nevertheless bound to contain numerous inaccuracies. Despite these currently unavoidable limitations, such figures are nevertheless useful as analytic tools for visualizing China's fleet age and type structure and its development over time.

As the position of the vertical dotted line shows, as of 2015, China operates ca. 26 destroyers belonging to 8 different ship classes, all of which—besides the 4 Russian *Sovremenny* class destroyers—were indigenously built in Chinese yards. Assuming a maximum service life of ca. 30–40 years for steel surface vessels, the figure illustrates that new building programs to counterbalance the impending retirement of the remaining *Type 051 "Lüda"* class vessels had to begin during

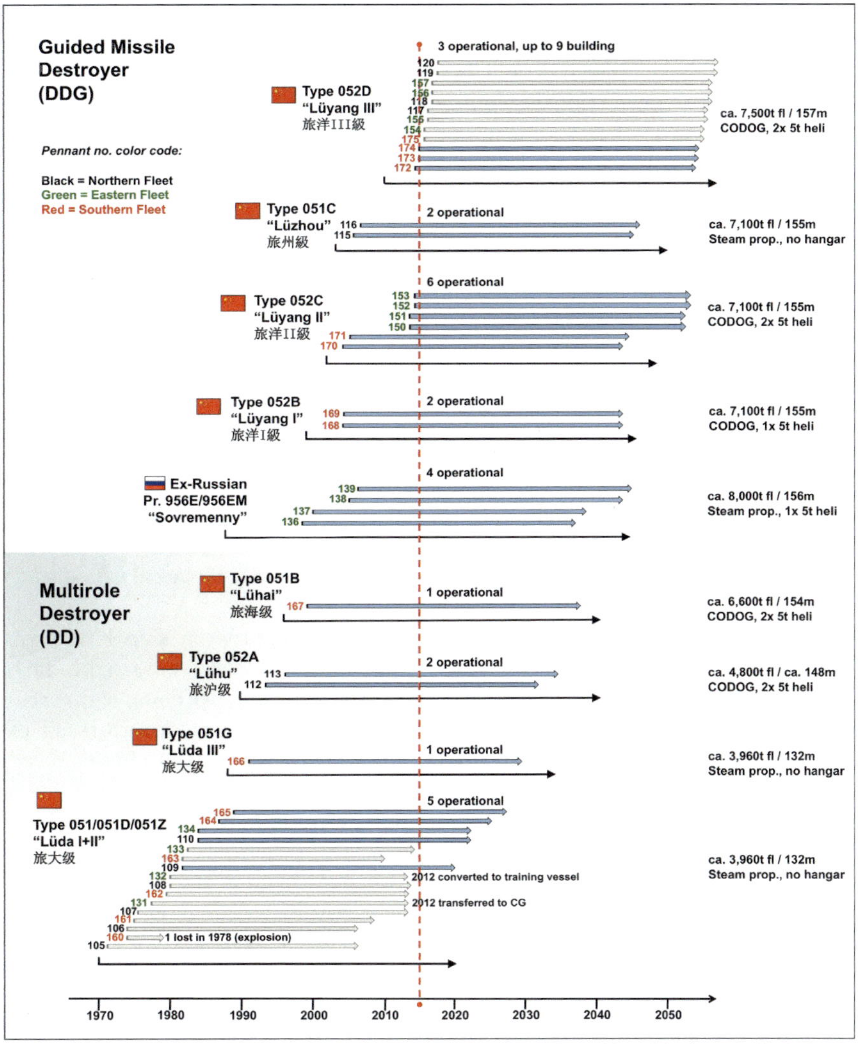

Source: Drawn using data from Wertheim (2007, 2013) and *Jane's Fighting Ships* (2014) cross-checked with Chinese Wikipedia entries

Fig. 5.6 Type and age structure of the PLA Navy destroyer fleet, 2015

the late 1980s in order to maintain fleet strength. This suggests that these programs were for replacement procurement rather than fleet enlargement. If the *Type 052D "Lüyang III"* class program progresses as expected, and the remaining *"Lüda"* class vessels all retire during the next few years, then the destroyer fleet could be expected to slightly grow from 26 to ca. 30 vessels altogether around 2020.

5.3.1.1 The Older Generation Destroyers

As a RAND study has noted, the old *Type 051 "Lüda"* class destroyers (as well as the contemporary *Type 053/053H "Jianghu"* class frigates)—which were built between 1972 and 1991 and until recently constituted the numerical backbone of the fleet—seem to possess somewhat limited survivability due to various hull construction deficiencies:

> Both suffer from poor welding with signs of premature failure, inoperable machinery, and overall poor hull workmanship. These deficiencies, in turn, seriously degraded their war-fighting ability (Medeiros et al. 2005: 143).

According to Wertheim, their design was based on the "unsuccessful" Soviet *Project 41 "Tallin"* class. They are steam powered and have no organic helicopters, except in one case (hull no. 105) where a heli deck and hangar were added during a 1987 upgrade. Further problems with these vessels cited in the RAND study include a lack of "damage-control facilities and basic safety features (such as fire-retardant systems, automatic firefighting systems, or watertight doors)" (Medeiros et al. 2005: 143–144). Originally, all *"Lüda"* class vessels "typified the anomaly of offensive ordnance lacking any combat direction system (CDS) for weapon control" (Bussert and Elleman 2011: 19). Attempts to remedy this weakness during the 1980s fell victim to the 1989 post-Tian'anmen sanctions.

Although all remaining *"Lüda"* class destroyers, except for one ship, have been equipped in the meantime for underway replenishment, they still lack air-conditioning, NBC warfare protection, and a central combat command space. Weapons and electronics outfitting varies greatly within this class as a result of the various types of extensive refits. Only a single hull (no. 105) also received advanced electronic systems such as the Italian *Alenia IPN-10* combat data system and was probably used as a "trials ship for equipment to be used in this and later classes" (Wertheim 2007: 118). Out of all the *"Lüda"* ships, only five seem to have had any fire-control radar systems installed. Moreover, some of the older destroyers of this class reportedly even lacked the necessary electrical power generation capability to operate all sensor and weapon systems simultaneously (Wertheim 2007: 118).

The most advanced of these older destroyers is hull no. 166, the only ship to receive the so-called *"Lüda III"* upgrade in 1991.[2] This ship became

[2] Although up to three other *Lüda* vessels (hull nos. 165, 109, and 110) received different kinds of upgrades at the same time as hull no. 166, and are thus often simply classified as *"Lüda III"* vessels, according to the most specialized study to date, this identification is *not* correct (cf. Bussert and Elleman 2011: 20). However, as of September 2014, the electronic edition of *Jane's Fighting Ships* does still identify all these four hulls as *"Lüda III"* vessels, whereas Wertheim (2013: 118) includes only the two hull nos. 165 and 166 in this category. The armament and sensor equipment listed as installed in these vessels varies greatly from source to source, and sometimes even within one and the same source.

> [...] a trial platform for systems that would later appear in the new gas turbine Luhu/Luhai-class guided missile destroyers. It featured a [French] Thomson-CSF [...] TAVITAC [...] combat direction system that made the Luda III the first Chinese vessel with a multithreat picture CDS display. The Luda III includes French communications and electronics to handle advanced weaponry, but the vessel seems lacking in means to pass data to fleet headquarters ashore using satellite communications or ultrahigh frequency (UHF) data links (Bussert and Elleman 2011: 21).

The French *TAVITAC* CDS has been used in China since the mid-1980s, when China obtained two *TAVITAC* systems with the matching *Exocet* missile technology in a legal export deal with France (Friedman 2006: 516). These two officially procured *TAVITAC* systems were installed on the *Type 051DT "Lüda II"* class destroyer *Kaifeng* (hull no. 109), which also received the matching French *Sea Tiger* air/surface search radar and *Crotale* naval missile systems, and on the *Type 052A "Lühu"* class destroyer no. 112, the *Harbin* (cf. Friedman 2006: 62).

French missiles and sensors also apparently left their mark on Chinese naval technology during the years that followed. As Bussert and Elleman note,

> [...] several of the upgraded DDGs have the [French] Crotale SAM system [...]. The system maintenance and operation needs include not only the launcher, loader, magazine, and missiles, but also associated Sea Tiger search and Castor missile guidance radars, and infrared localizer, and a television tracker (Bussert and Elleman 2011: 30).

It seems that after the 1989 sanctions took hold, and official dealings with France came to an end, the *"Lüda III"* upgrade of 1991 served as a "test bed" for integrating and debugging the French *TAVITAC* CDS and various other pieces of French and Russian equipment within the same hull (Bussert and Elleman 2011: 29).[3]

Although command and control in the older *Lüda* destroyer variants was accordingly limited and combat capability was strongly compromised, it may be concluded nevertheless that the remaining vessels are capable of foreign deployment due to added underway refueling facilities. They can, at any rate, serve as test and trials or training platforms, or, as in some cases, they might also be used later on as coast guard vessels in order to help China assert its maritime claims. Today, about a third of China's operational destroyers still belong to the technically obsolete *"Lüda"* class in its various upgraded variants. Given the country's past practice of retiring these vessels after a maximum 30 years of service life, they will probably all see decommissioning in the coming years.

[3] As an example of the integration problems, it may be noted that the French export version of *TAVITAC*-related systems such as *Vega* digital fire control and *Crotale* missiles use 60 Hz, while the Chinese systems likely use 50 Hz. According to Bussert and Elleman (2011: 44), "[t]his would require some sixty-hertz-to-fifty-hertz conversion in the combat direction system. Although a relatively simple conversion, equipment differences like this can easily undermine system interoperability."

5.3.1.2 Replacing the *"Lüdas"*: The Intermediate Destroyer Generations

Replacement procurement planning for the *Type 051 "Lüda"* class vessels was timely when it began during the 1980s. However, the political fallout from the 1989 Tian'anmen protests, which resulted in a U.S./EU arms embargo against China, had a destructive impact on these plans. More precisely, the very short time span during which advanced Western technologies such as combat direction systems, modern missiles, and advanced propulsion technologies were made available to China came to an abrupt end. This had severe consequences for destroyer designs such as the *Type 052A "Lühu"* class because of the sudden unavailability of key components and maintenance support. This is undoubtedly the major reason why the *Type 052A* class was discontinued after just two ships.

The *Type 052A* class has been termed "China's first modern surface combatant design" (Wertheim 2007: 117). However, grave survivability issues associated with the preceding *"Lüda"* destroyers seemed to persist in this class as well. According to reports cited in a RAND study of experts who were able to tour the *Type 052A* class destroyer *Harbin* (hull no. 112) in San Diego during late March 1997:

> The [*Type 052A*] Luhu's damage-control capabilities are limited to a room with an illuminated display board of the entire ship, but no actions can be taken from the room to address emergencies when they occur. In the weapon-control room, none of the operator's seats was bolted to the floor; during combat, operators would be thrown around (Medeiros et al. 2005: 144, fn. 54).

The main impact of this ship class on the current Chinese naval weapon systems may consist in the fact that it was the first to be conceived with a modern combat direction system. Although the Italian *IPN-10* CDS was apparently never officially sold to China, several experts agree that it is nevertheless installed on the *Type 052A* class, and also seems to have formed the basis for subsequent generations of Chinese CDS designs:

> According to a recent Chinese official account, the second-generation missile destroyers (Luhu class, the first of which was delivered in July 1994) were the first for which the CDS was designed before the ship. Work on a concept study of a new CDS (by the Seventh Academy, 701, 709, 716, and 724 Institutes) began in 1983. No. 701 Institute was designated primary organization for destroyer combat systems. Land tests of the new system began in December 1986. At about the same time, the connection to China was being very openly discussed at the Italian Mostra Navale show (1987), and it seemed likely that the Chinese would receive the MAGICS consoles with 286 chips (Friedman 2006: 62).

This conclusion is in line with existing visual evidence from photographs of combat information center (CIC) consoles aboard a *Type 052A "Lühu"* class destroyer:

> Published Chinese photographs of the CIC of a Luhu [*Type 052A*] and a shore training center show a system similar to that of the Italian IPN-10 or -20, with vertical consoles plus

> a three-operator horizontal (conference) console. [...] These data probably apply to *Qingdao* [hull no. 113], which is credited with ZKJ-6 (Friedman 2006: 62).

Regarding this analysis, Bussert and Elleman note that the destroyer in question must be the *Qingdao's* sister ship *Harbin* (no. 112), because photos of the console arrangement within its CIC taken during its visit to San Diego in March 1997 revealed an *IPN-10* combat direction system (Bussert and Elleman 2011: 46).

The CDS aboard the *Type 053HT "Jianghu III"* class frigates exported to Thailand in the early 1990s also closely resembled *IPN-10*. According to Friedman:

> In the Thai ships, it comprises a three-man console plus individual single-operator consoles for the surface and underwater pictures; for ASW and ASUW; for data link control; for EW control; and for decoy launcher control. The system fitted to the Thai ships was modified after delivery to incorporate a U.S. intership data link. Note that IPN-10 (which was also sold to India) does not incorporate a data link; the Italian navy bought its link terminals from Signaal. The Thai ships were delivered with a link back to a shore headquarters (Friedman 2006: 62).

Since no official sale of an *IPN-10* to China was ever documented, Bussert and Elleman seem to imply that the *IPN-10* observed on the Chinese vessels may have been illegally obtained or reverse engineered. According to their assessment, "[...] it is probable that there is no vendor training, production rights, or logistical support" for this complex system (Bussert and Elleman 2011: 30).

Establishing a clear timeline of all the various imported, reverse-engineered, and indigenous combat direction systems aboard Chinese destroyers seems to be a fairly hopeless task, given the fact that even specialized works on the subject such as Friedman (2006) and Bussert and Elleman (2011) are greatly at odds in not just some, but rather many instances.[4] The exact same vessels are variously described as carrying an *Italian IPN-10* or a French *TAVITAC* combat system, such as the *Type 051B "Lühai"* class destroyer *Shenzhen* and several others. Bussert and Elleman reveal their own hesitancy in the following way:

> Although reference books list various 'ZKJ-#' CDS listings on several ships, it is uncertain if they are variations of Type 88 [the oldest indigenous system], TAVITAC, IPN-10, or other Chinese or Russian design [sic]. [...] Earlier ZKJ systems should have lower series numbers, but the 1986 Jianghu system was ZKJ-4, and a Luda system later was ZKJ-2, which is unexpected (Bussert and Elleman 2011: 46).

In their own otherwise informative and knowledgeable account, textual descriptions of CDS aboard ships do not match the accompanying table in every instance, which adds considerably to the confusion.[5] Moreover, the supposed Chinese

[4] Cf. the diverging accounts put forward in Friedman (2006: 61–63) and Bussert and Elleman (2011: 44–45).

[5] In Bussert and Elleman's discussion of Chinese CDS, the text and table on pp. 44 and 45 are not quite in agreement e.g. with regard to the *"Lüda II"* upgrade of 1987, once associated with *TAVITAC* on p. 44 and then with *IPN-10* on p. 45. Because Friedman lays out the reasoning behind his identification of the various CDS more transparently, I largely follow his account here.

designations for combat direction systems in the accounts published by Bussert and Elleman and by Friedman are not in agreement with each other. Since Friedman's account, although older, is at least internally consistent and gives relatively transparent reasoning, I cite his list of known Chinese CDS designations in Fig. 5.7 for illustrative purposes.

The rather large degree of uncertainty surrounding the CDS configuration aboard the various destroyer classes notwithstanding, based on the data given in Fig. 5.7 as well as various other sources, it seems clear that modern Chinese CDS development has been derived primarily from two different European systems that China got hold of during the 1980s: The French *TAVITAC* and the Italian *IPN-10*, and their associated missile and search radar systems (French *Crotale* missile and *Sea Tiger* search radar, and the Italian *Aspide* missile).

Designation	Date	Characteristics, origin, other comments
ZKJ-1 a.k.a. Type 673-I	1984	Developed by nos. 706, 723, and 724 R and D laboratories; first test installation of naval version aboard DD 132 in 1984
ZKJ-II (ZKJ-2)	1986	Simplified version of *ZKJ-I* for frigates; first installed 1986 aboard *"Jianghu"* class FF (nos. 535 and 536)
ZKJ-3 a.k.a. CCS-3 (?)	1990	Export version of *ZKJ-2*; based on or identical with Italian *IPN-10* (which includes no data link); installed on FF exported to Thailand 1990–1991 (with added U.S. intership data link)
ZKJ-4	n.a.	May correspond with the Italian *IPN-20*; *"Lüda"* class vessels 165 and 166 received variants of this CDS
ZKJ-5	(1987)	Likely based on or identical with French *TAVITAC*, which was first installed in 1987 on Hull 109 (*"Lüda III"*) and perhaps on *Type 052A* class hull 112
ZKJ-6	n.a.	Likely based on either *IPN-10* or *-20* according to photographs of console arrangement; in service e.g. with *Type 052A* class DD *Qingdao* (hull 113)
ZKJ-7	2003	Likely based on Italian *IPN-10* or *-20* system according to photographic evidence; in service with *Type 051B "Lühai"* class DD *Shenzhen*
JRSCCS/JRNG	n.a.	Possibly related to the French Thales *TACTICOS* system
H/ZBJ1.2	2006	*Type 052C "Lüyang II"* class DDGs
Sapfir-U	2006	Russian CDS of the *Project 956EM Sovremenny* class, also installed aboard the *Type 051C "Lüzhou"* class DDGs

Source: Compiled from Friedman (2006: 61–63) and, with added data for later ship classes, from Bussert and Elleman (2011: 44–45). Note that the descriptions for the earlier ship classes in both sources cited are not in accord with each other

? = uncertain

Fig. 5.7 Chinese combat direction systems (ZKJ/CCS-3 series) according to Friedman (2006)

Thus, while the equipment introduced by the *Type 052A "Lühu"* class destroyers did represent a step ahead compared with that of the preceding class, their war-fighting capability in the event of conflict would in all probability still be compromised due to system integration challenges and the lack of vendor support after 1989. Nevertheless, since 2012, ships of this class have been deployed to the Gulf of Aden and made port visits in the Black Sea, which indicates their suitability for extended training missions.

5.3.1.3 Indigenous Destroyer Designs and Russian Imports

The following period saw a frenzy of new building and simultaneously pursued import programs with considerable overlap, resulting in an unusually high number of different destroyer classes now all in use. In 1996, the first contract was signed between China and Russia for the purchase of two ex-Soviet *Pr. 956E "Sovremenny"* class destroyers, with two more vessels of a more advanced *Pr. 956EM* class purchased later in 2002. In terms of carried combat systems, the import of these four destroyers represented an important watershed for China's naval capability.

Even though, as of 2014, these steam-powered Soviet-era ships are not exactly considered state-of-the-art, at the time of their import they nevertheless represented a marked improvement over China's indigenous ship designs. They further had the added advantage of being proven, reliable systems with established maintenance routines and training regimes, which Russia also provided to China. Moreover, since they were imported as complete, all-Russian weapon systems, they did not suffer from any of the system integration challenges which had made the indigenous Chinese designs of that era so problematic.

Concerning the major equipment introduced into the Chinese Navy through this import program, the "highly versatile" Russian *Mineral ME* (NATO nickname *"Band Stand"*) active/passive tracking fire control radar system for cruise missiles and main gun was clearly highly significant. Ever since its introduction, all subsequent major Chinese surface warships have continued to integrate this particular system. Reverse engineered copies can be found installed aboard the *Type 052B* and *Type 052C "Lüyang I* and *II"* DDGs and *Type 054A "Jiangkai II"* frigates (Friedman 2006: 272–273; see Figs. 5.8 and 5.9). On the first three *Sovremenny* ships, this system controls the supersonic (Mach 2.4) sea-skimming *SS-N-22* SSM, which Bussert and Elleman (2011: 34) describe as "a formidable weapon." This type of missile is extremely difficult to counter, and is in principle even capable of carrying a nuclear warhead, which China may or may not have added on its own. This missile later became the blueprint for a reverse-engineered indigenous variant, the *C-301*. Nonetheless, China received a more advanced weapon system with the last of the *Sovremenny* ships: On the fourth vessel, the *SS-N-22* was replaced with the even more powerful *SS-N-27 Sizzler 3M54E* SSM (Bussert and Elleman 2011: 35).

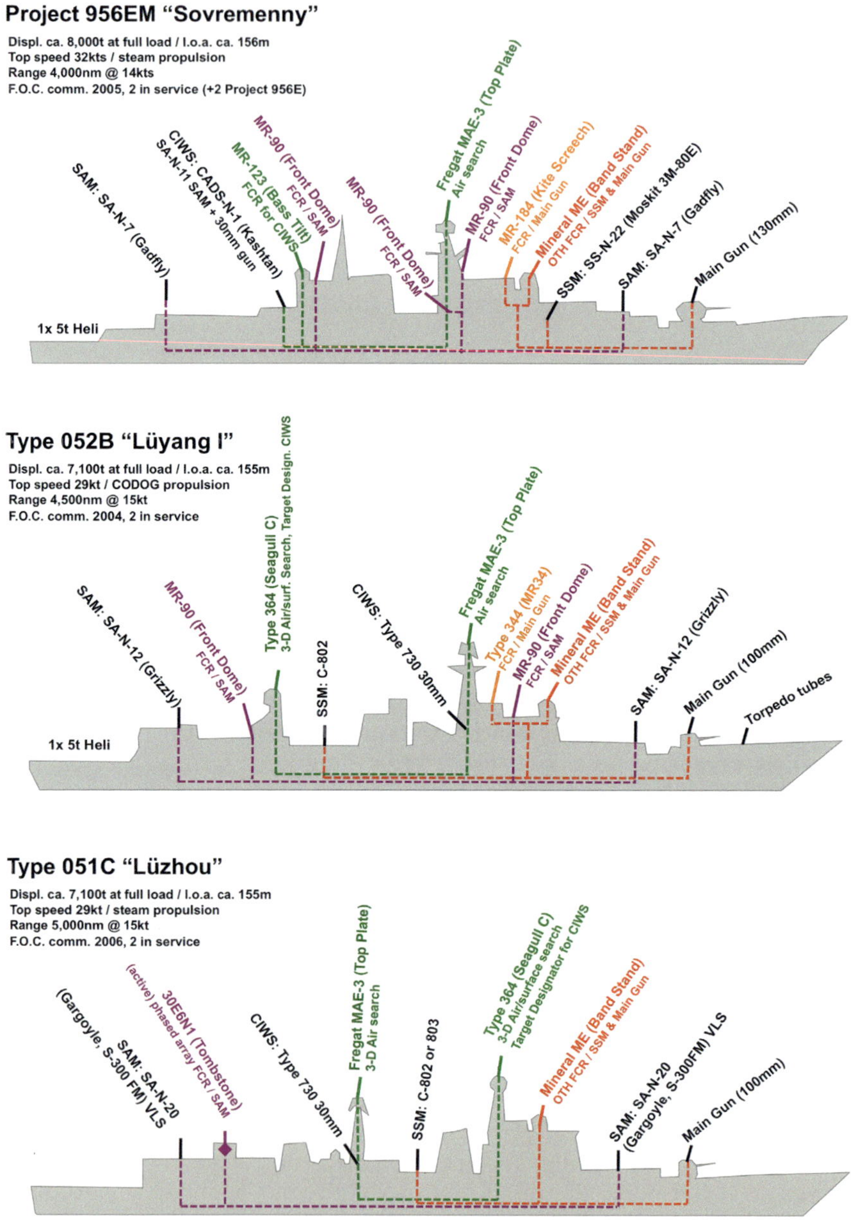

Source: Drawn using technical and visual data from *Jane's Fighting Ships* (2014), Wertheim (2013), Friedman (2006), and Bussert and Elleman (2011). Main functional chains shown

Fig. 5.8 Combat systems of China's mid-generation destroyers: main functional chains

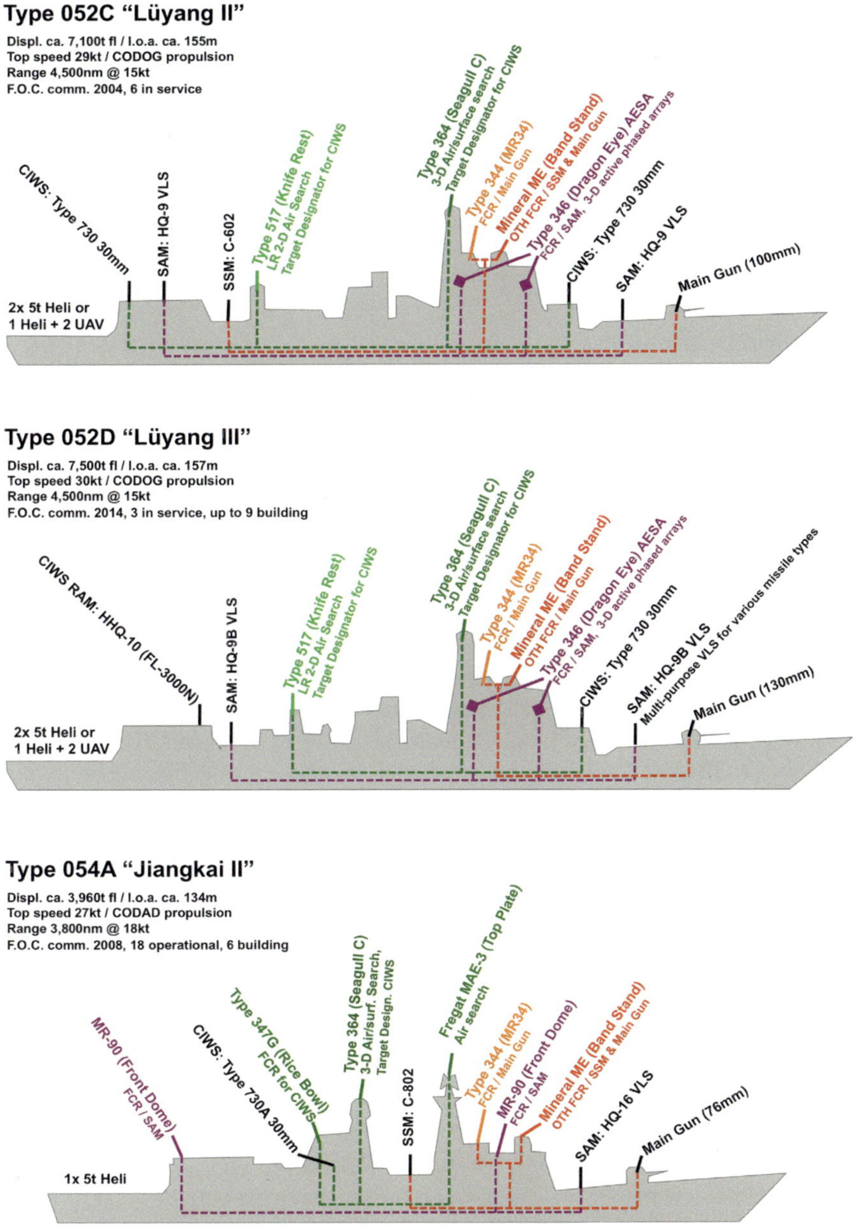

Source: Drawn using technical and visual data from *Jane's Fighting Ships* (2014), Wertheim (2013), Friedman (2006), and Bussert and Elleman (2011). Main functional chains shown

Fig. 5.9 Combat systems of China's latest surface vessel designs: main functional chains

In addition, the *Sovremennys* also came equipped with previously unparalleled long-range air-defense capabilities through their *SAN-7 Shtil* SAM and *MR-90 "Front Dome"* tracking radars. The ships have an S-band target designator/data link system (*"Light Bulb"* data links, *"Bell Strike"* video link to midlevel guidance aircraft, and the *Mineral ME "Band Stand"* radar track data link) with a 30-km horizon range. The passive range of the search radars reaches 450 km, while their active range goes up to 180 or even 250 km (according to some accounts). Fifty targets can be tracked passively, 30 actively. The data link system can handle 200 tracks (cf. Bussert and Elleman 2011: 34). According to Friedman's plausibly deduced estimation, *Mineral ME/"Band Stand"* probably uses L-band (cf. Friedman 2006: 272–273).[6]

The introduction of this comprehensive naval weapon control system into the Chinese fleet has clearly left its mark on all the modern major surface vessel designs China has produced since then, a fact that is plainly illustrated by Figs. 5.8 and 5.9 which show the major combat systems' main functional chains in comparison. It has also greatly enhanced China's potential level of naval combat proficiency. According to the most recently published major study of China's naval combat systems, the *Mineral ME "Band Stand" FCR's* "over-the-horizon (OTH) detection capabilities coupled with supersonic surface-to-surface missiles (SSMs) are the greatest threat to other major navies" (Bussert and Elleman 2011: 35). This over-the-horizon fire control/data link system can detect up to 200 contacts and designate nine as targets for *SS-N-9* or *SS-N-22* SSMs (Bussert and Elleman 2011: 37). The supersonic sea-skimming SSMs carried by the *Sovremennys* and controlled by the *Mineral ME "Band Stand"* FCR system have been dubbed "carrier killers," because currently only very limited countermeasures exist against such a threat. Moreover, as Bussert and Elleman note, there is an element of deliberate obscurity at play regarding the actual armament that is carried not only by the *Sovremenny* class destroyers, but other modern Chinese warships as well:

> Russia has sold China a wide range of missiles, including the P-270 (3M80 missile) and P-100 (3M82 missile), variously called SS-N-22, Moskit, and Sunburn. There are large differences in speed and range between these SSMs, however, and it would be difficult for their adversary to know which is being carried. Destroyers and submarines could fire a barrage of antiship weapons that spot aircraft carriers or escort ships and then drop back down to run along the water's surface before accelerating to supersonic speeds for the kill. The U.S. Navy Aegis weapon system may be challenged to provide effective defense against such formidable OTH low-level supersonic antiship weapons (Bussert and Elleman 2011: 36).

However, as another expert notes, the *Sovremennys* may in turn have difficulty defending themselves against quiet modern diesel submarines:

> Designed during the mid-1960s, the ships are primarily intended for surface-warfare tasks, including antiship, shore-bombardment, and antiair defense; the minimal ASW capability is primarily for self-defense (Wertheim 2013: 116).

[6] Bussert and Elleman (2011: 39) by contrast give I-Band.

Another noted drawback of this ship class is the fact that its 130 mm guns are "restricted to firing arcs of 40° each side of the centerline," and the SAM launchers "may be limited to launch angles within 30° of the centerline, a significant handicap" (Wertheim 2013: 117). These problems were effectively remedied with the introduction of improved vertical launch systems aboard the newer Chinese ship classes to be discussed below.

5.3.1.4 Possible Explanations for a Fragmented Build Strategy

From an industry standpoint, it seems odd that China—in addition to purchasing the Russian *Sovremennys* as an export order—started building almost simultaneously four different indigenous destroyer classes, no less at different shipyards. While some observers hail this fragmented build strategy as an ingenious move towards overcoming technical limitations by fostering experimentation and inter-shipyard competition (sometimes dubbed a "Hundred Flowers Strategy" of naval shipbuilding), this author's view of the motive behind it is somewhat more skeptical. This is supported by the fact that, in one case (the *Type 051B "Lühai"* class destroyer), only a prototype was built, without any follow-up vessels. Analysts generally agree that—in some respects at least—this vessel was an improvement over the preceding classes. Compared to the earlier classes, its hull is markedly larger and wider and thus more seaworthy:

> "It has a widened hull beam to enhance stability, armament-carrying capacity, and crew living space. In particular, the Luhai's larger size permits four quad launchers for C801/C802 anti-ship missiles, which is double the number, deployed on the [*Type 052A*] Luhu" (Medeiros et al. 2005: 144–145).

The *Type 051B "Lühai"* design was nevertheless discontinued after the first-of-class was built, which in all probability would not have been done had the performance of the ship been deemed satisfactory (cf. Bussert 2003). Three other destroyer classes that went into production nearly simultaneously with the *Sovremenny* procurement program were the *Type 052B "Lüyang I"*, of which only two ships have been built, the *Type 052C "Lüyang II"*, which is expected to result in a class of six ships, and the steam-powered *Type 051C "Lüzhou"*, a relatively old-fashioned design, which has only seen two vessels built and carries no organic helicopters. As late as 2000, no clear destroyer fleet development strategy had become visible yet to outside observers. This interpretation is supported by the fact that a follow-on procurement decision for further imports of two *Sovremenny* class destroyers from Russia was made as late as 2002. At the very least, this suggests that these Soviet-era vessels were still regarded at that time as technically superior to any of the indigenous Chinese destroyer classes that had been developed so far by the Chinese naval shipbuilding industry.

Judging from the general tendencies of naval shipbuilding programs in all Western countries as well as in numerous developing countries over the past few decades, a fragmented build strategy such as China's with added imports seems highly unusual, and especially so for a leading shipbuilding nation. Development costs for the first-of-class of any type of naval vessel are always disproportionally higher than the costs of follow-on vessels built from the same design, which means that the overall unit price can only be lowered if a sufficient number of units are built. Moreover, the usual technical shortcomings found in a typical first-of-class are often only remedied in the follow-on vessels, indicating the steepness of the learning curve normally to be expected in any new building program. (This is reflected by the fact that navies in the market for naval exports are often unwilling to serve as "test customers" for newly developed designs that have not been previously built.)[7] Historically, if naval shipbuilding programs have failed to make *any* use of economies of scale and technical learning curve effects, and thus been aborted after the first one or two ships, this has usually indicated either sudden budgetary problems (due, for instance, to extreme cost-overruns or unforeseen financial pressures faced by the national budget), severe technical issues deemed irremediable, or all of these at the same time.[8]

Another related aspect concerns the operational disadvantages of having too many different classes of the same vessel type within the fleet. The common view is that producing more vessels of the same class not only offers economy of scale and technical learning curve effects, but that it also greatly enhances interoperability within the fleet due to optimized training, logistics, and maintenance procedures. Different vessel classes almost automatically require highly specialized crew training, logistics, spare parts provisions, and maintenance regimes. In having so many different destroyer designs simultaneously in development, as well as opting to purchase the ex-Soviet *Sovremennys* at a time when other indigenous destroyer programs had already been launched, China's actions are not consistent with such general efficiency rules. Rather than ingenuity, this in all probability indicates persistent problems in key technical areas such as systems integration of naval electronics.

The fragmented build and design strategy observable in China during this phase may thus simply reflect the difficulties experienced in CDS development. At the same time, after various difficulties and setbacks, it seems that open architecture concepts have finally been decided upon. Commenting on the *Type 051B "Lühai"* and *052A "Lühu"* classes, Bussert and Elleman write as follows:

> The new destroyers' command and control (C2) draws from several sources. Integration problems with the French TAVITAC combat control system and various French- and Russian-style equipment was probably minimized on these vessels because the architecture was previously integrated and debugged on the Luda III test bed. Chinese combat systems

[7] This problem prompted French shipbuilder DCNS to take the unprecedented step of building a prototype *GOWIND* type patrol vessel (the *L'Adroit*) as a demonstration model, and offer the ship for lease to the French Navy in order to make the design more marketable (cf. Dunnigan 2010).

[8] An example of this would be the U.S.N.'s *Zumwalt* class destroyer program, which was originally planned to encompass 32 vessels, but was cancelled after only three ships because of obvious extreme cost overruns, technical deficiencies, and a shift in naval requirements towards non-traditional security issues.

> employ a common display console approach such as the U.S. UF-452 or newer UYQ-70 for various systems that portends display sharing and common bus architecture. The Chinese 2JK consoles have system codes, similar to the 2KJ-8 for fire control, 2KJ-5 for sonar, and 2KJ-3 for radar, which again indicates open-architecture mission displays (Bussert and Elleman 2011: 29).

With the *Type 052C "Lüyang II"* program, China seems to have finally settled on a design concept suitable as the basis for future serial production. It is founded on the preceding *Type 052B* design, which was still strongly influenced by Russian naval electronics. Figures 5.8 and 5.9 illustrate the gradual evolution of Chinese combat suites in the modern surface vessel designs.

5.3.1.5 The Modern Generation Destroyers: A Russian Imprint

The first two Type *052B "Lüyang I"* class destroyers were the first Chinese naval vessels to be modularly constructed and were therefore finished at a much faster pace than their predecessors. They thus represented an important step ahead in terms of modern, efficient production methods, as Medeiros et al. noted in 2005::

> This serial production of an indigenously designed vessel is a first in the PRC's naval history and a testament to improved project management. The four new 052B- and 052C-class vessels have been built or have been under construction within the past four years. By comparison, in the entire decade of the 1990s China only built a second Luhu (1993) and one Luhai (1997) destroyer (Medeiros et al. 2005: 146).

Even though the *Type 052B* class was discontinued after just two ships were built, and also despite relying strongly on Russian naval electronics, its design was quite remarkable:

> [It was] the first Chinese-designed and indigenously constructed ship with the OTH [*Mineral ME*] Band Stand SSM data link and [*Fregat MAE-3*] Top Plate three-dimensional air/surface search radar. [. . .] The ZJK-5 command and control (C2) may use an Ethernet local area network (LAN) architecture and the aviation 1553B data bus, giving a new level of systems interoperatibility. [. . .] The 052B DDGs [. . .] were a remarkable advance for China and the PLAN, and showed that China understood what its most serious naval limitations were and how to correct them (Bussert and Elleman 2011: 37).

The Russian surface-air-missiles (SAM) integrated in this class also offered a marked advance over earlier Chinese AAW capabilities, which were relatively weak. All in all, this small class of two vessels represented a new capability level and served as the basis for the even more effective *Type 052C* and *052D* classes that followed, and which have already replaced some of the Russian naval technologies with indigenous or reverse-engineered systems.

5.3.1.6 Introducing Modern Area-Defense Capabilities

The most interesting development was the critical advances made in weapons and sensor technology starting with the *Type 052C "Lüyang II"* class (cf. Fig. 5.9). The major departure from the combat system configuration of the *Type 052B* consists in the introduction of a phased-array radar system for SAM control. The roughly contemporaneous, but in many respects less advanced *Type 051C "Lüzhou"* class (which has no organic helicopter and uses steam propulsion), features a Russian *30N6EI "Tombstone"* phased-array radar antenna located at the aft, which controls the *SA-N-20* SAM vertical-launch system (cf. Fig. 5.8, which shows the main functional chains). Regarding this technology transfer, Bussert and Elleman write:

> China had purchased two advanced S-300-derived Russian area defense 30N6E1 phased-array Flap Lid antenna for 051C. The antenna is located just forward of the aft VLS cells [. . .]. These are the only VLS launchers and associated radar provided by Russia to the PLAN for any ship (Bussert and Elleman 2011: 43).

The two ships of this class, despite shortcomings resulting from the lack of helicopters, nevertheless have "Chinese HN-900 and Russian Mineral ME3 data link command and communications [which] provide the North Sea Fleet with the capability to conduct joint Chinese-Russian naval operations" (Bussert and Elleman 2011: 43).

The existing, even if limited, access to some advanced Western, Ukrainian, and Russian naval electronics has apparently allowed China's Nanjing Research Institute of Electronic Technology to develop an indigenous active electronically scanned phased-array radar system, which is variously called the *Type 346* or the *Type 348 "Dragon Eye."*[9] Friedman assumes that this system was probably derived from a Ukrainian C-band active phased-array radar produced by Kvant Design Bureau, which was sold to China by Ukraine in 2004 with the necessary design and technical assistance package (Friedman 2006: 222–223; cf. also Wertheim 2013: 115). Chinese publications described work on an indigenous S-Band radar like *AN/SPY-1*, and Bussert and Elleman (2011: 38) consequently suppose this to be the operating band of *"Dragon Eye."* Friedman, however, notes that

> [. . .] other manufacturers of active-array radars are finding it difficult to package sufficient power in S-band Tx/Rx modules small enough to fit the sort of array involved. It seems more likely that Type 346 operates in X- or in C-band, like APAR or the Russian Tombstone (Friedman 2006: 223).

Starting with the first *Type 052C "Lüyang II"* class vessel, which was commissioned in 2004, Chinese destroyers have since begun to integrate this indigenous phased-array radar system for controlling the *HQ-9* or *HQ-9B* SAM and their 48-cell vertical launch system. The antennas look "strikingly similar"[10] to the U.S. *AN/SPY-1* radar antennas (Bussert and Elleman 2011: 38). In addition to this new multi-function radar, the very capable and proven Russian *Mineral ME "Band*

---

[9] While Friedman (2006) uses the designation *Type 346*, Bussert and Elleman (2011) call it *Type 348*. The latter name is a little ambiguous, given that *Type 348* is also used as the designation for the *MR-35* fire control radar installed aboard the *Type 037/2 "Houjian"* class FAC. Since the matter stills appears to be unresolved as of mid-2014, I follow Friedman's convention for the reason given.

[10] For a close-up photograph of two of the four *Type 346* panels, see Friedman (2006: 222).

Stand" fire control system for cruise missiles and main gun was retained, while the indigenously designed *Type 364 "Seagull-C"* 3-D air/surface search radar for CIWS control has replaced the Russian *Fregat MAE-3 "Top Plate"* search radar on the main mast in all newer destroyers, and is typically combined with the vintage Soviet-era *Type 517 "Knife Rest"* long-range 2-D air search radar located further aft (cf. Fig. 5.9). *Fregat*, though, remained on the main mast of the latest *Type 054A "Jiangkai II"* frigate class.

Introducing an indigenous active phased-array radar system (AESA) for SAM control is of course consistent with the trend in most advanced naval shipbuilding nations. This is because such powerful—and extremely costly—multi-function radars, which are able to search, detect, and track a high number of targets while simultaneously providing command guidance and terminal guidance as well as target illumination to missiles, are the cornerstone of modern combat systems designed to counter massive saturation attacks which would otherwise overwhelm traditional ship self-defense systems that rely on mechanical scanning. Western equivalents include the U.S. *AN/SPY-1* family of radar systems, which is at the center of the *AEGIS* CDS and has also been adopted by the South Korean, Japanese, Australian, Norwegian, and Spanish Navies; the Dutch *APAR*, which is integrated aboard the *LCF (De Zeven Provinciën)* class, the German *F124 "Sachsen"* class, and the Danish *Iver Huitfeldt* class; the British *SAMPSON* as part of the *Sea Viper* air defense system aboard the *Daring* (*Type 45*) destroyer class; and the Franco-Italian *EMPAR* radar in service aboard the *Horizon* class, the *FREMM* class, and the Italian aircraft carrier *Cavour*. Among the named radars, *AN/SPY-1* and *EMPAR* are passive electronically scanned arrays, while the newer *APAR* and *SAMPSON* are active electronically scanned arrays (AESA), representing a step ahead in technology with markedly enhanced performance and reduced power consumption.

The follow-up of the *Type 052C* class, the *Type 052D "Lüyang III,"* was reported to have received a modified variant of the *"Dragon Eye"* AESA consisting of larger antennas, possibly pointing to a lower frequency radar band (Bussert 2013).

5.3.1.7 Potential Technical Issues with the Latest Destroyer Classes

Command, control, and communications technology represents a major challenge not only because of the need to build the physical systems and to resolve their very high energy consumption aboard a warship. Another difficult hurdle lies in developing efficient data links in order to create a true joint battlespace with other participants who access the same data for a shared situational awareness. Yet another problem concerns creating the necessary software for interpreting the data and exchanging it with the various weapons, effectors, and sensors of not just one ship, but of several other vessels and aircraft, and providing consoles for coherently displaying the data. The basic technical requirements on board a single ship include "shipboard LAN and common display consoles shared by sensors and a computer/control station." Creating the joint battlespace additionally requires encrypted high-speed data links (Bussert and Elleman 2011: 40–41).

Compared with all earlier classes, the *Type 052C "Lüyang II"* class can be considered the first Chinese warship design with a true air-defense capability that is theoretically comparable to the *AEGIS*-equipped ships in service with the U.S. Navy, South Korea, Japan, and Australia (Wertheim 2013: 115). However, the actual performance of the Chinese systems relative to their better known Western equivalents still awaits independent evaluation. As Grubb and Collins noted in 2008, while the "integration of these three sub-systems [a phased-array radar antenna, a SAM and VLS, and data links] into a comprehensive long-range area air-defense system is a notable achievement, and may indicate a move towards improved PLAN blue water capability," they also pointed out that

> [...] little is currently known as to the actual capabilities or operational effectiveness of the *Luyang II*'s systems, and one might plausibly interpret the purchase of advanced *Sovremenny*-class destroyers and Kilo-class submarines from Russia as indicators of continued limitations in indigenous capabilities for integrating the most complex sets of warship systems (Grubb and Collins 2008: 47, italics in the original).

As of late 2014, this state of affairs has not markedly changed. This is not surprising given that the potential difficulties not only of developing, but also operating such a complex "system of systems" can hardly be overstated. As Bussert and Elleman warn, by taking note of historical examples it is easy to infer just how challenging this task really is: "Soviet problems indicate it might be very difficult to get the Chinese system to work properly" (Bussert and Elleman 2011: 38). As a matter of fact, the first Soviet attempt to create a phased-array radar, the *Sky Watch/Mars-Passat* (first observed aboard the aircraft carrier *Admiral Gorshkov* in 1988, and later also aboard the *Admiral Kuznetzov*), appears to have been a complete failure. According to Friedman:

> This radar was apparently unsuccessful. The evidence is that the superstructure of the second full-deck carrier, the incomplete *Varyag*, was adapted instead to a conventional mechanically scanned radar (Friedman 2006: 277, italics in the original).

The *Varyag* in question here is of course the same ship hull that China later acquired from Ukraine and then outfitted and re-christened the *Liaoning*.

Bussert and Elleman likewise note that during sea operations, the Soviet *Sky Watch* radar apparently "failed to detect and engage incoming ship or aircraft targets" and was therefore "replaced by mechanical scanning Top Sail/Top Pair radars on the next Soviet carrier, the *Tblisi*" (Bussert and Elleman 2011: 38, italics in the original). In line with the above observations, a Chinese analysis in 2003 concluded that in direct comparison with U.S. AAW capabilities, Russian systems were still clearly deficient (cf. Chen 2003: 21).

The development of the U.S *AN/SPY-1* family of radar systems was likewise a lengthy and cumbersome process. Development of the first U.S prototype phased-array radar system, the *AN/SPG-59* (put into service in 1964) was ultimately discontinued due to unsatisfactory performance. Its successor, the *AN/SPY-1*, suffered from initial challenges as well. One particularly egregious example of the problems that were encountered during early operational history was the tragic misidentification by the *USS. Vincennes* of an Iranian civilian airplane as a military target on July 3, 1988, resulting in 290 casualties.

The potential problems that must be faced and solved during design, building, integrating, and using such systems for creating a joint battlespace are thus quite numerous. Not only can they not be remedied simply by acquiring new technology or software piecemeal, but they seem to be experienced by all the leading navies. In order to get an idea of a plausible timeline, it is useful to recall the following in regard to the U.S. Navy:

> The Aegis technology was on the USS *Norton Sound* [...] in 1974 for nine years of development testing prior to the first installation on the USS *Ticonderoga* [...] more than twenty-five years ago. Since 1983 more than fifty [...] Aegis warships have been built. Integration with various indigenous ship guns and missiles and other sensors, as well as other ships' data management and weapons, is a challenge. Developing the software for signal processing and tracking a hundred air, surface, and submarine targets could take even longer for China (Bussert and Elleman 2011: 38, italics in the original).

Similar software-related problems seem to have plagued nearly every comparable system in the world, with the latest AESA systems such as the European *APAR* being no exception. As for current U.S. active phased-array radar development and related software troubles, it is noteworthy that as of September 2014, more than seven years after its introduction, Raytheon's newly developed *AN/APG-79* AESA radar for *Super Hornet* fighter airplanes reportedly still suffers from serious software problems, with no final date given for the resolution of the issues.[11]

In light of the universal track record of the world's leading builders of phased-array radar systems, it is reasonable to assume that the first Chinese ships carrying such systems will also likely experience "long-term software headaches." All in all, integrating a new phased-array radar, data links, and a new type of VLS launcher is going to be a formidable challenge for China. It is, however, equally reasonable to assume that China will eventually overcome these difficulties, and that its navy will ultimately reach a new level of command and control capability comparable to other leading navies (Bussert and Elleman 2011: 41).

As for data links, the U.S. and their NATO partners currently use *Link 16,* a wide-band ultra-high frequency (UHF) digitally encrypted data bus, which has about ten times the speed of the previous *Link 11*. A Soviet equivalent to *Link 11* was the *"Bell Crown,"* which was replaced with the *"Bell Thumb"* in 1993 (Bussert and Elleman 2011: 40).

In China, the Russian *Mineral ME3 "Band Stand"* data link command and communications technology filled this capability gap. The modern Chinese data link is called *HN-900:*

> China's newest frigates, updated Ludas, Luyang DDGs, and Houbei catamarans have the Chinese tactical data link system designated as HN-900. The HN-900 probably includes some [...] foreign data link technologies. Luyang DDGs have the Russian Light Bulb data

[11] Cf. the discussion in 'Raytheon's APG-79 AESA Radars' (2014).

> link, right above the DDG 170 helicopter hangar, and Band Stand provides coordinated operations with Russian Navy and Chinese data links (Bussert and Elleman 2011: 41–42).

There are several indications of China's difficulties in solving the problems associated with creating the technical requirements for a joint battlespace indigenously, most notably, through reports of the country's spying activities and foreign purchases.

As Bussert and Elleman note:

> The Chinese Beijing Readsoft Corporation copied the MIL-STD 1553, under the designation GJB289A. It was initially used on PLANAF J-10A Chinese avionics and PLA systems, and was adapted to 052C weapon equipments, which supports interoperability of forces (Bussert and Elleman 2011: 41).

Another incident pointing in the same direction concerns a case of *AEGIS* battle management data theft, when, "in November 2005 four Chinese contractors were charged in Los Angeles with stealing military technology" (Bussert and Elleman 2011: 38). Yet another sign that China lacks domestic data link design capability were reports indicating that the Chinese tried to get the Russian *TSIMSS-1* digital data link installed during negotiations for the purchase of Russian *Su-30 MK2* aircraft. In order to exchange data with such airplanes, the modern Chinese destroyers would need the equivalent data link for the *Sukhoi* fighter planes as well (Bussert and Elleman 2011: 40).

It is important to note that in the interval between gaining access to foreign technologies and knowledge and the development in China of a coordinated engagement capability, it is also to be expected that there would be a relatively long period of reduced detecting and tracking capability of the combat systems aboard the *Type 052C* class. The strategic consequences of this for China would be grave. As Bussert and Elleman note,

> [...] until coordinated engagement capability matures, these two Luyang II ships may be limited to 1940s-era radar tasks of detecting and tracking air and surface targets for just their own ship's weapons, not for a truly joint battlespace (Bussert and Elleman 2011: 38–39).

The mission-critical importance of the *Mineral ME "Band Stand"* OTH targeting and data link technology for essentially all modern Chinese surface vessels can hardly be overstated. Furthermore, the resulting dependency on Russian support for these and other weapon systems (which was at first physically limited to a support enclave at Dinghai), well explain China's attempts to reverse-engineer these systems despite Russian protests (cf. Bussert and Elleman 2011: 38).

5.3.1.8 A Next-Generation Destroyer Design: The *Type 052D "Lüyang III"* Class

Starting with the currently under construction *Type 052D "Lüyang III"* class, which is largely based on the *Type 052C*, serial production seems to have begun in earnest, with up to 10 vessels of this type expected to enter service during the next few years. Technologically, they represent marked progress. They will also likely form the backbone of the expected 3 to 4 carrier battle groups envisaged by China's naval planners, together with the new *Type 054A "Jiangkai II"* class frigates. As *Jane's Fighting Ships* reports in its 2013 electronic edition:

> The principal features [...] include a development of the Dragon Eye [Type 346] phased array radar incorporated in the forward superstructure and a new Vertical Launching System [...] capable of housing air-to-surface, surface-to-surface, and anti-submarine missiles. Reports suggest that there is a total of 64 rectangular cells (32 forward, 32 aft). Armament forward includes a single-barrelled 130 mm gun and a Type 730 CIWS. Aft there is a flight deck and centre-line hangar on which there is a short-range missile system (FL-3000).

The new multi-purpose VLS is a notable improvement, for it is able to handle a range of different missile types. Another milestone is the improved ship self-defense system against saturation attacks, achieved through a CIWS that includes a launch system for *HHQ-10* fire-and-forget Rolling Airframe Missiles (RAM) aft (cf. Fig. 5.9).

Like all modern Chinese surface vessels, the *Type 052D* class uses the same Russian *Mineral ME "Band Stand"* OTH fire control system for SSM and main gun. The *Dragon Eye* phased arrays on this class are larger than those on the preceding *Type 052C* class destroyers, suggesting a lower frequency (Bussert 2013).

Also in the preceding class, the search radar on the main mast is the Type *364 "Seagull C"* 3-D air/surface search radar, which is again combined with the vintage *Type 517* 2-D air search radar further aft to provide target data for the CIWS. The CDS for the new *Type 052D* class has been named the *H/ZBJ-1*. According to Bussert, "[w]ith the PLA having new missiles, launchers and sensors, it would be reasonable to expect a next-generation CDS complex," but details are still speculative (Bussert 2013).

The sensor configuration is thus very similar to the established design. As for data links, Bussert expects *"Light Bulb"* data link radomes for this class, and a new Chinese integrated datalink system, namely the

> [...] joint service integrated datalink system (JSIDLS), that was certified in June 2012 and is comparable to the U.S. Navy's successful Link 16 for Aegis. The 052D would share a new automated maintenance support link from combat systems to the support auxiliaries, and even home base, on fault isolation and repair part ordering that was being used in Somali deployments on 054A frigates. A July 2009 PLA Daily article quoted Xu Jianmao, an instructor at the National Defense University, saying 'ship equipment support soft power,' such as a software information management system, is quantified and is in use. This alleviates the popular conception that China relies on other nations for repairs. The common HN-900 intership datalink on all current warships requiring Naval Tactical Data System-like capability, probably will be replaced on 052D with a new naval common tactical data link (NCTDL), which is the next generation two-way encrypted datalink that can support new electro-optic and laser pod UAVs (Bussert 2013).

Since many details regarding the combat suite aboard the *Type 052D* class remain unknown, it is too early to make an assessment of its potential capability once all integration problems have been solved. Given its new multi-purpose VLS and the active phased-array radar system, Bussert holds that its capability potential from the hardware point of view is even greater than the decades-old *AEGIS-centered* technology. However, it is certainly true that, when taken as a complete, proven, and functional system, *AEGIS* ships will nevertheless still likely outperform this new vessel, mainly because

> [t]he associated Aegis software remains decades more advanced than the Chinese software, especially the new antiballistic missile and cooperative engagement capability missions capabilities (Bussert 2013).

Overall, there has clearly been marked progress in indigenous Chinese naval weapon systems development over the past decade, and Chinese naval architects seem to have arrived at a convincing combat systems configuration consisting of vintage systems like the *Type 517* early warning radar antenna, legally imported or reverse-engineered Russian systems like the *Mineral ME "Band Stand,"* and cutting-edge systems such as the *Type 346 "Dragon Eye"* AESA, which combine with a variety of imported and indigenously developed missile systems. Just how well this system of systems actually performs in comparison to the various tried and proven Western equivalents, however, will only be known when actual and verified performance data becomes more readily available.

5.3.1.9 Overall Characteristics of Technical Innovation in the Destroyer Fleet

Judging from current European and American naval shipbuilding trends, the main realms of innovation in naval shipbuilding today are: a) survivability, which includes signature reduction features ("stealth") through hull form innovations such as x-shaped cross sections or encapsulated radar masts, as well as literal survivability in case of damage, e.g. through systems redundancy or so-called "two-island-concepts"; b) efficient and reliable propulsion concepts; and c) state-of-the-art combat systems able to handle the most advanced threats. These critical areas also define to a large extent a warship's strategic value as a weapon system. In China's case, the first two criteria are relatively easy to assess.

China's first indigenously built destroyers were still highly deficient both in design and construction, which impaired both their seakeeping and their survivability. These problems were addressed in the newer classes, and it appears that the performance of the Type 052C *"Lüyang II"* class was sufficient for it to become the basis of larger vessel classes developed from this design. It seems that, especially since the *Type 052C* and *Type 052D* destroyer classes, China's naval shipbuilding has made significant progress in building more seaworthy and more stealthy surface vessels with adequate provisions for at least one organic helicopter per ship, a crucial feature for over-the-horizon targeting and anti-submarine warfare (ASW). However, large destroyers in the leading Western navies typically carry two large (10 t) helicopters. Such helis can carry a heavier payload and thus be equipped to function as an independent weapon system. They can also be used as a shared asset between several ships within the same fleet. So far, however, China's naval helicopters have all been much smaller, 5 t aircraft, and are thus not as capable as the helicopters carried, for instance, by Japan's large "Helicopter carrying destroyers" (e.g. the *Hyuga* class) or by South Korea's *Dokdo* class LPX. A state-of-the-art helicopter capability for modern AAW destroyers would normally include provisions for two ship-borne 10 t helis, with at least one of them equipped for submarine search and the other for combat missions or search-and-rescue. This is all the more relevant given that industry sources interviewed by me uniformly expressed the opinion that ASW helicopters are today far more effective in submarine search missions than surface ships with bow-mounted sonars, which are apparently only rarely able to detect modern quiet submarines under real-life conditions.[12]

As Bussert and Elleman explain, the *AEGIS* concept would usually include two shipborne helicopters. One of these would be equipped with distant reconnaissance/targeting capabilities, which would then "pass target data to a consort shooter that would launch its weapons" (Bussert and Elleman 2011: 40). For this kind of coordinated action, efficient and secure data links are a necessity. The same goes for data exchange between ships and shore-based long-range fighter aircraft. Here, Chinese naval aviation equipment and mission capability is still considered somewhat deficient in comparison to regional and international navies of similar overall capability.

Regarding propulsion concepts, it is remarkable that even some of the newer vessel classes (notably the *Sovremennys* and the *Type 051C "Lüzhou"* destroyers) still use steam propulsion. Although somewhat outdated, this propulsion concept is not necessarily a bad choice in strategic terms, since gas turbines are very expensive in procurement, maintenance and fuel consumption, and yet, as numerous related problems experienced by Western surface ships have illustrated, also not necessarily reliable or easy to maintain. All other modern Chinese destroyer classes however rely on a CODOG arrangement comprising gas turbines and diesel engines. This places China in the uncomfortable position of having to import or license-build gas turbines as well as naval diesels, neither of which can be produced independently yet in China. So far, the gas turbine supply has been sourced from licensed manufacture of a Ukrainian design. As already described at length in Sect. 4.6, China's dependence on importing critical components such as propulsion is a potential area of strategic vulnerability.

Evaluating the third criterion of innovation in naval electronics and the degree of reliance on foreign technology transfers is fraught with much more analytic

[12] I was given detailed descriptions of failed submarine detections from several former active service members who had participated in various national and international ASW helicopter training missions. In their experience, in the case of modern, quiet diesel submarines, systems such as bow-mounted sonars or towed arrays were often frustratingly insufficient for accomplishing the task in practice. See also Scott (2011a) who states that even "a few relatively unsophisticated submarines can conduct sea denial".

difficulty. Overall, as discussed in the previous sections, China's destroyer fleet shows the distinct marks of a turbulent history in the field of cooperative arms production. In the beginning, the Soviet influence on Chinese fleet development was immense and its influence is still felt today, especially in the realm of electronics. Bussert and Elleman refer here to the "Soviet technology legacy"[13]:

> Most of the PLA Navy's new vessels [after 1949] [. . .] were provided by the Soviet Union during the 1950s and 1960s, either as actual equipment or in the form of plans, tools, and production facilities (Bussert and Elleman 2011: 2).

Even though direct Soviet assistance in Chinese shipbuilding ceased with the Sino-Soviet split, there has nevertheless been surprising continuity in terms of the naval equipment that is used, even in vessels built during the 1980s and later: "The 1989 inventory of PRC naval electronics indicates that through the 1980s, very little basic equipment had changed since the 1950s" (Bussert and Elleman 2011: 2).

Some vintage radar systems, like the *Type 517 "Knife Rest"* early warning radar dating from the early 1950s, are even commonly installed in newly built vessels (Bussert and Elleman 2011: 4–5). On the *Type 052C* and *052D ("Lüyang II and III")* class destroyers, this long-range 2-D air search radar system is combined with the modern *Type 364 "Seagull C"* 3-D air search radar to provide target data to the CIWS. One interpretation would be that the continued reliance on such highly outmoded systems implies that little technological innovation has taken place in this particular field. On the other hand, it is also possible that these types of radar systems offer advantages in detecting modern stealth fighters. Finding ways of detecting American stealth fighters such as the *F-22 Raptors* stationed at Guam has been a priority for Chinese military planners for some time (Minnick 2014).

Another continuing technical legacy is the ubiquitous presence of the Russian *Mineral ME "Band Stand"* weapon control system, which has been installed on all modern Chinese destroyers and frigates since the arrival of the *Sovremennys*. As described in Sect. 4.6, the alleged Chinese reverse engineering of this system has already caused problems between the Chinese and the Russians (cf. 'Frictions between Russia and China' 2009: 15). Assuming such reports are true, incidents like these confirm that this imported weapons control system is of critical importance for virtually all modern Chinese warships. This would help explain the Chinese attempts to replicate it, even though such a course of action would risk endangering good relations with Russian suppliers.

Likewise, China's very short period of access to advanced Western technologies between 1980 and 1989 had a surprisingly significant impact on some of the currently active destroyer classes. Notably, the two *Type 052A "Lühu"* class destroyers were greatly affected by the arms embargo, for they were originally designed to incorporate a mixture of U.S., German, French, and Italian systems—specifically, U.S. gas turbines, German diesel engines, the French *TAVITAC* combat system plus *Sea Tiger* air search radar and *Crotale* missiles, and Italian sonar. This

[13] Cf. Li (2009: 122–125) for a detailed description of the various aspects of Soviet influence on China's military development after 1949.

combination, plus "the lack of spare parts and maintenance packages since 1989, has made operations and maintenance a logistical nightmare" (Shambaugh 2004: 268). A 2005 RAND study similarly concluded that China's dependence on foreign systems for some of its destroyer classes has had a number of strategic drawbacks:

> This high degree of reliance on foreign goods creates major challenges for systems integration and, given the inconsistent availability of certain weapon systems, complicates serial production of some platforms. In particular, Chinese combatants lack long-range air-defense systems, modern anti–submarine warfare (ASW) weapons, and advanced electronic warfare capabilities needed to outfit its new ships (Medeiros et al. 2005: 153).

In the realm of AAW weapons and sensors, the most critical components of military ships in terms of combat capability, China continues to be restricted from having access to state-of-the-art components from both the U.S. and the EU. Consequently, China appears to rely mainly on systems purchased from France or Italy before (or in some cases despite) the embargo, or on reverse-engineered indigenous developments that are based on those systems. China also has access—either legitimately or through reverse-engineering—to the proven and reliable, if not quite state-of-the-art, Russian and Ukrainian sensor systems.

The need to integrate systems of Russian, French, and indigenous origin within one and the same destroyer hull can be an enormous technical challenge. Ultimately, this problem can only be alleviated by developing more indigenous systems, an expensive and time-consuming course of action fraught with difficulties. China has managed to develop some missile systems based on reversed-engineered Russian and French technology, most famously the *C-801* and *C-802* sea-skimming cruise missiles and the *HQ-61*, *HQ-7*, and *LY-60N* air-defense missiles: "The C-801 [...] is very similar to the French Exocet missile with its radar homing, 165-kilogram warhead, and sea-skimmer flight profile" (Bussert and Elleman 2011: 26). At the same time, the *HQ-7* (export name: *FM-90*) appears to be "an unauthorized copy of the French Crotale Modulaire short-range shipboard SAM system. [...] Imported Crotale Mudulaire systems are aboard some Luda-class and Luhu-class destroyers (Wertheim 2013: 106)." These French systems were probably imported legitimately before the embargo along with the French *TAVITAC* combat direction system first used aboard some *"Lüda"* class destroyers. They have since been copied and installed on the single Type 051B *"Lühai"* destroyer, as described in the preceding sections. The *HQ-9* missile system (export designation *FT-2000*) is likely based on a Russian land missile system, the *Ryf-M (S-300/SA-N-6)*, of which two were sold to China (Friedman 2006: 569; cf. also Wertheim 2013: 114).

Around 2005, China's indigenous naval weapon systems developments were still judged to be far behind the contemporary industrial standard:

> [T]he capabilities of most of China's current naval SAM and SSM systems and much of its naval electronics are limited and not equivalent to U.S. capabilities or those of other Asian militaries. The limited range and accuracy of Chinese SSMs and SAMs create serious problems for air-defense and anti-submarine warfare. Many of these systems also do not operate with over-the-horizon targeting, further degrading their already-limited capabilities (Medeiros et al. 2005: 139).

These problems have been largely remedied with the introduction of modern Russian OTH targeting and weapons control systems, as well as very capable SSMs obtained through the imported *Sovremenny* class destroyers (and *Kilo* class submarines). These systems greatly influenced the combat system configuration of all subsequent surface ships. Until the newly developed combat suites of the latest destroyer and frigate classes are fully debugged and operational, the imported Russian-built destroyers will probably remain the most capable fully operational naval weapon systems in the Chinese inventory. Since they were developed specifically to target *AEGIS* carrier battle groups, and there are as yet only very limited countermeasures against their SSM, they remain a worrisome feature of the Chinese fleet for potential strategic opponents such as the U.S. and other major regional navies operating *AEGIS*-equipped ships (cf. Shambaugh 2004: 267; Bussert and Elleman 2011: 35).

Due to China's relatively high degree of secrecy in weapons performance data and its still limited co-operation with Western navies in large-scale fleet exercises, it must be concluded at the time of this writing that the most modern among Chinese surface combatants, the *Type 052C* and *052D* class destroyers, so far effectively remain black boxes when it comes to the capability of their naval electronics. As Kevin Pollpeter of the *University of California Institute on Global Conflict and Cooperation* (IGCC) similarly notes: "Overall, China's hull design [capability] is quickly improving but what's going into the ships is a bit of a black hole [. . .] We don't know how good the systems are" (quoted in Jean 2013).

Unfortunately, there is currently no way of addressing this information gap in the absence of voluntary transparency on the part of China's navy, or, alternatively—as unappealing as it may sound—actual naval combat situations that would allow for an evaluation of the precision of Chinese missile strikes or the efficiency of Chinese countermeasures against saturation attacks under battle conditions. Along similar lines, Bussert and Elleman note:

> Of course, final marks can really only be determined in actual combat with enemy forces, so an exact assessment of the new Chinese platforms might remain elusive for some time to come (Bussert and Elleman 2011: 41).

As enlightening as such an event would certainly prove from the point of view of naval research, such a scenario can hardly be hoped for.

5.3.1.10 Smaller Surface Combatants

In addition to the destroyer classes already analyzed, China's surface fleet encompasses a relatively large number of frigate- and corvette-sized vessels. As of late 2014, about 70 such units are in service, with around 10 of these being the obsolete or nearly obsolete *Type 053/053H/053H1 "Jianghu I and II"* classes. All frigate and corvette classes in service are depicted according to type and age structure in Fig. 5.10.

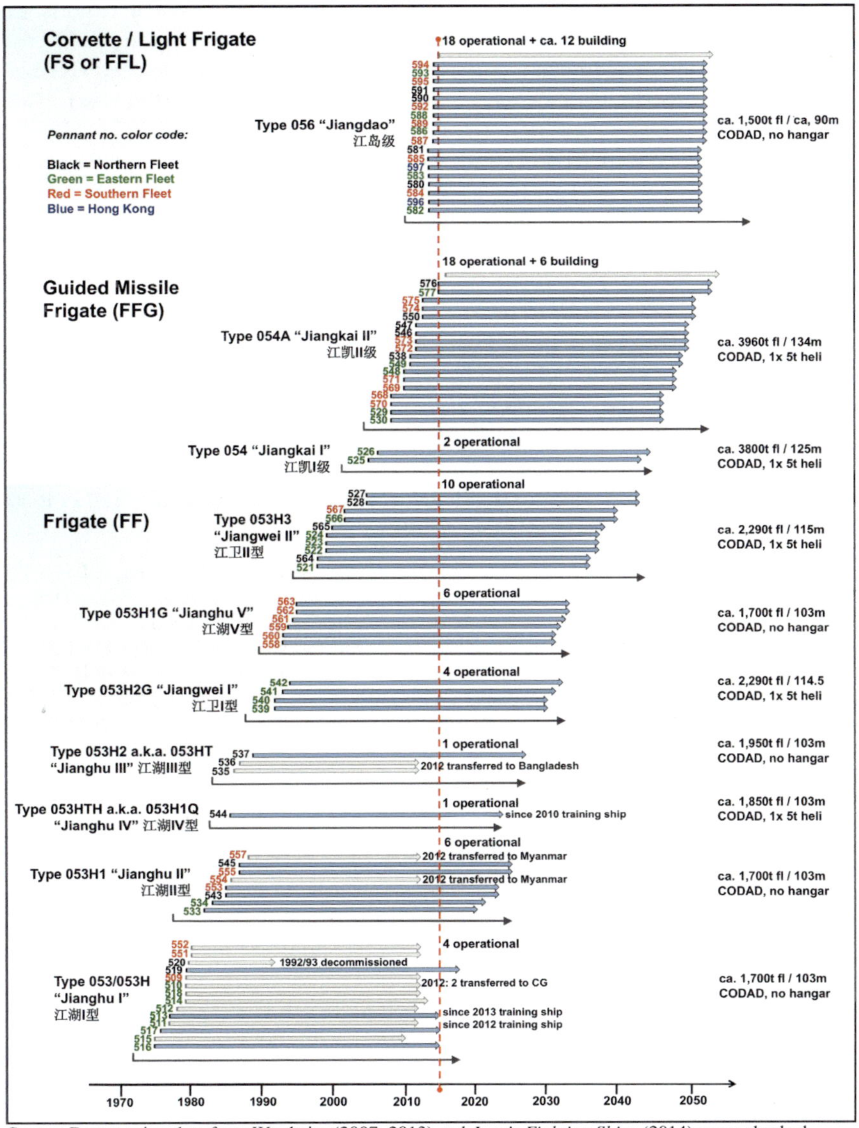

Source: Drawn using data from Wertheim (2007, 2013) and *Jane's Fighting Ships* (2014) cross-checked with Chinese Wikipedia entries

Fig. 5.10 Type and age structure of the PLA Navy frigate and corvette fleets, 2015

The development of these smaller combatants followed a roughly similar historical trajectory to that of the destroyer classes described in the preceding section, and therefore does not need to be described in as much detail. Some of the hull construction and combat system deficiencies of the earliest frigate classes were already discussed in the section dealing with China's early naval exports to Thailand and Bangladesh (cf. Sect. 4.6.5). It is therefore only necessary to briefly discuss here the modern frigate and corvette classes, which were deemed suitable for serial production by China's navy—the *Type 054A "Jiangkai II"* class frigate and the *Type 056 "Jiangdao"* light frigate/corvette.

The weapons and sensor outfit of the *Type 054A* class frigate (ca. 4,000 t at full load), which is the design that has been selected by the Chinese navy for serial production and will thus form the backbone of the future frigate fleet, is shown together with the modern destroyer designs in Fig. 5.9 above. It closely resembles the configuration of the *Type 052B "Lüyang I"* class destroyer, which was developed a few years earlier and uses the same Russian *Fregat MAE "Top Plate"* main search radar, combined with the *Mineral ME "Band Stand"* weapons control and data link system (cf. Fig. 5.8). Unlike the smaller (ca. 1,500 t at full load) new corvette design, the *Type 056 "Jiangdao"* class, which has no hangar, the *Type 054A* class frigate is designed for one embarked 5 t helicopter.

Interestingly, the *Mineral ME "Band Stand"* systems aboard these vessels, as well as other electronics, seem to be illegal, reverse-engineered copies rather than legal imports. As Bussert and Elleman report:

> Upon closer examination of photos, Russian trade officials reported that four major 054A systems were illegal Chinese copies of systems provided on the Sovremennyys. The illegally copied systems included the Mineral ME tracking radar, the Fregate [sic] M2EM 3-D radar, the MR90 SAM tracking radar, and the modern MGK-335 MF hull sonar. It would seem likely that the Chinese had non-Russian [by implication: most likely Ukrainian] assistance to reverse engineer such sophisticated systems in only four or five years. This has had a negative impact on future Russian sales of complex weapon systems to China (Bussert and Elleman 2011: 56).

Notwithstanding the well-known problems usually associated with integrating illegally-acquired, reverse-engineered technology, Bussert and Elleman come to the somewhat surprising conclusion that the *Type 054* class frigate "compares favorably with the French *La Fayette* or the German/Spanish/Dutch 124 FFG" (Bussert and Elleman 2011: 55). Although a comparison with the somewhat dated *La Fayette* frigate, of which the first of class was laid down already in 1990, may theoretically be in order, the "favorable" comparison of the *Type 054* with the German *F124 "Sachsen"* class and the Dutch *LCF* is nonetheless surprising, and one that most naval architects would beg to differ with, not least because the much larger and indefinitely more capable *F124 and LCF* are in fact hardly comparable with the *Type 054* at all. Despite their misleading "frigate" denomination, they have the capability of a modern air warfare destroyer due to a highly advanced combat suite centered around the *APAR* AESA multi-function radar, American *SM-2* missiles, and a *Mk 41 Mod 10* VLS with installed NATO *Links 11* and *16*. (It is worth noting that the Spanish variant also mentioned here by Bussert and Elleman, the *Álvaro de Bazán* class, actually ended up a rather different ship, carrying *AN/SPY-1* and becoming the first European *AEGIS* vessel). Moreover, the *F124/LCF* design is much larger than the *Type 054* (almost 10 m longer, with a full load displacement of 5,700 t compared with the former's less than 4,000 t), has a more advanced

CODAG propulsion plant compared with CODAD, and can operate and embark two large (12 t) helicopters, compared with the *Type 054*, which can only carry one embarked 5 t heli. There is thus very little actual basis for comparison regarding either the mission profile or the capability of these ship classes, and it would be highly unusual to assume that the Chinese *Type 054* class frigate could somehow perform the same tasks as the *F124*—even if all its systems worked flawlessly. *The F124/LCF* rather belongs in the same category as the large modern Chinese DDGs, such as the *Types 052C* and *052D*. In time, these could indeed begin to perform comparable area defense tasks if all initial system integration challenges can be overcome.

The new *Type 056 "Jiangdao"* class missile corvette design—which is apparently destined to replace the aging Fast Attack Craft and could possibly comprise a whole family of corvette-sized vessels ranging from lightly armed OPVs to heavily armed light frigates using the same basic ship platform—is apparently based on a similar design to the OPVs exported to Thailand and Nigeria (cf. Sect. 4.6.5). Its full load displacement is ca. 1,500 t and, according to most descriptions, it appears to have no provisions for a hangar and therefore no organic helicopter capability, although Wertheim (2013: 124) does list a hangar in its specifications (cf. *Jane's Fighting Ships*, el. ed. 2014). It seems to be armed with a *C-803* SSM and a *FL-3000N* SAM system.

Judging from the outer hull lines, some of the older Chinese frigate and OPV designs still seemed to be derived directly from their Soviet-era predecessors. Regarding hull forms and the survivability of the smaller surface ship types—e.g. the frigates exported by China to Thailand, Bangladesh, and Pakistan, as well as smaller OPV designs offered to various customers in South America during the past 10 years—European shipbuilding industry sources interviewed by me in 2011 indicated that these ships still tend to be rather slim and low in displacement in relation to length overall. While this does enhance speed, it also has a negative impact on sea keeping and survivability in the case of hull damage. According to these sources, deck height aboard those Chinese ships also still tends to be somewhat lower than is customary in Western ships.[14]

Among the smaller surface combatants, another notable development with potential strategic consequences and thus meriting closer inspection is the 43 m *Type 022 "Houbei"* class missile catamaran that China has been building in large numbers since around 2005 (not shown in Fig. 5.10). With a full load displacement of only ca. 250 t and very shallow draft, this FAC can reach a top speed of ca. 36 knots using waterjet propulsion. These catamarans are heavily armed with eight *C-803* SSM missile launchers each. As already described in Sect. 4.6.4, the hull design was originally derived from the civilian *AMD 350* catamaran, which was supplied to China by the Australian design company AMD Marine Consulting.

[14] One shipbuilding executive interviewed by me had personally seen a Chinese technical description of a ca. 90 m, ca. 1,300 t patrol vessel design that was offered to at least two non-Asian developing countries in recent years. His opinion was based on the data contained therein.

Large numbers of these hulls are currently being built in license (cf. 'AMD Building Catamarans' 2008). Wertheim (2013: 124–125) estimated that 68 units had been built by 2013. These small and fast vessels are ideal for littoral warfare in confined and shallow waters such as the East China and Yellow Sea areas, and could play a critical role in asymmetric warfare scenarios, e.g. during possible conflicts over Taiwan, the Senkaku/Diaoyutai islands, or in the South China Sea (cf. Moss 2011a). Interestingly, they seem to be painted mostly in the colors of the Chinese Marine Corps (Bussert and Elleman 2011: 90).

5.3.2 Naval Power Projection Weapon Systems: Carriers and Nuclear Submarines

The present subsection describes the PLAN's attempts to build a fleet of offensive power-projection vessels. Aircraft carrier and amphibious assault ship development will be discussed first before turning towards nuclear-powered ballistic missile submarines.

5.3.2.1 China's Aircraft Carrier Program

Ever since buying the decommissioned Australian aircraft carrier *HMAS Melbourne* in 1984, China has been studying retired carriers. China's ambition to obtain carriers gained momentum in 1998, when China, via a nominally privately owned business based in Macao, purchased the incomplete, rusting hull of the ex-Soviet carrier *Varyag* from Ukraine. The *Varyag* hull is an *Adm. Kuznetsov* class carrier of ca. 304 m in length and a displacement of almost 60,000 t which had been laid down already in 1985. It was delivered to China only 70 % complete, without propulsion, electronics, sensors or weapon systems (cf. Hardy 2011). After a lengthy transfer and study process—which remarkably included the unprecedented building of a full-sized concrete carrier mockup at Wuhan, purportedly for studying system integration issues[15]—finishing and outfitting of the *Varyag* began in earnest in 2005. During this year, the *Varyag* ultimately received a steam propulsion plant. Seven years later, the ship, now re-christened the *Liaoning*, was finally commissioned into service in September 2012. Most realistic assessments assume that the primary function of the *Liaoning* will be training and research missions for a future indigenously built carrier force rather than actual combat missions. Due to its strategic role, the *Liaoning*, like China's strategic nuclear forces, is under the direct

[15] An article in the December 2009 issue of *Kanwa Asian Defence* entitled 'Structure of 'Wuhan aircraft carrier'' contains clear images and interesting technical observations. *Kanwa* experts were able to inspect the concrete structure and draw conclusions regarding the likely future carrier design.

command of the Central Military Commission (CMC) rather than PLAN command (Cole 2011). An overview of China's emerging carrier and amphibious assault fleet is given in Fig. 5.11.

The first official announcement of China's intention to build indigenous carriers was made in May 2010 in the *2010 Ocean Development Report* published by China's State Oceanic Administration. Most foreign observers did not become aware of it until much later in the year (LaGrone 2010). There is no doubt that China's major naval shipyards are now able to design and construct a carrier hull. It is similarly known from official statements and matching satellite imagery that China is currently in the process of building up to three indigenous carrier hulls at the Jiangnan Group's Changxing Island shipyard near Shanghai, which has large enough docks to accommodate such ships (Hardy et al. 2013). While the *Liaoning* can only be expected to be fully operational (including air wing) sometime after 2015, the second carrier, according to plausible estimates, could be operational by around 2020 (Saw 2008: 38).

It is however important to note that the hull design and construction of even a very large ship such as a carrier is not by any means the most difficult technological challenge to overcome. Compared to densely integrated high-end vessels, such as AAW destroyers, guided-missile frigates, or submarines, carriers mainly consist of vast quantities of empty space. Aside from building the hull, the most crucial challenge therefore lies in mastering aircraft-launching technologies, and then in training and operating the air wing. Since a carrier is basically a floating air base, the number, size, and reach of the aircraft being carried, as well as the range, speed, and endurance of the carrier itself, are decisive for the carrier's strategic value. The maximum number of aircraft that can be carried and operated simultaneously is dependent upon flight deck and hangar size. Medium-sized carriers can range

Class	Length o.a./displ. full load	Propulsion	Main radar	Main armament/CIWS	Total crew/air wing	CIWS	Aviation
Ex-Adm. Kuznetsov CVL	306.45 m/59,100 ton	Steam (4 turbines, 4 props)	1× *Top Plate* 3D air search; 1× *Type 346* phased-array radar	4× 24 cell FL-3000N SAM	ca. 2,000/ca. 600	2× 30 mm Type 730, 2× RBU-6000 ASW RL	ca. 20 J-15 or SU-33
Type 081 LHD	ca. 210 m/ca. 20,000 ton	CODAD?	n.a.	n.a.	n.a.	n.a.	ca. 8 helis
Type 071 LPD	210 m/17,600 ton	CODAD (4 diesel, 2 props)	*Type 364* air/surf. search; *Type 360* air/surf. search	HQ-7 SAM?	120 crew + 400–800 landing troops	4× 30 mm AK-630	4× Z-8 helis

Source: Compiled from Wertheim (2013) and *Jane's Fighting Ships* (2014)

Fig. 5.11 China's carrier and amphibious assault fleet—comparison of main data

between 40,000 t and 60,000 t full load displacement, but the most advanced U.S. "big deck" carriers have displacements in excess of 70,000 t:

> Aircraft carriers come in a wide variety of sizes, costs, and capabilities, from small vessels capable of operating a handful of vertical takeoff and landing (VTOL) aircraft to super-carriers with complements approaching a hundred large conventional aircraft. In the middle of that range, more or less, is the Russian *Kuznetsov* class (Li and Weuve 2011: 213).

The most advanced aircraft carriers in naval service today are the nuclear-powered big-deck carriers of the U.S. Navy. Due to a nuclear propulsion plant, they have unlimited range and can reach very high maximum speeds of more than 30 knots. Apart from the U.S., France is the only other country currently operating a single nuclear carrier. The *Charles de Gaulle* displaces ca. 43,000 t fl and was fitted with two U.S. steam catapults (Wertheim 2013: 196). A small ex-French aircraft carrier, the *Sao Paulo,* was purchased in operational state by Brazil in 2000. It is also fitted with steam catapults, but uses steam propulsion (Wertheim 2013: 54). All other carrier-operating navies have so far relied exclusively on conventional propulsion, usually gas or steam turbines.

The installations necessary for operating an air wing ultimately determine the actual strategic value of a carrier, and also make its building so difficult. The level of technology available for launching fighter, transport, and reconnaissance aircraft from a carrier is directly related to the maximum take-off weight, and therefore limits the choice of aircraft. This, in turn, has a direct consequence on their maximum operational range and capability. The Catapult-Assisted Take-Off But Arrested Recovery (CATOBAR) technology in use aboard all currently operational U.S. carriers uses steam catapults to generate enough thrust for launching the aircraft, which means the aircraft are not dependent upon the carrier's own movement for generating enough wind-over-deck to assist the take-off. Moreover, this technology allows for the launching of much larger and heavier aircraft than any other option, which enhances their potential reach and armament (Li and Weuve 2011: 216–217).

A further advance in future aircraft-launching technologies will be the Electro-Magnetic Aircraft Launch System (EMALS) developed by General Atomics, which is being integrated in the future U.S. *Gerald R. Ford* class carriers currently under construction. According to reports dating from May 2013, U.S. representatives have shown the EMALS system to the Indian navy. Despite its status as a critical technology, the system could eventually be transferred to India within the framework of an enhanced U.S.-Indian strategic arms co-operation, specifically, for use on India's planned second indigenously built *Vikrant* class aircraft carrier, the *Vishal*, which is expected to enter service sometime after 2020 (cf. 'Indian Navy seeks EMALS' 2013).

At present, all other carrier-operating navies rely on less advanced aircraft launching technologies, such as the Straight Take-Off But Arrested Recovery (STOBAR) or Short Take-off, Vertical Landing (STOVL) systems. Compared with CATOBAR launching, which is possible even when the carrier is not moving, "STOBAR aircraft carriers should maintain a speed of 20k–30k to generate wind-over-deck to support the mission" ('Indian Navy seeks EMALS' 2013). Such carriers typically require a ski jump design that allows the aircraft to gain enough height when taking off, and use the same type of arresting gear for recovering the aircraft as CATOBAR carriers. STOVL designs use aircraft which are capable of vertical landing and therefore require no arresting gear. Such aircraft could "theoretically also take-off vertically, but the performance penalty for doing so is high" (Li and Weuve 2011: 217). A ski jump is thus usually part of a STOVL design in order to enhance both payload and range of the aircraft. Yet another option for launching aircraft is the sole use of Vertical Take-Off and Landing (VTOL) aircraft, which require no additional thrust to take off. Such planes can theoretically take off even from helicopter carriers such as the Japanese *Hyuga* class or the Korean *Dokdo* class, an arrangement which could in effect convert such vessels from helicopter carriers to small aircraft carriers of limited capability.

All options other than CATOBAR or EMALS have the strategic disadvantage of limiting the choice of aircraft carried to smaller, muss less capable models. The available technology for launching the aircraft is thus a key factor for evaluating the strategic value of a carrier:

> A STOBAR design is generally less capable and much simpler to build and maintain than a CATOBAR design but may still be a large, fast ship. STOBAR is less appropriate for the strike role, so a decision to forgo catapults may indicate intent to not perform the strike mission (Li and Weuve 2011: 217).

This is because STOBAR puts limits on the maximum take-off weight while

> [. . .] increasing the amount of fuel consumed during takeoff. This restricts and reduces the fuel and weapons payload that an aircraft can carry, thereby reducing its range, loitering time, and strike capabilities. STOBAR is also more affected by wind, tide, rolling, and pitching. Furthermore, it needs more flight deck space for takeoff and landing (Li and Weuve 2011: 221).

However, as Medeiros et al. noted in 2005, CATOBAR may not be an option for China thus far because

> China lacks the capability to build either large-capacity aircraft-lift elevators or steam catapults for the movement and launching of aircraft; so a Chinese carrier would have to rely on a ski-jump design. Thus, a Chinese carrier would not resemble in any way, shape, or form a U.S. 'big-deck' carrier (Medeiros et al. 2005: 150).

This assessment still holds true today. The main problem is not simply the lack of technical knowledge, which could perhaps be acquired or developed. It also concerns gaining access to high-precision tools and critical materials, such as *HY-100* military high strength steel. So far, the latter can only be produced by the U.S., and it had to be imported e.g. by India for its carrier *Vikramaditya,* an entirely Russian-built carrier that was also refurbished in Russia (Sheldon-Duplaix 2011: 13). It seems unlikely that an export license for *HY-100* would be granted to China under the prevailing political conditions.

Current indications suggest that China's indigenously developed carrier would use a similar STOBAR ski-jump design as the *Liaoning*, even though the indigenously built carriers could end up having a larger flight deck (Hardy et al. 2013).

For the future air wing, this means that its "fixed-wing aircraft [. . .] are essentially limited to air superiority—fleet air defense or offensive air superiority—operations, or relatively short-range strike missions" (Li and Weuve 2011: 218).

Before China can be considered a full-fledged carrier-operating naval force, not only would all remaining system integration challenges of the *Liaoning* need to be overcome on the vessel itself. A functioning air wing would need to be in place, including fast fighter aircraft and fully trained pilots. China is at a disadvantage here compared with India because Indian fighter pilots receive flight training on U.S. carriers, and Indian naval forces regularly participate in large-scale exercises with a number of advanced navies including the U.S. Navy. These factors are quite decisive. As Erickson and Wilson warn:

> If the experience of other navies is any measure, the Chinese also need to realize that getting carrier operations right will involve the loss of expensive aircraft and hard-to-replace pilots. In 1954 alone [...] the U.S. Navy lost nearly eight hundred aircraft. [. . .] the PLANAF has traditionally been poorly funded and its pilots have only a fraction of the flying hours that their peers in the United States, Japan, and even India have. These factors will make China's mastery of carrier aviation even more costly in human terms. Quantum leaps forward are required not only in sea-based aviation and midair refueling but also in PLAN doctrine and antisubmarine warfare (ASW) as well as in PLANAF service culture if China's aerial power-projection capabilities are to be improved dramatically (Erickson and Wilson 2007: 243).

Although the type and number of aircraft ultimately to be carried by the *Liaoning* and the future indigenous carrier class is as yet unknown, and all related data remains speculative, the *Kuznetsov* class carrier air wing in the Russian navy usually encompasses "12 Sukhoi Su-33 Aircraft, five Sukhoi Su-25UTG/UBG aircraft and up to 20 Kamov Ka-27 helicopters" (Saw 2008: 37). This could give a rough indication of the ultimate size of a future Chinese carrier air wing. As Andrew and Kopp (2012) explain, the *Admiral Kuznetsov* "is without doubt an ASW platform with sufficient fighter capability to protect the carrier, and its ASW payload." So far, it is not clear which role the Chinese Kuznetsov-class carrier *Liaoning* would perform. The above cited Australian study discusses the various possibilities as follows:

> The actual strategic intent underpinning the PLAN development of a CV fleet will only be determined once the composition of the PLAN shipboard air wing is known. If the air wing is optimised for air defence and ASW, then clearly the fleet will be intended to protect PLAN SSBN bastions in the Yellow Sea, and possibly South China Sea. If the air wing is oriented toward air superiority and strike operations, then the PLAN CVs would qualify as traditional CVAs or 'attack carriers', intended for blue water sea control operations and distant or regional interventions. A balanced mix in PLAN CVW composition would yield less clear conclusions, but would also present difficulties given the smaller displacement of the Varyag class, compared to US Navy CVNs, and resulting smaller air wing size. There are pragmatic limits to how many roles any given CVW asset can perform well. [...] At this time the final systems fit, propulsion fit and CVW composition of the PLAN Varyag is not known, other than the intended use of the J-15B Flanker D. Therefore any assessments of the vessel's role and strategic impact would be at best speculative (Andrew and Kopp 2012).

In 2008, China was reported to have started a program to train 50 pilots for carrier operations at the Dalian Naval Academy. Given the large complement needed to operate a carrier, however, many more highly trained specialists (not limited to pilots) will be needed in the long run: "It should be noted that [. . .] a [Russian] Project 1143.5 aircraft carrier has a crew of 1,960, plus the air group with 626 people and a flag staff of 40 people" (Saw 2008: 36). China has built a full-scale land-based carrier test and training site at its Wuhan Naval Research Facility, which was modeled after the Ukrainian *NITKA* carrier flight training center (Scott 2011b; Johnson and Hardy 2013). After the sea trials period of the *Liaoning* has been completed, most experts believe that a lengthy period will be needed for conducting flight training and joint exercises before a realistic and demonstrable capacity can finally begin to emerge.

After establishing the air wing, effectively integrating the carrier into a battle group will prove to be its own organizational and logistic challenge. As one analyst noted:

> Carriers are difficult enough to build, but they are altogether more complex to operate in a manner that makes proper use of their potential. It will take the PLAN a very long time to learn this particular craft, of which it has no prior experience whatsoever (Moss 2011a).

With little outside help, China will likely face greater hurdles than, for instance, India or Brazil, which have already operated carriers for many years and do not face as many restrictions in military co-operation with more advanced powers.

Protecting a high-end naval asset such as a carrier against all kinds of attacks—symmetric as well as asymmetric—will itself soon become a strategic challenge, and forming several carrier battle groups will require larger numbers of AAW capable surface ships than China can currently spare. China will also need to deploy frigates and submarines for carrier protection, as well as naval auxiliaries for logistic support. Eventually, much more than the simple replacement procurement currently observable in the fleet will be needed to supply these additional capabilities without compromising coastal defense and EEZ patrols. The naval shipbuilding industrial base will also need to be significantly upgraded even beyond its current state in order to provide enough multi-purpose warships within the necessary timeframe. About 10 years ago, an expert concluded that in order to form just one carrier battle group, China's navy at that time would have needed to deploy "virtually every advanced vessel in the fleet" (Shambaugh 2004: 271). With enhanced programs for large surface combatants underway, several of the newly built vessels belonging to the most modern destroyer and frigate classes are probably destined for service in this role.

Given the confined and shallow nature of China's territorial waters, and the relatively close proximity to China's coastline of several at least equally capable and potentially hostile navies, China is bound to find that protecting its carriers against attacks is a daunting challenge. As Erickson and Wilson note:

> Without major improvements in ASW [. . .], any Chinese CV [aircraft carrier] would be an easy target for a diesel-electric or nuclear-powered attack submarine (SS/SSN). Chinese

> ASW capabilities, while slowly improving, cannot yet be counted on to provide a reasonable degree of security in open waters (Erickson and Wilson 2007: 243).

In light of the assumed "extreme vulnerability of Chinese aircraft carriers in the event of a conflict—in particular to submarines that will increasingly proliferate in the South China Sea" and the risk of losing or damaging them, some analysts assume that they may never be deployed in actual combat scenarios (Moss 2011b). As a news report aptly summed up: "The military utility of China's carrier program is questionable—at least in the context of any future showdown with America" ('China's aircraft carrier: 58,500 tonnes of coincidence' 2012). Another reason for this is that U.S. fixed-wing AEW aircraft such as the *E-2C* (which "can patrol for up to 6 hours, monitor a sea area of 12.50 million square km, track 2,000 targets, and direct engagement against 40 targets simultaneously," and have "a detection range of 741 km for surface targets, 556 km for aircraft, and 270 km for missiles") have effectively established an air-defense perimeter of ca. 300 km around the carrier, which exceeds the range of most anti-ship cruise missiles. This means that unless China can match these capabilities, "China's carrier group would be reduced to a vulnerable 'sitting duck', particularly if it engages the highly stealthy U.S. combat aircraft" (Li and Weuve 2011: 221).

The influential Chinese military leader Liu Huaqing, usually considered to have been the "father" of China's carrier program, wrote in his memoirs: "The aircraft carrier is a symbol of a country's Comprehensive National Power, and it lies at the core of a navy's ability to carry out combined operations at sea" (Liu 2011: 481, my transl.). Due to its extreme symbolic value in the perception of the Chinese populace, as well as its very high cost,[16] losing a carrier to either accident or attack would likely be seen as no less politically disastrous by China's leaders than it would be by their counterparts in the United States. Following interviews with several naval analysts, Moss comments as follows:

> Fighting wars is probably not what China's carriers are principally designed for. As the commotion over the as yet non-existent carriers has already demonstrated, the idea of an aircraft carrier is a powerful thing: a symbolic capability that transforms China from a developing country into a world power. Acquiring this status is more precious for China's leaders than any effective long-range strike capability, both as an expression of China's international stature and as a domestic form of legitimisation for the Communist Party's continuing unopposed rule (Moss 2011a).

It has to be concluded that the added, and yet so far purely theoretical, strategic capability of a Chinese carrier therefore automatically gives rise to questions of strategic vulnerability. In the view of this author, it is possible that the nascent carrier fleet will chiefly be deployed to maintain and strengthen the Chinese foothold in the South China Sea. China's opponents in that particular arena have

[16] In 2011, Li and Weuve estimated that the total hardware procurement cost for one *Liaoning* class carrier battle group, including ships and aircraft, would likely amount to ca. US$10 billion, with an added estimated cost for maintenance, repair, training and fuel of ca. US$200 million per year (cf. Li and Weuve 2011: 213).

far less advanced navies than either the U.S. Navy or the Japan Maritime Self Defense Force (JMSDF), and their submarine capabilities are for the most part still nascent. Deploying one or several floating airbases to that region would unquestionably enhance Chinese ambitions to control the Spratly and Paracel archipelagos.

5.3.2.2 The Emerging Amphibious Assault Fleet

In addition to the *Liaoning* and the future indigenous carriers under construction, China has also decided to build large amphibious assault vessels for its surface fleet. These belong to the *Type 071 "Yuzhao"* class and the future *Type 081* class landing helicopter dock (LHD) type (cf. Fig. 5.11).

The *Type 071* class has a full load displacement of ca. 18,000–20,000 t and can carry up to four heavy-lift military transport helicopters such as the *Z-8*, the Chinese version of the French *SA 321 Super Frelon* in service with the PLAN, which has a maximum take-off weight of 13 t. The *Type 071* class can also dock four landing craft and transport 400–800 additional troops. It is powered by a CODAD propulsion plant, which yields a top speed of 22 knots. The *Type 071* class vessels currently do not seem to be armed with SAM, although these could be retrofitted later.[17] A class of up to six such ships is ultimately expected, with three already in service as of 2013 (Wertheim 2013: 128). As for the additionally planned new *Type 081* class, a similar sized vessel design intended to carry up to eight helicopters which apparently resembles the French *Mistral* class, only speculative data is available at present.

While Moss (2011a) expects the new amphibious assault ships to serve mainly in humanitarian and disaster relief missions, they could also conceivably be used to transport troops and equipment, as well as to bring in large numbers of helicopters and landing craft in the event of a territorial conflict. Vessels of a similar type are already well established in the Japanese Maritime Self-Defense Force, which has been operating so-called "Large helicopter-carrying destroyers" (the *Hyuga* and the new building 30,000 t fl *Izumo* classes) for years. Those ships can carry up to eight large helicopters, which would mostly operate in an anti-submarine warfare role. The Japanese helicopter carriers are clearly more capable than the existing Chinese designs, as they have a much more powerful COGAG propulsion plant which yields top speeds of more than 30 knots, and are heavily armed with a *Mk 41* VLS, *Evolved Sea Sparrow Missiles* (ESSM), *ASROC* ASW missiles, and a *Phalanx* CIWS. In South Korea, the *Dokdo* class LPD performs a similar role: In size, it resembles the Chinese *Type 071* design, but can carry up to ten helicopters. Like the Japanese

[17] During a port visit to Hamburg of the *Type 071* class vessel *Changbaishan* on 23 January 2015 I had the opportunity to take part in a tour. No missiles were installed, and the ship carried only two helicopters in a hangar large enough to accommodate four. The quality of welding and construction overall seemed rather high. Safety standards on the bridge, however, seemed below the international norm, with seats installed for two officers only.

helicopter carriers, it has the potential to be alternatively converted into a small carrier for STOVL aircraft. Russia has currently no indigenous LHD design of its own, which explains why it made the unprecedented decision in June 2011 to procure two *Mistral* class vessels from France, a NATO member state. France approved the export deal despite protests from U.S. and other NATO members. In the wake of the Crimea crisis in 2014, the suitability of the transfer of the first finished vessel to Russia has been called into question and was stalled by the French government. As of the time of this writing, the transfer of the two *Mistral* class ships to Russia remains uncertain (cf. Harress 2014).

Vessels of the LHD or LPD types can generally fulfill very different types of missions, ranging from humanitarian disaster relief to anti-submarine warfare to supporting offensive landing operations. Given the fact that all three East Asian nations operating such vessels are currently involved in conflicts over island territories, the trend towards building such large multi-purpose vessels for sea-basing is likely to continue, or even intensify.

Having reviewed the development of offensive power-projection vessels within the surface fleet, the following section will now discuss the issue of strategic deterrence, and provide an overview of China's nuclear submarines.

5.3.2.3 China's Nuclear Submarine Fleet

China's submarine fleet is among the world's largest, which causes it to attract considerable attention. The following figures depicting the fleet-age structure of the nuclear and conventional submarine fleets assume a shorter maximum life span than for surface vessels of only ca. 30 years, depending on the actual days spent at sea.

China's nuclear submarine fleet is comprised of four different classes with 7–11 boats currently in operational status (cf. Fig. 5.12). If the current building programs of the *Type 094 "Jin"* class ballistic missile submarine (SSBN) and the *Type 093 "Shang"* class nuclear attack submarine (SSN) continue to progress as projected, with the older classes going out of service, China's nuclear submarine fleet could rise to ca. 12 boats by 2020. Very little is known about the actual status of all Chinese nuclear submarine programs, however. Technical data on all these vessels remains speculative, and estimations vary greatly from source to source.

The remaining boats of the oldest *Type 091 "Han"* class SSN have been dubbed "sharks without teeth" as well as "Whiskey class with a reactor" by Chinese and Western observers alike. This is because, apart from the usual Chinese problem of high internal radiation levels, they seem to suffer from an inability to fire missiles while submerged, a defect which of course massively impairs their strategic value (Wertheim 2007: 111).

The relative loudness of the *"Han"* class was evident in November 2004. When a boat of this class passed between the Japanese islands of Miyako and Ishigaki, it caused an outrage among Japanese officials. Even so, it was apparently not a problem for the Japanese naval forces to track the submarine's entire route

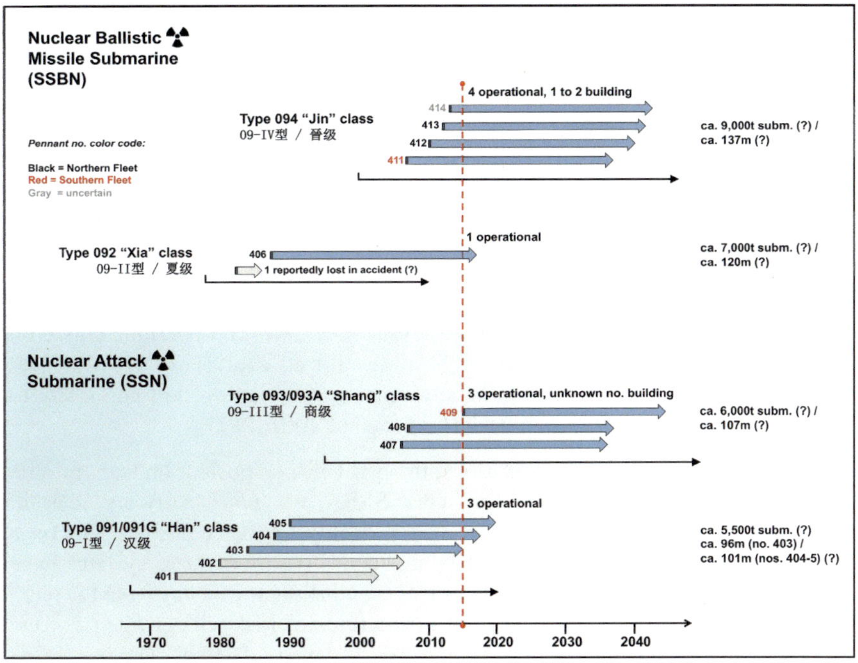

Source: Drawn using data from Wertheim (2007, 2013) and *Jane's Fighting Ships* (2014) cross-checked with Chinese Wikipedia entries

Fig. 5.12 Type and age structure of PLA Navy nuclear submarine fleet, 2015

(Holmes and Yoshihara 2008: 63). Despite the very long time the *"Han"* class took from completion to commissioning due to initial problems with the reactor, these boats still represented in the end an impressive technological accomplishment.

The first vessel of the *Type 093 "Shang"* class SSN, which were destined to replace the ageing *"Han"* boats, was designed during the 1990s with extensive Russian assistance and commissioned into service in 2006. The design is likely based on the double-hulled Soviet *Victor III* class designed during the late 1970s (Shambaugh 2004: 272). The major task of an SSN is anti-submarine warfare, including the tracking of enemy SSBN. Little is known about the current level of Chinese ASW capabilities, but it is important to note that such operations involve complex interactions between technology, crew training, and the ability to integrate multiple cues from different source into one coherent picture. Since upholding such capabilities requires constant training, the seemingly low activity rate of the entire Chinese submarine fleet noted by analysts such as Kristensen (2007) can thus be assumed to have had detrimental consequences for the overall level of Chinese ASW skills at least until very recently.

As Bell notes, the only task of a SSBN consists in providing a credible sea-based nuclear deterrent. "Because of their basic mission: to remain hidden in all situations, the most important characteristic of [...] ballistic missile submarines is their emitted noise" (Bell 2009: 31). Therefore, the strategic value of a SSBN will to a large degree depend on its ability to avoid detection. This, however, covers only part of the required capabilities. The former commanding officer of the *USS George Bancroft* (SSBN 629), Robert G. Loewenthal, summarizes the requirements to be fulfilled by a SSBN at all times as follows:

"SSBNs are required

- to remain undetected by anyone, friend or potential foe, essentially in a wartime posture;
- to remain able to receive radio messages continuously;
- to maintain navigation accuracy [. . .]; and
- to take the crew through missile launch sequences, with no previous notice on orders received any time day or night, either for an exercise or for a tactical launch. The requirement was to have all preparations completed and ready to launch in less than fifteen minutes" (Loewenthal 2007: 287).

If measured against such expectations, the first Chinese nuclear ballistic missile submarine design, the *Type 092 "Xia"* class SSBN, was not exactly a success in strategic terms. According to Bell, it was probably "never considered to be a survivable deterrent. The product of 'trial and error' (many onboard systems have been upgraded and tested at sea) the Type 092 spends little time away from the pier" (Bell 2009: 35). Only a single boat of this class remains in service, while a second vessel seems to have been lost in an accident during the mid-1980s. The remaining boat has apparently never sailed beyond China's home waters and has thus effectively not conducted any deterrent patrol activities whatsoever (cf. Kristensen 2007). According to reports, it suffers from high levels of on-board noise. It is soon to reach the end of its maximum service life, and was apparently mainly used for missile trials (Shambaugh 2004: 272).

The replacement for the *Type 092 "Xia"* class SSBN, the *Type 094 "Jin"* class, is expected to be much more capable overall. Similar to other nuclear submarines of Chinese design, this class also experienced reactor problems at first (Shambaugh 2004: 272). Four boats are reportedly operational, but are expected to remain without ballistic missiles until the ongoing trials of the intercontinental-range *JL-2* ballistic missile system are completed. This missile is predicted to have a range of 7,000–8,000 km and could reach targets in Alaska or India from positions in the Yellow or South China Seas (Fisher 2013). However, due to the missile's unknown operational status as of October 2014, it is not entirely clear whether China already has a fully functioning sea-based nuclear deterrent capability in place. Given the Cold War era SSBN requirements listed above, providing the necessary technologies is obviously only a minor part of the conditions that need to be fulfilled to reach this goal. Rigorous crew training, regular exercises, and extended global patrols would need to be continuously demonstrated in order for China to match the criteria set by the U.S. Navy as explained by Loewenthal (2007: 287).

Compared to other types of warships, reliable information about technical performance data on Chinese nuclear submarines is even harder to come by through open sources. In order to arrive at plausible estimations regarding the maximum level of quieting reached by a submarine design, photographic evidence can be used at least to some degree. Bell describes a method of 'visual qualitative analysis comparison' for making noise level estimations of the *Type 094* class:

> The proper approach involves breaking down the images into separate hydrodynamic design categories. By looking at obvious design factors, including shape, skin friction (sail/surface), flood openings, and propellers a better assessment can be made. [...] In addition, utilizing estimated speed to complement these factors would help narrow the sound estimate. Overall, many design features found on acquired technologies from advanced submarine builders, such as the French and Russians, should be considered in use on the Type 094 (Bell 2009: 31).

When discussing these features separately, Bell concludes that "[t]he additional height needed for the JL-2 missile certainly imposes noise penalties." The shape of the submarine is not optimal as a result of a need to accommodate the missile (Bell 2009: 32). All in all, Bell expects the *Type 094* class to be markedly more noisy than e.g. the U.S. *Ohio* class SSBN[18]:

> Overall, based on visual qualitative analysis comparison, the Type 094 is likely much louder than the super quiet Ohio. It has a large sail, deviates from the ideal shape, and includes vents. An advanced propeller will mitigate, but not eradicate these problems. However, this is not to say that the Type 094 is going to be a loud platform overall; it is likely a small step ahead of the Delta III SSBN. It is important to consider that the average speed of an SSBN on station is less than five knots; comparisons made in terms of attack boats, which travel in excess of 15 knots have limited applicability. According to reports, the Delta III registered between 125-130 dB at 4-8 knots, the likely speed of an SSBN on patrol. Considering modern propeller design, this correlates with certain Chinese reports, which argue that the Type 094's acoustic signature was 120 dB (Bell 2009: 34–35).

Given the practical difficulties of tracking SSBNs en route, the *Type 094* class is thus likely to be a relatively capable and quiet submarine that will effectively provide China with a survivable sea-based nuclear deterrent if adequate training and mission competency can be achieved. Moreover, according to Bell, the submarine was "released at the opportune moment in United States anti-submarine warfare (ASW) decline" (Bell 2009: 35). After the end of the Cold War, ASW skills in many countries including the U.S. gradually deteriorated due to decreasing emphasis and practice. Citing Christopher McConnaughy, Chief of Submarine-Launched Ballistic Missile Quality Assurance at the United States Strategic

[18] Regarding a classification of noise levels, Bell notes: "According to E.V. Miasnikov, Senior Research Scientist [. . .] at the Moscow Institute of Physics and Technology, a very quiet submarine registers about 100 dB, a quiet submarine about 120 dB, and a loud submarine about 140 dB. If the Type 094 puts out 120 dB at sea, it will be very difficult to track. When the limitations of using one platform (SSNs) to track the Type 094 are considered, the United States Navy will have to make adjustments" (Bell 2009: 35).

Command, Bell notes that nuclear attack submarines are the only platforms capable of continuously tracking SSBN at sea. He further adds:

> Once at sea there are a limited number of platforms capable of finding and tracking an SSBN on patrol. There is no guarantee that these platforms will actually find the SSBN, only a probability. The more assets and efforts used, the higher the probability of detection and tracking (Bell 2009: 36).

Effective tracking, moreover, requires the integrated use of cues from very diverse sources of information such as "satellite imagery, antisubmarine aircraft, and fixed, passive underwater acoustic arrays" (Bell 2009: 37).

Under these circumstances, it is easy to see why the Chinese leadership decided to build a submarine base at Hainan island that provides nuclear submarines with direct access to deeper waters from underwater tunnels drilled into the rock. It is also apparent why China did not publicly announce the building of that base before it was discovered in satellite images published by *Jane's* in 2008 (cf. 'Secret Sanya' 2008). The strategic advantages of such a base for China's nuclear submarines are obvious: "[I]f China bases the Type 094 from this Island, in port satellite imagery becomes impossible" (Bell 2009: 38).

Russian assistance in Chinese nuclear submarine development is widely assumed, but not officially acknowledged by both sides. Fisher (2007: 141) cites a report by *Jane's Defense Weekly* dating from August 13, 1997 which contends that "Russia's Rubin submarine-design bureau began assisting China's SSN development in 1995." In a CIA report made to the U.S. Congress in 2003, which Fisher also cites, Russia was likewise named as "the main supplier of technology and equipment to India's and China's naval nuclear propulsion programs" (quoted from Fisher 2007: 141). The further co-operation of China with other Russian partner organizations in nuclear submarine development is thus possible, but remains speculative.

5.3.3 China's Conventional Submarine Fleet

As can be seen from Fig. 5.13, China currently operates up to 53 diesel-electric submarines. This means that together with its up to 11 nuclear boats, the PLA Navy has currently well over 60 at least theoretically operational submarines. As discussed at the beginning of this chapter, this is a very high number in an international comparison. In this regard, China is roughly on par with Russia and second only to the USA, which operates 72 submarines, all nuclear powered. By comparison, as of 2014, the other two BRIC countries—India and Brazil—have only 16 and 5 conventional submarines, respectively, while Japan has decided to gradually increase its maximum number from 16 to 20 in 2010 (see Fig. 3.9). China's total number of boats is not likely to shrink further in the near future due to the fact that the 12 Russian *Kilo* class and 13 *Type 039 "Song"* class boats were built almost simultaneously, and also at the same time as some of the *Type 035 "Ming"* and the latest *Type 041 "Yuan"* class boats, resulting in a fleet with a fairly homogenous age structure.

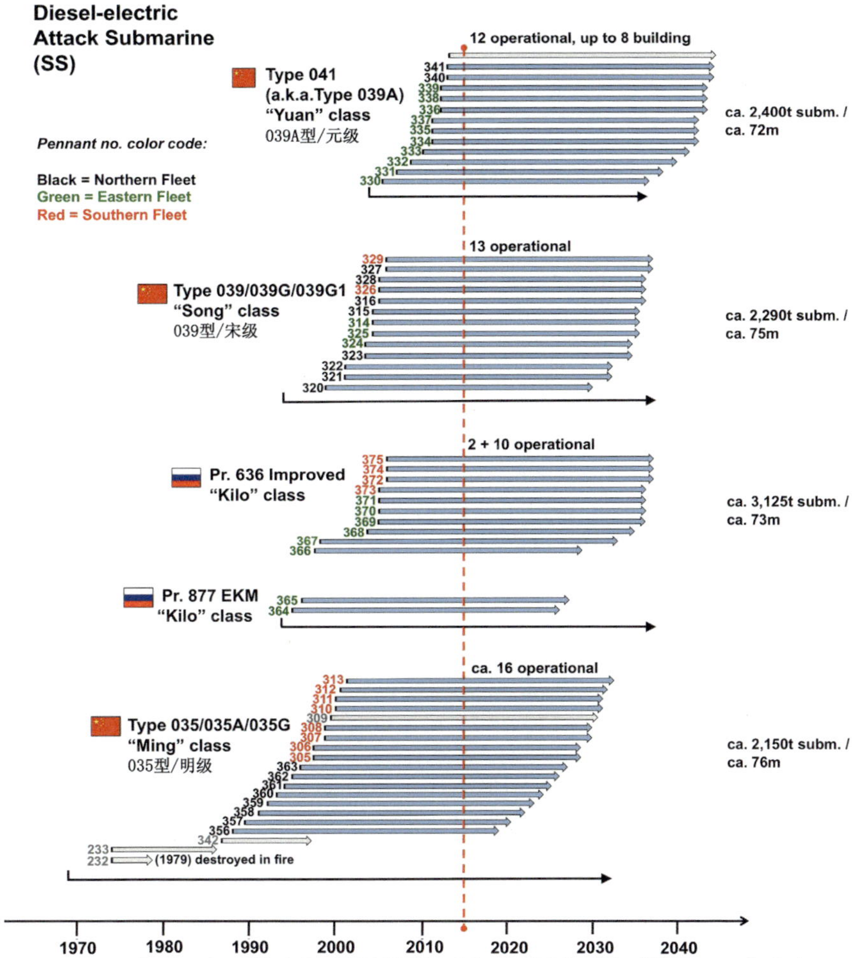

Source: Drawn using data from Wertheim (2007, 2013) and *Jane's Fighting Ships* (2014) cross-checked with Chinese Wikipedia entries

Fig. 5.13 Type and age structure of the PLA Navy conventional submarine fleet, 2015

5.3.3.1 Historical Development and Technical Issues of the Conventional Submarine Fleet

Despite its imposing size, the Chinese conventional submarine fleet loses some of its luster when technical and operational limitations are taken into account. Accidents due to technical malfunctions seem to have occurred more frequently than in other navies, including the suffocation of the entire crew aboard a *Type 035 "Ming"* class diesel submarine in 2003 (cf. Fig. 5.14).

Date	Vessel type	Incident
1978	*Type 051 "Lüda"* class destroyer	Ship lost in explosion
ca. 1975–1979	*Type 035 "Ming"* class diesel submarine	Vessel lost in fire
1985	*Type 092 "Xia"* SSBN	Vessel lost in accident
1985	*Type 033 "Romeo"* class diesel submarine	Vessel sunk
1993/1994	*Type 033 "Romeo"* class diesel submarine	Vessel lost in accident
2003	*Type 035 "Ming"* class diesel submarine	Technical malfunction, entire crew of 70 suffocated
2005	*Type 035 "Ming"* class diesel submarine	Onboard fire while submerged

Source: Compiled from Wertheim (2007, 2013) and *Jane's Fighting Ships* (2014)

Fig. 5.14 Presumed accidents due to technical malfunctions of PLA Navy vessels

This was merely the worst among the known incidents in a series of problems with *Ming* class subs, which seem to have experienced fires and malfunctions on various occasions, indicating that safety standards aboard these boats are still below state-of-the-art. Moreover, the listed incidents may only constitute the tip of the iceberg, since China as a rule "does not publicize accidents" (Bussert and Elleman 2011: 19). While Howarth (2006: 18) is certainly correct in stating that submarine operations are "naturally hazardous," and accidents are therefore bound to happen to any navy, a cursory review of publicized submarine accidents of other navies since 2000 shows that almost all publicly known U.S.N. submarine accidents were collisions or groundings, and therefore due to human error, rather than technical malfunctions. Russian-built submarines by contrast seem to have been much more prone to equipment failures or technical malfunctions, which often resulted in human casualties (cf. Fig. 5.15). The last comparable incident for the U.S.N. occurred in April 1988, when the diesel-electric submarine *USS Bonefish* suffered a battery compartment fire after seawater leakage. Three crew were killed, and the boat had to be scrapped.

Even though the list provided here is likely far from complete since SSBN accidents are almost never publicized and a high number of unreported minor accidents can be assumed to have occurred in all submarine-operating navies, it is still possible to infer that Russian-built submarines have proven far more accident-prone due to technical malfunctions than modern Western ones, which mostly experienced accidents due to crew errors, and that Russian submarine accidents and malfunctions have claimed on average far more human lives.

The oldest boats still in service with the PLAN are the *Type 035 "Ming"* class submarines mentioned above. Submerged, these ca. 76 m long subs displace about 2,100 t and can reach a maximum speed of ca. 18 knots. Their design was modeled after the Soviet-era *Romeo* class and improved upon in various batches.

The follow-on class destined to replace the ageing *Type 035* class boats, the *Type 039/039G "Song"* class, represented a significant step ahead according to most analysts, both in design and regarding its industrial production rate:

Date	Country	Accident/malfunction description	Error type
5/2000	Russia	SSN *Kursk* (K-141) is sunk by torpedo malfunction while submerged. Entire crew of 118 killed.	Technical (explosion)
2/2001	USA	SSN *USS Greeneville* (SSN-772) collides with and sinks a Japanese fishing vessel. Nine Japanese crew aboard fishing vessel killed.	Human (Collision)
8/2001	USA	SSN *USS Greeneville* runs aground in Saipan, sustaining heavy damage. No casualties.	Human (grounding)
1/2002	USA	SSN *USS Greeneville* collides with amphibious transport dock ship *USS Ogden* (LPD-5), damaging both vessels. No casualties.	Human (collision)
5/2002	USA	Diesel-electric research sub *USS Dolphin* is partially flooded due to torpedo shield door gasket failure. Crew abandons ship but manages to salvage the boat. No casualties.	Technical (partial flooding)
11/2002	USA	SSN *USS Oklahoma City* (SSN-723) collides with LNG tanker near Gibraltar and sustains damage. No casualties.	Human (collision)
11/2002	UK	SSN *HMS Trafalgar* (S107) runs aground near Isle of Skye, with vessel damaged. Three crew injured.	Human (grounding)
2/2003	Australia	Diesel-electric sub *HMAS Dechaineux* (SSG 76) is partially flooded when diving due to pressure hose malfunction. No casualties.	Technical (partial flooding)
8/2003	Russia	Decommissioned SSN *K-159* sinks in Barents Sea while being towed to scrapping site. Nine out of ten crew killed.	Technical (flooding)
10/2003	USA	SSN *USS Hartford* (SSN-768) runs aground at Sardinia, sustaining heavy damage. No casualties.	Human (grounding)
10/2004	Canada	Diesel-electric sub *HMCS Chicoutimi* suffers fire after partial flooding while surfaced, due to opened hatches and insufficient insulation of cables. Sub salvaged. One crew killed.	Technical (fire)
08/2005	USA	SSN *USS San Francisco* (SSN-711) collides with undersea mountain off Guam, sustaining heavy damage. One crew killed, ca. 60 injured.	Human (grounding)
09/2005	USA	SSN *USS Philadelphia* (SSN-690) collides with Turkish merchant ship off Bahrain, sustaining and causing light damage. No casualties.	Human (collision)
09/2006	Russia	SSN *Daniil Moskovsky* (K-414) suffers onboard fire due to electrical fire in wiring while anchored near Rybachiy peninsula. Two crew killed.	Technical (fire)
01/2007	USA	SSN *USS Newport News* (SSN-750) collides with Japanese tanker while submerged in the Strait of Hormuz, damaging both vessels. No casualties.	Human (collision)

Fig. 5.15 (continued)

Date	Country	Accident/malfunction description	Error type
03/2007	UK	SSN *HMS Tireless* (S88) suffers explosion while submerged in the arctic, with emergency surfacing through pack ice. Two crew killed, one injured.	Technical (explosion)
05/2008	UK	SSN *HMS Superb* (S109) runs aground in the Red Sea near Suez Canal. Vessel damaged and retired ahead of schedule. No casualties.	Human (grounding)
11/2008	Russia	SSN *Nerpa* (K-152) suffers gas leak due to malfunctioning fire extinguishing system while submerged on trials off Vladivostok. 21 killed, 20 injured.	Technical (gas leak)
02/2009	France/UK	Two SSBN, the French *Le Triomphant* (S616) and the British *HMS Vanguard* (S28), collide in Eastern Atlantic, with damage to both vessels and both boats salvaged. No casualties.	Human (collision)
03/2009	USA	SSN *USS Hartford* collides with Amphibious transport dock ship *USS New Orleans* (LPD-18) while crossing Strait of Hormuz submerged, with both vessels damaged. Fifteen *New Orleans* crew injured.	Human (collision)
02/2010	India	Russian-built SSK *INS Sindhurakshak* (S63) suffers onboard fire after explosion in battery compartment due to faulty battery valve that leaked hydrogen. One crew killed, two injured.	Technical (fire)
10/2010	UK	SSN *HMS Astute* (S119) runs aground off Isle of Skye. Vessel sustained damage. No casualties.	Human (grounding)
06/2011	Canada	Diesel-electric sub *HMCS Corner Brook* (SSK 878) runs aground off Vancouver, with heavy damage sustained. Two crew injured.	Human (grounding)
10/2012	USA	SSN *USS Montpelier* (SSN-765) collided with cruiser *USS San Jacinto* (CG-56) off Florida, with the latter sustaining damage. No casualties.	Human (collision)
08/2013	India	Russian-built SSK *INS Sindhurakshak* (S63) suffers onboard fire while berthed and is damaged beyond repair. Ca. 18 crew killed.	Technical (fire)

Source: Compiled from various news reports, cross-checked with Wikipedia entries

Fig. 5.15 Non-Chinese publicized submarine accidents since 2000

> The Song [*Type 039/039G/039G1*] class has a hydrodynamically sleek (tear-drop) profile, possesses new cylindrical environmental sensors, and relies on German engines for propulsion. Most significantly, the Song is much quieter because it is fitted with an asymmetrical seven-blade skew propeller, and the Song uses anechoic rubber dampening tiles on the hull and shock absorbency for the engine to reduce its acoustic signature. The Song may also be able to launch cruise missiles when submerged, another design advance for China's conventional submarines (Medeiros et al. 2005: 148).

However, Bussert and Elleman note that this submarine design suffered from similar systems integration challenges to those experienced by surface vessels already discussed, which may have impaired its strategic value:

> The first hull of the Song class started sea trials in 1995, but the integration of Chinese, Russian, and imported systems such as the French TSM 2225 sonar and the German diesel engine is blamed for serious system design and operational problems on the lead boat, and there were reports of high noise levels while operating (Bussert and Elleman 2011: 67).

These problems led to a major redesign of the class, which was launched after several years. As Fisher describes, this new design variant was judged much more favorably by outside observers based on existing photographic evidence:

> Pictures [. . .] released by the PLA in 2001 showed that it lacks the distinctive step sail of the first 039, thereby improving hull-noise dynamics. The Type 039 resembles the French Agosta-90B-class conventional submarine but there is no open reporting that indicates that France provided any assistance (Fisher 2007: 153).

It is widely assumed, however, that Pakistan, China's strategic ally in co-operative arms production, has shared foreign-imported arms technology illegally with China on multiple occasions (cf. Sect. 4.6.5), and Pakistan has also officially imported *Agosta* class submarines from France, including one equipped with a *MESMA* air-independent propulsion system (AIP). Fisher thus concludes that

> [. . .] it is possible that Pakistan would share advanced French submarine technologies with China. [. . .] Onward transfer of this technology could improve future versions of the Song and Yuan, or help China to develop countermeasures to French and other Western submarines (Fisher 2007: 154).

So far, 13 *Type 039 "Song"* class boats have been built. They are ca. 75 m long and are estimated to displace ca. 2,250 t submerged, and to have a maximum diving depth of ca. 300 m (Wertheim 2013: 111–112). Their armament is not exactly known, but could include a new indigenously developed wire-guided torpedo and anti-ship missiles based on the *YJ-81* or *YJ-82*. Chinese television coverage broadcast in 2004 gave evidence of "extensive use of digital systems in the control room, which suggests the utilization of more modern automatic combat control systems" (Fisher 2007: 153).

Advances with domestic submarine designs notwithstanding, the PLA Navy nevertheless opted to purchase altogether 12 Russian-built *Kilo* class submarines on the export market during the mid-1990s, when the indigenous "Song" class boats were already being built. The first two *Kilos* belong to the older *Project 877EKM* type, while the subsequent ten boats are an improved export version designated *Project 636*. As with the purchase of *Sovremenny* class destroyers discussed earlier in this chapter, this submarine procurement decision seems to point to the less than satisfactory performance of the indigenous Chinese submarine designs at that point in time. According to an interview that Jean published in 2013 with Kevin Pollpeter, this is indeed the most likely explanation:

> If China's shipbuilding industries are doing well, what would prompt the military to go out and buy Russian submarines? One answer is that they are still not satisfied with domestic-built submarines," Pollpeter said. Another possibility is that China believes it requires subsurface equipment faster than its own defence industry can build it to deal with regional threats, he added (Jean 2013).

Although the Soviet-era *Kilo* design is not exactly considered state-of-the-art as of today, analysts generally agree that these boats are relatively well armed, quiet, and capable. They are ca. 73 m long and displace between 3,000 and 4,000 t submerged, depending on the respective batch. They can reach a maximum submerged speed of 17 knots and have a maximum diving depth of 300 m. The ten *"Improved Kilo"* boats (pennant nos. 366–375) are armed with *3M-54E1 (Klub-S)* supersonic anti-ship missiles in addition to torpedoes. The apparent difference in capability between the two batches is remarkable. According to Wertheim, the *Project 636* "[o]ffers vastly improved sound quieting over the Project 877 series and a greater degree of automation. [. . .] The outer hull is coated with 0.8-m-square rubber anechoic tiles, attached with glue over special studs" (Wertheim 2013: 113).

Their relatively advanced technology, however, seems to pose challenges of its own regarding maintenance and crew training, and several of the *Kilos* appear to have been at sea only infrequently. The specialized training requirements of the crews seem to have posed yet another problem, for the Kilos require "very different training regimens from what Chinese sailors are accustomed to, and hence the crews are not interchangeable with other submarines in the fleet" (Shambaugh 2004: 273). In the past, boats have also had to be sent to Russia for repairs. Maintenance problems are also associated with these foreign purchases. As Collins and Grubb note: "The Chinese literature includes accounts of sailors tracing out systems by hand on new Kilo-class submarines due to a lack of technical documentation" (Collins and Grubb 2008: 32).

Regarding their strategic value and capability as a weapon system, Shambaugh estimated that the Improved *Kilo* class subs are roughly as quiet as the U.S.N. *Los Angeles* class SSN, and comparable in capability to the older Japanese *Yushio* class SSK, which has since been decommissioned (Shambaugh 2004: 273).

The most advanced indigenous Chinese diesel submarine design so far is the *Type 041 "Yuan"* class, which is also sometimes designated as *Type 039A*. The first-of-class was launched in 2004. This 72 m long boat displaces ca. 2,400 t submerged and has a maximum speed of ca. 23 knots (Wertheim 2013: 110). As of 2014, about 13 boats appear to have been built, with a class of ca. 20 ultimately expected.

Wertheim assumes that the design is heavily influenced by the Russian *Kilo* class and comparable to it in terms of capability (Wertheim 2013: 111). Fisher (2007: 152), however, sees a stronger resemblance to the Russian *Project 677 Lada/Amur-1650* class. It is possible that the *Type 041* is powered by an AIP system, as has been claimed by Chinese sources, though the exact type of AIP is not clear.[19] China has developed a smaller export variant of the *Type 041* (minus the fuel cell), and, according to reports, is in talks with both Pakistan and Bangladesh for delivery of this export variant (cf. Bokhari 2014).

[19] For an overview of the various technical solutions for achieving AIP—the German Siemens polymer electrolyte membrane fuel cell, the French *MESMA*, and the Swedish Stirling engine based system—see Scott (2011a). AIP equipped submarines have also been developed in Japan (*Souryu* class, which uses a Stirling engine) and Russia (*Lada* class).

One noteworthy development is China's new *Type 032* test submarine, which is not included in Fig. 5.13. This very large diesel submarine (ca. 92 m length, ca. 6,600 t displacement submerged), first seen in pictures in 2011, was apparently commissioned in 2012 and is capable of launching nuclear missiles (Fisher 2013). It is likely a trial platform for the new *JL-2* ballistic missile, which is destined for integration in the new *Type 094 "Jin"* class SSBN. Fisher (2013) suggests that this type of conventionally powered ballistic missile submarines could become a cost-efficient alternative to SSBNs. However, given the requirements for a credible sea-based nuclear deterrent outlined by Loewenthal (2007), such a replacement strategy does not appear all too likely. Nevertheless, the *Type 032* is certainly an interesting new system that should be more closely watched and analyzed.

5.3.3.2 Low Patrol Rates of the Chinese Submarine Fleet

Apart from high radiation levels and reactor problems aboard some of the earlier nuclear-powered submarines, other technical problems often discussed in connection with Chinese submarine technology have included relatively high noise levels and relatively slow maximum speeds. In 2006, Howarth reported the results of a study which effectively concluded that China's mostly "old, noisy and slow" submarines were still outclassed by the fleets of most other submarine-operating navies in Northeast Asia, including the Japanese, in qualitative, if not quantitative terms (Howarth 2006: 16). Since then, South Korea has received German state-of-the-art *Type 209* and *Type 214* submarine designs from HDW. The altogether nine new *Type 214* subs are equipped with air-independent propulsion and therefore extremely quiet and capable compared to even the most modern Chinese submarines. According to Fisher, "Germany's Type 212 and 214 SSKs are perhaps the world's most advanced, using new fuel-cell based AIP systems, and the latest combat system technologies" (Fisher 2007: 146). At the end of the day, a small number of such highly capable boats can prove to be much more capable than a large fleet of relatively ineffective submarines.

Operational problems with the Chinese submarine fleet are furthermore indicated by the seemingly low degree of activity reported for many Chinese subs until a decade ago. As U.S. nuclear weapons expert Hans Kristensen wrote in 2007, declassified U.S. Navy information showed that China's submarines have consistently had a very low "patrol rate" since the early 1980s, and on average conducted less than two patrols per year between 1981 and 2006 (see Fig. 5.16). Not only was the fleet as a whole "surprisingly inactive," but the single *"Xia"* class nuclear submarine appears to have never conducted even a single "deterrent patrol."

Year	81	82	83	84	85	86	87	88	89	90	91	92	93	94	95	96	97	98	99	00	01	02	03	04	05	06
No.*	1	0	2	2	2	1	1	5	2	0	1	1	0	1	1	1	2	3	2	6	3	4	3	3	0	2

*Number of "patrols" conducted by Chinese submarines during each year
Source: Declassified U.S. Navy data reported by Kristensen (2007)

Fig. 5.16 No. of Chinese submarine patrols per year, 1981–2006

The low average number of patrols per year is surprising, especially given the huge size of the fleet. Moreover, as Kristensen points out,

> In four years (1982, 1990, 1993 and 2005), no patrols were conducted at all. Over the 25-year period, the trend is that patrols have only increased from one per year to approximately 2.8 patrols per year (Kristensen 2007).

Note that the term "patrol," as it is used here, does not necessarily imply that the subs were otherwise never at sea. While it remains unclear how exactly "patrol" is to be defined in this context, it seems to apply to extended offshore patrols, in contrast to shorter operations near the coastline. Nevertheless, according to Kristensen, the data "implies an almost dormant submarine fleet," at least until 2006. Moreover, if the data is indeed reliable, its very existence further implies that the U.S. Navy had no difficulty during the covered time frame in tracking almost any kind of submarine activity conducted in Chinese territorial waters, a rather telling fact.

If the low patrol rate during 1985–2006 is indeed accurate, then even if the frequency has somewhat picked up since then, as it appears to have done, this would still imply the crew's relatively low training status and degree of readiness, and thus the compromised operational capability of the entire submarine fleet. As Kristensen concludes from the data indicated above:

> The implications of the low patrol rate are significant. The total operational experience for the entire Chinese submarine force is only 49 patrols in 25 years, corresponding to each submarine conducting an average of one patrol every third year. As a result, Chinese submarine crews appear to have relatively little operational experience and consequently limited skills in operating their boats safely and competently. It suggests that the tactical skills that would be needed for the Chinese submarine force to operate effectively in a war may be limited. China continues—at least for now—to use its submarine force as a coastal defense force (Kristensen 2007).

It is not possible to blame the bad technical state of older or obsolete submarine classes for this situation, because, as Kristensen points out, "during the period between 2000 and 2006, when China acquired a dozen new Kilo and Song class submarines, the number of patrols *declined* from six to two (with no patrols at all in 2005)" (Kristensen 2007, emphasis in the original). The problem seems to be systemic, and thus organizationally rather than technologically determined.

The reasons behind this phenomenon are not quite clear, though bathymetry might be a factor. As already discussed in Sect. 2.3, China's confined and shallow territorial waters may be relatively difficult terrain for conducting the full spectrum of extended submarine operations, especially when it comes to large submarines, because of the dangers of accidental collisions and groundings.[20]

---

[20] Confined and shallow waters are generally difficult terrain for larger submarines, as a classic study of Soviet submarine incursions into Swedish territorial waters explains: "Even the older Soviet Golf-class missile submarines would for the most part not be able to transit many of the archipelago channels even if they travelled on the surface—and certainly not if submerged" (Leitenberg 1987: 141). China's territorial waters, and especially the Yellow Sea, are comparably shallow as the Baltic Sea.

In recent years, the increased submarine activity in the South China Sea and the building of an underwater submarine base on Hainan island perhaps signal a paradigm shift towards the greater out-of-area patrol activity of Chinese submarines, especially China's nuclear-powered submarine fleet. As of 2014, China has started to deploy submarines to support the Anti-Piracy mission off the Gulf of Aden, and has publicly announced its intention to further expand its naval presence in that region. In September 2014, a PLAN submarine (reportedly a "*Song*" class vessel) for the first time made a port call in Colombo, Sri Lanka (Rahmat 2015). More extensive Chinese submarine activity in the Indian and the Pacific Oceans can thus be expected in the future.

5.3.4 Summary of Technical Aspects Influencing Combat Proficiency and Total Capability

In the past, one major critical weakness of the PLA Navy was the absence of an area-defense AAW capability. Until the introduction of the *Type 052C "Lüyang II"* class destroyers, Chinese destroyers were effectively limited to a point-defense capability, and thus unable to defend other ships in addition to themselves against attacks (Cole 2010: 193). The introduction of modern AAW capabilities has theoretically ameliorated this major shortcoming. However, so far, the actual performance of these new systems has not been demonstrated to the outside world. Nevertheless, it is easy to see how this capability area will remain a focal point of research for the PLAN, because

> [t]he development of a sea-based AAD [advanced air defense] capability is critical to the PLA(N)'s aspirations of operating in 'distant seas.' It allows PLA(N) combatants to operate outside of shore-based air defense more confidently, with long range SAMs capable of engaging air targets outside of air-to-surface weapons range. It also increases the efficiency of the surface force by allowing a single ship to focus on air defense for an entire group (ONI 2009: 18).

Whereas the older destroyer and frigate classes still operational today—i.e. the various *Lüda* and *Jianghu* class vessels—had no SAMs installed whatsoever, the Russian *Sovremennys* came equipped with a *SA-N-7* point-defense system, though without modern AAW capability. The French *Crotale (HQ-7)* SAM installed on some of the intermediate surface vessel generations offered point-defense capabilities only. Moreover, even though a U.S. Navy study (ONI 2009: 18) notes that the range and potency of China's SAMs has increased considerably over the past 10 years—from only ca. 7 nm (*HQ-7*) to ca. 80 nm (*SA-N-20*), with the first vertical launched SAM with a range of ca. 55 nm (*HQ-9*) appearing on the *Type 052C "Lüyang II"* class—Cole is likely correct when he states that

> [e]ven the [*Type 52B*] Luyang I and [*Type 51C*] Luzhou classes commissioned with the [*Type 52C*] Luyang IIs in the middle of this decade, while armed with very potent AAW missiles, lack a true area-defense capability (Cole 2010: 193).

This means, in effect, that in the absence of more capable platforms, such vessels would be difficult to defend against saturation attacks.

The other critical weakness often observed in today's PLAN is comparably low ASW capability. There is no "silver bullet" available to address this problem, however, since:

> Detecting submarines, especially from a surface ship, is a very difficult process, and the PLAN does not appear to be taking advantage of available ASW technology, especially in the field of passive detection—[even though] some of it forty years old (Cole 2010: 193).

Although more ASW capable helicopters and towed arrays have been added to the fleet, and modern surface vessels now typically embark at least one 5 t helicopter, these measures are probably not sufficient for making a marked improvement towards a true integrated ASW effort against modern, quiet, AIP equipped diesel subs such as those in service with e.g. the South Korean Navy or the JMSDF. Most significantly,

> China's Navy lacks significant airborne ASW resources, with only a dozen old aircraft assigned to the ASW mission; there is no open-source knowledge that China has deployed sea-bottom listening arrays in it coastal waters. A serious ASW effort would encompass all these systems, properly integrated and supported by timely operational and technical intelligence (Cole 2010: 194).

Rigorous training and joint exercises would be imperative to ensure that the necessary level of expertise among ASW operators remains constant.

Another area of concern for the PLAN is without doubt its still relatively high degree of dependence on outside sources of technical support, especially when it comes to certain mission-critical systems (Johnson 2013b). From China's point of view, the likely elimination of Ukraine as a bargaining lever/alternative source of technical know-how in areas where Russia has been reluctant to share technology with China will likely exacerbate this situation. Efforts to minimize the de-facto dependency on Russia, and to reduce the resulting strategic vulnerability, can therefore be expected to continue. In light of the arms embargo and the lack of alternative arms production partners, this means that China can be expected to continue using reverse-engineering and illegal intelligence gathering methods in an effort to become a second-tier or even first-tier arms producing country in its own right, including in the most mission-critical technology areas. The near and mid-term prospects of China succeeding in this endeavor, however, must be judged dimly after a careful weighing of the publicly available evidence. As very recent information that was released on a planned large-scale arms procurement deal with Russia for 24 *Su-35* fighter aircraft and four *Lada* class submarines has shown,[21] and as many other procurement choices made by China's naval planners in the past imply, China has still not reached the necessary level of development for being independent when it comes to naval electronics design, the critical area for achieving a higher level of capability in network-centric warfare, area defense, and AAW.

---

[21] For details, see Foster (2012), Johnson (2013a), and Sect. 4.6.3.

Moreover, according to estimations given to me by various professionals from Western defense industries, neither Ukrainian nor Russian defense electronics are regarded as being on par with the leading Western systems, implying an even greater capability gap yet to bridge for China.

All in all, China today seems either dependent on or at least heavily influenced by Russian technology imports and licenses in the area of helicopters (Kamov), Fire Control Radar (*Mineral ME*), SSM, and fighter airplanes; French technology in the realm of medium-frequency sonars (*SJD 8/9*), helicopters (*Z-8*), diesel engines (Pielstick) and perhaps, CDS design; Ukrainian technology for gas turbines; German technology for diesel engines (MTU); and, according to Bussert and Elleman, Dutch technology for the *Goalkeeper*-like CIWS (2011: 46). In aircraft development, the same authors note that "nearly all export models, and many aircraft flown within China, feature imported avionics suites," which again points to insufficient electronics development (Bussert and Elleman 2011: 101).

Since even the most recently built Chinese vessels remain combinations of technologies from various sources of origin—some legally, some likely illegally obtained—and thus very likely lack proper documentation and support, this in all likelihood presents the PLAN with difficulties in training and maintenance for such systems (Bussert and Elleman 2011: 46).

5.4 The PLAN's Operational Capability

As already explained in Chap. 3, technological factors can only serve to establish the maximum level of combat proficiency theoretically achievable by a navy. The existence of certain types of naval hardware, however, is not in itself a sufficient condition for assuming that the related combat capability also indeed exists. Crew training, maintenance support, repair facilities, logistics, and regular and realistic training exercises are essential to ensure that systems can be efficiently operated and maintained even under duress. The remaining part of Chap. 5 will therefore discuss these criteria.

5.4.1 Primary Function, Maximum Mission Capability, and Mission Flexibility

According to Johnson et al. (2009: 44), apart from coastal defense, the primary mission of the PLA Navy is preparing and assisting in an attack on Taiwan. Specifically, the Navy would be charged with preparing to invade and/or blockade the island in conjunction with the other services. Another, equally important mission includes the safeguarding and defense of China's territorial claims in the East China and South China Seas. The Taiwan issue, as well as the island conflicts, thus figure quite strongly in Chinese naval planning, which is also reflected in the procurement decisions made for naval equipment, as well as in the attempts to fortify islands in the contested South China Sea areas in order to install military infrastructure there (Fig. 5.17).

The emphasis on these kinds of primary missions is also evidenced by the predominance of cruise missiles among the armament of all naval units, even submarines. Reporting on an interview with William Murray of the U.S. Navy War College in 2011, Moss states:

> Unusually, the PLAN also appears to be harnessing its submarine fleet in order to boost its ASCM [anti-ship cruise missile] options even further, in spite of the many other anti-ship missile platforms already at its disposal on land, at sea and in the air. 'The Chinese concept of a navy is that all ships are there to shoot cruise missiles,' said Murray. 'I also have an outlier premise that they do not see submarines as torpedo shooters but as cruise missile-shooting platforms.' This is the PLA's 'active defence' doctrine in action: the attacking of the enemy with as many different, possibly asymmetric, platforms as possible (Moss 2011a).

If and when the experimental *DF-21D* anti-ship ballistic missile (ASBM) system reaches operational capability, the PLA Second Artillery Force might be able to threaten foreign carrier battle groups from land-based launchers within a range of 1,500–2,000 km. According to skeptical observers, however, manifold technological challenges need yet be overcome to actually be in a position to hit a moving target from that distance. Even though rumors of a successful Chinese ASBM test against an unmoving target in the Gobi desert were floated on the Internet in 2013, to date the veracity of that information has not been ascertained. Moreover, even if true, this would not necessarily imply that the weapon system will be operational anytime soon in light of the immense challenges of accurate over-the-horizon targeting in an actual combat scenario in the middle of the ocean (cf. Cole 2013). Moreover, as Lennox (2011) points out in this context, "a direct attack on a US carrier group would invite a massive retaliation that the Chinese leadership would want to avoid." There are thus powerful political constraints in place which could in themselves be sufficiently strong to prevent the ASBM's actual use. Along similar

PLA Army	PLA Navy	PLA Air Force	Second Artillery Force
■ Prepare to invade and occupy Taiwan ■ Defend China's land borders ■ If necessary, defend China's government against a domestic uprising	■ Prepare to invade or blockade Taiwan ■ Protect China's coastline ■ Defend China's maritime territorial claims	■ Prepare to support an invasion or blockade of Taiwan ■ Prepare to execute a coercive campaign against Taiwan ■ Support the defense of china's coastline and land borders	■ Deter a nuclear or other regime-threatening attack against China ■ Prepare to support an invasion or blockade of Taiwan ■ Prepare to execute a coercive campaign against Taiwan ■ Support the defense of China's coastline and land borders

Source: Compiled from Johnson et al. (2009: 44)

Fig. 5.17 Major missions of the PLA

lines, Cole suspects an element of deliberate deception at play when it comes to leaked information on supposed Chinese advances in ASBM technology, deception having been a crucial element of classical Chinese strategic thinking ever since *Sunzi's Art of War* (and later, of Leninist political culture as well). The current uncertainty surrounding the ASBM's operational status can thus be assumed to suit the Chinese leadership's interests rather well, and is therefore likely to persist during the next few years.

In terms of operational range, it is noteworthy that during the past few years, China's navy has seen a marked increase in international activity, attesting to an ever wider reach of the PLAN. Given the fact that it was already founded in 1949, the PLA Navy started to perform overseas cruises relatively late. The first one did not occur until 1985, when two ships visited Pakistan, Sri Lanka and Bangladesh. Other notable milestones include a successful cruise of four months in 1997, when a small flotilla visited Hawaii and San Diego. Finally, in 2002, two ships embarked on a round-the-world voyage (Bussert and Elleman 2011: 30). While its continuous participation in the international anti-piracy operations off the Gulf of Aden since late 2008 has undoubtedly presented China with very valuable learning opportunities, this does not signify a sudden improvement of mission competency in a high-threat environment. Instead, these operations are mostly an opportunity to practice certain aspects of non-traditional security missions, including the logistic challenges of sustained out-of-area deployment.

Another important milestone for China's navy was the successful evacuation operation of Chinese nationals from Libya in February and March 2011. China sent a frigate and military transportation aircraft, along with private chartered aircraft, to the Mediterranean Sea in order to evacuate Chinese personnel from the warzone. However, as Shambaugh notes, the evacuation operation itself was performed "entirely by leasing commercial planes, ferries, and boats in neighboring countries" and thus mostly an "impressive display of financial firepower" (Shambaugh 2013: 285).

Due to the sustained modernization effort of its naval forces over the past three decades, China has, in the view of a recent analysis, succeeded in building a "credible force for deterrence, power projection, expeditionary operations, and surveillance and reconnaissance in the EEZ." The mission types that the PLAN is now thought to be capable of include "littoral defense, sea control in adjacent sea areas (around mainland and South China Sea), [and] limited naval-air dominance (in the absence of an aircraft carrier force)" (Sakhuja 2011: 72). This assessment may be overly optimistic at the moment. As a study published by the U.S. National Defense University has argued, "most of the tasks performed and lessons gained from out of area operations are not directly transferrable to either a Taiwan contingency or a notional out of area [contingency]" (Yung and Rustici 2010: 3). At the very least it is true that the PLA Navy, and the PLA as a whole, has had relatively little recent combat experience, as a news article pointed out:

> The last time it fought a real enemy was in the war against Vietnam in 1979, when it got a bloody nose. In contrast, a decade of conflict has honed American forces to a new pitch of professionalism. There must be some doubt that the PLA could put into practice the

complex joint operations it is being increasingly called upon to perform ('China's military rise' 2012).

In the past few decades, the lack of experience, coupled with China's then relative technological backwardness and a general embarrassment of its military's shortcomings, were perhaps the main reasons why China has restricted contact with other navies until very recently.

In view of the PLAN's early performance during out-of-area missions off the Horn of Africa, the study by Yung and Rustici has concluded in 2010 that there were still limits to China's actual naval competency in more complicated mission scenarios:

> The PLAN still has some ways to go before it can operate effectively out of area. At present, it can effectively replenish at sea, conduct intra–task force resupply, perform long-distance navigation, conduct formation-keeping with competent seamanship, and operate in all weather conditions. The PLAN cannot currently conduct a full-scale joint forcible entry operation, maintain maritime superiority out of area, conduct multicarrier or carrier strike group operations, or provide comprehensive protection against threats to an out of area task force (antiaircraft warfare, ASW, and antisurface warfare) (Yung and Rustici 2010: 2).

As of 2014, this assessment appears still plausible according to the available data. In terms of the system of naval combat proficiency levels described in Sect. 3.2, this would correspond with a maximum capability Level of VI—"Anti-submarine warfare with submarines." This is on par with Japan, Australia, or South Korea, but still well behind the U.S., which has demonstrably reached Level VIII—"Multimission air control, limited sea control, and deep strike." China also remains slightly behind India, Russia, the UK, France, Italy (and also, Brazil and Spain) in naval combat proficiency, as these powers have reached Level VII—"Naval strike and limited air control" at least theoretically. This situation could change once the Chinese carrier capability is fully developed.

5.4.2 *Logistics and Maintenance Issues*

The ability to deploy and sustain naval forces, and thus make effective use of the theoretical capability of a navy, hinges on logistics, maintenance, and crew training and motivation. These are very decisive factors for assessing a navy's total strength, and may be just as, or even more, critical than the quality of the military hardware. Capability gaps in these fields will automatically have far-reaching consequences. This subchapter will therefore discuss a few aspects of logistics and maintenance, while the following subchapter will then analyze the main challenges connected with naval training in China.

Maintenance and logistic support, particularly concerning the ability to perform repairs in a timely and efficient manner, are crucial for keeping ships in operational state. Given the unusually heterogeneous class structure of China's surface fleet (described earlier in this chapter), which complicates maintenance procedures and spare parts management, it is no small feat that, as of 2010, "China's Navy [...] appears capable of supporting its operating forces despite the complexity of its disparate platforms and systems. The three fleets include, for example, sixteen different destroyer and frigate classes" (Cole 2010: 132).

The high preponderance of foreign, especially Russian, high technology in mission-critical areas has detrimental effects on logistics and maintenance as well. In this area, most observers agree that China continues to be relatively strongly dependent on outside sources of support, pointing to a potential area of strategic vulnerability. Commenting on the older *Lüda, Lühu* and *Lühai* class destroyers, Bussert and Elleman note that

> [e]ven though these ships are indigenously constructed, and are evolutionary in upgrades, there are doubts if the crews can maintain, repair, and properly operate the complex systems at sea (Bussert and Elleman 2011: 31).

This problem is even more severe with regard to the ship classes that were imported as complete weapon systems from Russia—the *Sovremenny* class destroyers and *Kilo* class submarines. According to convincing reports, it appears that

> China is still largely dependent on Russian advisers for training and operations. Maintenance and repair of foreign equipment can be particularly difficult, and China's Sovremennyys rely on Russian technicians for maintenance. China was forced to return two Russian-made Type 2D-42 diesel generators to the Elektrosila plant in Russia for repairs (Bussert and Elleman 2011: 33).

Information published by *Kanwa Asian Defense* and cited by Bussert and Elleman further suggests that a large infrastructure of Russian technical advisors is in effect clustered in two isolated "support cocoons" near the East Sea Fleet headquarters at Ningbo, which has become a necessity for maintaining the imported vessels:

> The Sovremennyys largely remain in the Russian support cocoon at the isolated port of Dinghai, on Zhoushan Island, rather than at a large city fleet base. Russia repeatedly urges China to have overhauls conducted by its Zvezdochka Factory, which has facilities and documentation for 956E and EM as well as Kilo submarines. China repeatedly refuses and is trying to establish 956 maintenance capability at Bohai Shipyard, and Russia has so far trained thirty-five technical staff. Every piece of equipment was totally new to the PLAN except for the Palm Frond navigation radar, the RBU-1000 ASW launcher, the Kite Screech and FC radars, the SA-N-7 Shtil SAM, and the Ka-28 ASW helicopter. [...] The modern Kilo submarines have a similar Soviet technical support enclave at the nearby harbor of Xiangshan, while Kilo submarine crews are trained and supported there (Bussert and Elleman 2011: 33).

At present, it seems that the extensive Russian technical assistance remains necessary, even indispensable for these two imported ship classes. The risk of such an arrangement lies in a heightened degree of dependency on the support of another nation—in this case Russia—which could at any time be reduced or withheld for political reasons; and, in a reduced level of fleet integration, if certain ships need to be kept at a distance from other vessels due to their specialized support needs. According to the study cited above, it is "unclear whether the

Chinese could keep these ships and submarines in good operational order" if they were fending for themselves (Bussert and Elleman 2011: 33).

Another test for logistics and maintenance are extended overseas missions. Such missions are a relatively new capability that has been developed by the PLA Navy, as the list of related milestones given in Fig. 5.1 shows.[22]

Given its limited experience before 2009, China's continuous active participation off the Gulf of Aden in the multinational anti-piracy effort in the meantime has added considerably to the PLAN's out-of-area experience, and certainly holds valuable lessons for foreign analysts and Chinese navy planners alike. For China, this was the first chance to gather actual experience in conducting extended blue-water missions; in fact, it was "the first time the country's vessels have operated out of territory since the founding of the People's Republic 60 years earlier" (Lanteigne 2013: 291). According to a Japanese assessment, "[t]his demonstrates that the Chinese Navy is improving its capacity to execute naval operations in increasingly distant waters" (Japan Ministry of Defense 2011: 85).

However, the experience collected at the Gulf of Aden has also highlighted some key areas where, despite the impressive progress, Chinese logistics and maintenance capabilities still have not been developed to the level of a true blue-water navy. For example,

> [i]n the absence of a nearby facility or military base, that task force had difficulty maintaining its ships; the ships had difficulty maintaining supplies of fresh vegetables, fruits, and potable water; and personnel did not have access to comprehensive medical care (Yung and Rustici 2010: 1).

This is of course to be expected in the absence of a network of overseas bases. As already outlined in Sect. 2.3.2, one of the pillars of the U.S. Navy's, and to a lesser degree the British and French navies', global role is an extensive network of overseas bases. Nonetheless, compared with the named examples, and despite any putative "string of pearls" strategy for building a Chinese network of bases in the future, Yung and Rusticy have pointed out the reality that so far,

> China [...] has no network of facilities and bases to maintain and repair its ships. The possession or absence of such a network may ultimately be the best indication of China's future intentions. If China lacks such a support network, it will have great difficulty engaging in major combat operations (MCOs) far from its shores (Yung and Rustici 2010: 2).

The present author tends to agree with this view. While Chinese officials have made it clear that they intend to continue sending naval vessels (including submarines) into the Indian Ocean, and to gradually expand China's naval reach into that region (cf. Rahmat 2015), thus far it seems unlikely that building and maintaining a permanent overseas military base network extending as far as the Indian Ocean will be high on the Chinese agenda for the foreseeable future, or indeed at any point, as long as other, much more pressing concerns in China's

[22] For an insightful analysis of China's early Gulf of Aden mission experiences including detailed information on the ships deployed, see Erickson (2010).

immediate vicinity (especially in the South China and East China Sea areas) are not satisfactorily addressed. However, China will likely need more effective basic logistic support for its naval presence in the Indian Ocean, and could thus be expected to forge stronger ties with friendly nations in the area to get better access to supply and maintenance installations. So far, China seems to have signed port agreements with Djibouti, Oman and Yemen (cf. Lanteigne 2013: 303).

5.4.3 *Naval Training and Human Resources Issues*

The quality of military training is another critical area for assessing fleet capability. In the case of China, this factor is relatively difficult to evaluate, for, as Cole notes: "No area of inquiry into China's Navy is less clear than how that force manages and trains its personnel" (Cole 2010: xvi).

What is known is that the manning and support of China's ca. 200 major warships, including submarines, is done by approximately 107,000 personnel. Cole writes:

> China has a personnel-to-ship ratio of roughly 535, about the same as the naval forces of Korea (525), Taiwan (480), and, Japan (470), and not much more than the U.S. Pacific Fleet's 430, indicating that the PLAN is utilizing its manpower efficiently (Cole 2010: 118).

The ratio of officers, on the other hand, is significantly higher than in other forces. According to a RAND study, "[r]oughly one-third of PLA personnel are officers, whereas about 15 percent of U.S. military personnel are officers. The PLA is attempting to reduce this proportion by increasing the size and capability of its NCO corps, but this process is still under way" (Johnson et al. 2009: 32). NCO by now "account for between 65 to 80 percent of all enlisted personnel on board ships and serve as squad leaders, mess officers, conscript trainers, and technical experts" (ONI 2009: 32).

Naval officer training is done by nine naval academies, with the Naval Command Academy in Nanjing and the Naval Engineering University in Wuhan at the top of this system. In 1985, a National Defense University was founded in Beijing. The various naval academies every year graduate ca. 5,000 sailors, marines, and officers. The Navy recruits ca. 1,000 college graduates from civilian universities per year, and as much as 25 % of total PLAN strength may be "uniformed civilian employees" (Cole 2010: 122–128).

5.4.3.1 Recruitment Challenges

One factor that is likely to increasingly frustrate Chinese military planners is the perceived unattractiveness of military service in the eyes of young and well-educated people compared with more lucrative careers attainable elsewhere. In fact, this is an almost universal trend in many countries of the world. According to Krishnan:

> People from the wealthier and more educated classes of Western societies increasingly evade military service [...] This clearly shows the general unattractiveness of military service. Most Western militaries experience great difficulties retaining and recruiting personnel. [...] Recruitment shortfalls are one of the main reasons why all Western armed forces are increasingly becoming smaller. [...] Military service is unattractive because it generally means a lot of sacrifice in regard to individual freedom and also hard work for comparatively little payment (Krishnan 2008: 20–21).

The problem of attracting qualified personnel in China is being exacerbated by the increasing demands of the ongoing "informationization" strategy. Until recently, about two-thirds of China's conscripts stemmed from poor rural backgrounds:

> Although the PLA provides stipends for conscripts and their families, these are minimal allowances that are not attractive to individuals from more affluent areas of the country. Therefore, [...] the core of the enlisted force continues to be made up of young men from poor rural areas. [...] most rural conscripts are intermediate school graduates with low degrees of technical proficiency (Cole 2010: 120).

In 2002, the official recruitment quota for urban conscripts was increased to 33.2 % from previously 26.5 % in an effort to enhance overall qualification levels (Li 2009: 279). However, due to attractive alternatives available in China's fast-growing economy, "even among rural residents, the PLA can no longer count on being able to recruit the most-capable and most-motivated young men and women" (Johnson et al. 2009: 34).

This is a problem because naval systems, which require qualified, motivated personnel with high learning potential, are getting consistently more difficult to operate and maintain. Moreover, once such recruits have been trained, the PLAN often finds it difficult to retain them long-term "because China's rapidly growing economy offers attractive alternatives to sea duty and military discipline, particularly for the highly trained technicians the PLAN most wants to retain" (Cole 2010: 124). To address this problem, the PLAN decided to professionalize the NCO corps, reduce the minimal service time for navy conscripts from four to two years, and introduce a new rank structure in order to offer more attractive career options for talented recruits.

Two further strategies have been employed to meet the shortage of qualified personnel: reducing the percentage of recruits from ca. 80 to 65 %, and offering significant pay increases. In some ranks, payment rose 100 % during 1999–2000, and 2009 saw another large rise (Cole 2010: 119).

The shortage of qualified personnel is especially critical in areas that place high demands on the recruits, such as submarine service, a phenomenon that is also well known to some Western navies (Cole 2010: 143). In fact, this seems to be a rather universal problem, for most nations have had to confront a shortage of qualified military personnel. Due to a lack of qualified candidates, physical fitness standards of the U.S. Marine Corps, for instance, have had to be reduced during the past few years. Similar complaints about supposedly physically "unsuitable" or even "effeminate" candidates have been voiced by Chinese officials. Citing PLA opinions from an article which appeared in the Chinese *Xueshi shibao* (Study Times), *Taipei Times* journalist Hsu Shao-hsuan reports:

> Most of the PLA's recruits have never been to battle, and the children from one child families born in the 1980s and 1990s account for a majority of recruits, the article said. The growing femininity of men is directly impacting the quality of the PLA, making it a 'socially disturbing' phenomenon, the article said (Hsu 2011).

Leaving aside such questionable labels as "effeminate," the demographic change inflicted on China's populace through the one-child policy could indeed indirectly undermine the fighting spirit of the people, a problem that the PLA leadership seems aware of (cf. ONI 2009: 31–32). According to Li, an official PLA study from Shenyang noted:

> Some of the only-child solders were less cooperative with peers and more egocentric than solders with siblings. In some units, their performance in personal drills and detachment training was good, but their performance in tactics coordination training was poor. Some were reluctant to participate in high-risk training because they were afraid of injury (Li 2009: 283).

Worries over the supposed "softness" of the Chinese have been voiced and widely echoed in other quarters of Chinese society as well. In his influential 2004 autobiographic novel *Lang tuteng* ("Wolf Totem"), author Jiang Rong credits the purported lack of a "wolf character" (狼性) in traditional Chinese culture and national character, which he contrasts unfavorably with the supposed wildness found in the "Northern Barbarians," as the main reason for China's perceived backwardness, history of "humiliation" by superior powers, and propensity for accepting authoritarian rule. This line of thinking that focuses on the Chinese mentality seems quite entrenched in popular as well as scholarly Chinese discourse, despite its highly problematic aspects. However, rather than uncritically celebrating the "wolf character," Jiang Rong also warns that

> "the wolf character in the nature of humankind is essentially similar to a thermonuclear reaction; its energy is enormous, and its destructiveness is also enormous; well controlled, it can become a benefit to mankind; badly controlled, it can destroy the planet" (Jiang 2004: 401, my transl.).

The unwillingness of one-child parents to have their only child serve in the military has likewise been noted as a detrimental factor for attracting high-quality recruits (ONI 2009: 31–32). Demographic factors like this could indeed be a useful indicator of the general level of societal support for warfare. According to German genocide research expert Gunnar Heinsohn, who conducted a large-scale comparison of demographic factors correlating with outbreaks of violence around the world, the level of societal support for warfare, as well as the likelihood of large numbers of young men volunteering for military adventurism, seems to be strongly correlated with the number of economically "superfluous" male children born per woman. Heinsohn calls this phenomenon the "Youth Bulge" (cf. Heinsohn 2003: 13–28). In many cases he studied, there was a clear correlation between the demographic Youth Bulge phenomenon and outbreaks of genocidal or ethnic violence. In wealthy countries with very low birth rates, on the other hand, general societal support for warfare and military adventurism, as well as the willingness of young men to join the armed forces, has markedly dropped. Applying this finding to China would mean that China's current social climate may not favor military adventurism, neither now nor in the long run, because of its increasingly skewed demographics. This is especially relevant when assessing the chances that China will actually become a "global military power" on par with the U.S. Notwithstanding nationalist sentiments voiced among some parts of the Chinese populace, it does not indeed seem very likely that many Chinese families would deem it worthwhile to risk losing not merely their only son, but their only child, for some mission of national glory in distant lands. Some degree of societal support would nonetheless be a precondition for China's leaders to start engaging more substantially in a strategy of increased out-of-area operations.

5.4.3.2 Challenges of Naval Training

Naval training in China poses several major challenges. First, if the full potential, creativity, and initiative of its officers and troops is going to be realized, China has to overcome remnants of the inherent limitations imposed by the old Soviet-style/Leninist military leadership model. Second, for historical reasons detailed in the sections above, the technological gap between the equipment of different generations is much larger in China than in most countries. The need to integrate highly diverse equipment represents a considerable hurdle for achieving interoperability not just within the three fleets, but across all services, including when jointly exercising with Ground Forces, Air Force, and the Second Artillery Force. Third, the strategy of "informationization" within the PLA is still an ongoing process that adds considerably to the educational demands placed upon officers and recruits alike. Since China's conscripts are still overwhelmingly from rural areas, and given the huge regional discrepancies in China's overall economic development, the educational levels of recruits can be expected on average to be somewhat behind those of the general workforce. The fact that the PLA as a whole, including most of its leadership, has very little or no actual combat experience adds another dimension to the difficulty of training. These areas of difficulty will be addressed in more detail below.

After its foundation in 1949, the PLAN first adhered to Soviet training models and even used translated versions of the respective Soviet guidelines and manuals.[23] The main difference between the Soviet training model and the training concept developed by the U.S. has been summed up as follows: "The U.S. model maximizes the amount of training conducted under operational conditions—at sea and in the air. The Soviet/Russian model maximizes the amount of training

[23] For an in-depth description of the intensity and scope of Soviet influence on China's military training, see Li (2009: 122–125).

conducted ashore and in port" (Cole 2010: 130). This distinction is reflected, for instance, in the amount of time a ship spends at sea, or the number of hours a pilot spends flying per year. However, according to Cole, "the increasing availability of sophisticated shore-based trainers that accurately simulate real-time operational conditions has to some degree lessened the difference between the two general training models." In recent years, the PLAN has shifted its training concept more towards the U.S. approach by spending more time on training when underway. The benefits are fairly obvious, since "[o]perating effectively at sea is physically demanding and requires actual training at sea. This is especially true for proficiency in complex, integrated joint operations" (Cole 2010: 130–131).

Training for "joint operations" involving the cross-service participation of units acting in unison thus requires realistic multiunit operational training scenarios, which however are still not conducted by the PLAN as regularly as by other leading navies.

One factor that could be considered detrimental for developing a highly professional naval personnel force are the prevailing Leninist command structures within the PLA, wherein political commissars of equal rank to the unit commander are appointed to each unit for the purpose of joint decision-making. In the past, this even amounted to the de-facto supremacy of the political commissar:

> By regulation, both the commander and the political commissar are responsible for their unit. In the early years of the PLAN the commissar had to approve the commanding officer's orders. This became a particular point of contention following the Korean War. [...] The commissar's importance as military decision maker has declined as increasingly complex systems and technology have come to characterize the Navy; twenty-first century political officers must become professionally competent if they are to affect the military decision-making process in a positive way (Cole 2010: 138–139).

As Howarth points out, this dual command structure within units could interfere, for instance, with the decision-making autonomy of submarine commanders, thus casting doubt on the operational value of China's submarine fleet in case of combat (Howarth 2006: 18–19). The insistence on political loyalty to the Party Central rather than the state leads to a situation where the leadership nevertheless still requires "its personnel to be both 'red' and 'expert'" (Cole 2010: xvi). Political education, even if it is now somewhat reduced in scope compared with earlier periods, thus continues to play a large role in naval training. This likely implies a somewhat reduced level of professionalism among the officer corps, while pointing to the difficulty of modernizing a Leninist military (Cole 2010: 128).

A comparison with past practices in the Soviet Navy, which operated under similar premises and was in effect the blueprint for the Chinese system, can offer a glimpse of what this could mean in practice. Harshly commenting in 1986 on the general dearth of the Soviet Navy's military success during the entire period of its existence, Tsouras writes that

> The concept of military professionalism is inextricably bound up with the degree of personal initiative a culture tolerates or encourages. Western and Japanese military institutions have consistently shown a positive degree of initiative, but nothing could be more

> alien [...] to modern Soviet political culture, with its preoccupation with suspicion and scapegoating (Tsouras 1986: 22).

While the problem with such a culture, the same author notes, is not necessarily readily apparent under normal conditions, it can become a critical factor under duress:

> In the Soviet mentality, if things go wrong, someone is always to blame. Even the Soviet use of the word 'initiative' carries a different meaning. It is used in the sense of doing your own job well, never in the sense of modifying or going beyond orders as the situation dictates. [...] Such institutional defects, which are rooted in the culture, are not always readily apparent in peacetime activities, and this fact tends to give a misleading impression of the navy's ability to perform in an emergency or under wartime conditions (Tsouras 1986: 23).

These problematic aspects were pointed out by other analysts as well. Commenting on training levels of the Soviet navy in the mid-1980s, Miller and Papp observed:

> How a ship would actually fight in wartime is unknown. However, there have been incidents of shipboard fires that have gotten completely out of control just after a ship has received excellent marks for damage control and fire-fighting ability. Another problem is that while ships are deployed, they spend a large amount of time at anchor in international waters. It is not certain whether this situation is the result of a need to conserve fuel or of problems with maintaining equipment and a need to minimize wear and possible breakdown. Either way, time spent at anchor is time not spent under way, and underway training for days and weeks on end is a major factor in having well-trained crews (Miller and Papp 1986: 43).

The same analysts also note that the "level of performance of some ships also indicates poor training, as collisions, fires, and other emergencies have occurred at rates much higher than in the U.S. Navy" (Miller and Papp 1986: 44).

Although it is not clear whether there are indeed any parallels between these historical Soviet examples and the present-day Chinese navy, some indications point in a similar direction, such as the above-described "near-dormancy" of large portions of the submarine fleet (observable only very recently), the officially as yet unexplained suffocation of an entire submarine crew in 2003, and the recent high-level and large-scale corruption investigations (or, rather, purges) directed against retired high-ranking military leaders, which even included retired CMC vice chairman Xu Caihou himself. Such campaigns may well be the result of political conflict within the leadership rather than solely an effort to fight corruption.

Another area of difficulty when training for future challenges is posed by increasingly complex technical equipment and the Chinese leadership's strategic aim of transforming the PLA into a fully informationized force capable of conducting joint operations by the mid-twenty first century.

As the former U.S. military attaché in Beijing, Lt. Col. Dennis J. Blasko, recently pointed out:

> C2 of joint operations has been one of the most frequently explored training subjects for several years. Headquarters from brigade to military region levels are experimenting with methods of integrating new capabilities from all services, reserve forces, and civilian

> entities into joint operations facilitated by modern communications and computer systems. Due to the introduction of new weapons and reconnaissance and surveillance technologies, commanders now have more options to accomplish various tasks than in prior decades and must learn to optimise and choose the most appropriate and effective means to accomplish their missions (Blasko 2013).

However, despite increased practice, Blasko reports that Chinese "internal military newspapers consistently point to a shortage of commanders and headquarters staff proficient in joint and informationised operations" (Blasko 2013).

The so-called "Revolution in military affairs" (RMA) has had a decisive influence on China's approach to naval training. The PLA changed its entire military education system following the new conviction that the future of warfare lies in informationization, which requires distinctly different training than traditional, mechanical warfare. The basic meaning of RMA is that "war by machine is being superseded by war conducted on the basis of controlling information" (Cole 2010: 126). However, even though the "RMA is often mentioned [in Chinese military writings] [...] its impact on military education and training is not clear." While concepts such as RMA and network-centric warfare are being widely discussed in Chinese military circles, most Chinese articles, according to Cole, "simply repeat buzzwords without offering realistic links between concept and operational practice" (Cole 2010: 133).

It seems clear, nevertheless, that the trend towards informationization has had a great impact on Chinese military thinking and doctrine. It is worth noting the relatively wide scope of the term "information warfare" as defined by Chinese military thinkers:

> In addition to traditional information operations, such as electronic warfare, destruction of enemy communication and information systems, and computer-network operations, information warfare includes what the PLA calls the *three warfares*: psychological warfare, media warfare, and legal warfare (Johnson et al. 2009: 51, italics in the original).

These "three warfares" are further defined as follows:

> Psychological warfare largely entails traditional techniques, such as using propaganda leaflets and loudspeakers. Media warfare refers to public-information campaigns designed to mobilize China's populace, demoralize the enemy's populace, and educate Chinese military personnel on the importance of minimizing civilian casualties and other types of collateral damage. Legal warfare refers to efforts to argue that China's military actions are justified under international law and to ensure that Chinese forces adhere to internationally recognized standards of conduct (Johnson et al. 2009: 51–52).

Chinese strategists have intensively studied U.S. operations over the past few decades and consistently stressed the need to develop the capability to operate under conditions of information warfare and in a "complicated electromagnetic environment."

In this context, it is instructive to pay attention to the way Chinese military thinkers study historical case studies in order to extract lessons for China in the absence of actual Chinese naval combat experience. In an interesting review of Chinese analyses of the 1982 Falklands conflict between Britain and Argentina, Yung concludes that Chinese military planners have drawn several lessons from this scenario which could be applied to a possible conflict over Taiwan or other island territories, and sees clear evidence of this in Chinese procurement patterns, doctrine, strategy, training, and logistics (cf. Yung 2011: 90–104). Other foreign conflicts that Chinese military experts closely watched and analyzed include the Gulf Wars, the recent Afghanistan campaigns, the air warfare campaigns conducted during the Kosovo conflict, among others (cf. Scobell et al. 2011).

The actual training needs for China's three fleets are likely rather varied, since each fleet has relatively dissimilar geographic, strategic, and operational conditions to cope with (Cole 2010: 134). A general trend common to all fleets, however, consists in the need for more realistic training scenarios. As Johnson et al. point out:

> In the past, training was conducted largely by small units of a single type [. . .] operating independently under benign condition of familiar terrain, daylight, good weather, and either no adversary or an adversary whose actions were scripted and known. Now, training routinely occurs on unfamiliar terrain, at night, in bad weather, under intense electronic-warfare and cyberwarfare conditions, and against unscripted opposing forces. It frequently involves combined-arms or joint operations of relatively large scale (Johnson et al. 2009: 50–51).

This need has thus apparently been identified and addressed. Naval training has been both intensified and enlarged in scope, with more realistic training scenarios being played out. According to a report by PLAN commander-in-chief Wu Shengli, cited by Blasko, the Chinese navy conducted over 30 "combat operations group campaign exercises at sea" during 2000–2009. Blasko concludes:

> These numbers averaged out to about one major exercise for each of the three fleets per year during that timeframe. In the four years since, the navy has more than doubled the number of exercises, taking its surface forces further out to sea, with seven reported in 2012, and at least that many on track for 2013. New ships and aircraft train together to improve area air defense, anti-submarine, anti-surface, and amphibious capabilities, sometimes in formations combining assets from two or more fleets. Likewise, extended submarine patrols are reportedly on the increase. At the same time, coastal defense operations integrating air, surface, and subsurface forces are practised as new ground-based anti-ship cruise missiles and fast attack missile patrol craft are added to the navy inventory. The navy's extended anti-piracy deployments in the Gulf of Aden have tested long-distance operations, logistics, and C2 capabilities for all three fleets since task forces were first deployed in December 2008 (Blasko 2013).

However, compared to other countries' levels of transparency, it is relatively difficult in China's case to both identify and analyze all of China's naval exercises (Cole 2010: 134). The most publicly visible series of live firing exercises was perhaps the one timed to coincide with elections on Taiwan in 1995 and 1996. Apparently done with a view to intimidating Taiwanese voters, this triggered the so-called "Taiwan missile crisis."[24] However, according to Cole, the PLAN does not appear to have engaged in many complex interfleet exercises until 2010 (Cole 2010: 135).

[24] For details on the background of this crisis, see Shirk (2007: 187–190). For details on the exercises performed at that time and the U.S. military's reaction, see Li (2009: 284–290).

Another chance for China to acquire skills are joint naval exercises with other navies. Remarkably, the PLA Navy had comparably little out-of-area exercise experience until only very recently. In a marked departure from its former practice, the PLA started to regularly participate in international exercises a little over a decade ago. Since 2003, China has "conducted an average of three or four exercises annually with other nations, and about half of these exercises have been conducted outside of China's borders and nearby waters" (Johnson et al. 2009: 40).

Until 2013, the PLA as a whole had reportedly taken part in more than 60 such international exercises involving about 30 countries. Bussert and Elleman list an exercise in 2005 that concerned two vessels with the Pakistani and Indian navies and another exercise in 2007 where a destroyer and an auxiliary visited Australia and New Zealand and participated in exercises with both navies. Li Mingjiang gives a more detailed overview of the Chinese regional naval exercises with other powers until 2010 (cf. Li 2010). In addition to such activities, China had participated in about 20 UN peacekeeping operations until 2009, most of them since the year 2000, and has deployed over 9,000 personnel since then (Johnson et al. 2009: 39–40).

However, most of these international exercises and missions have been on a small scale and focused primarily on "non-traditional" security missions such as anti-terrorism, search and rescue, or disaster relief. For instance, since 1997, China regularly participated in an annual series of "Sea and Air Rescue Exercises" (SAREX) by the Hong Kong Maritime Police together with U.S. Coast Guard forces (Cole 2010: 136). However, in a few cases, larger scale exercises were performed, with the largest by 2014 being "Peace Mission 2005," a joint Chinese-Russian exercise involving about 18,000 soldiers (Blasko 2013). In 2014, China was officially invited for the first time by the U.S. to join the biennial RIMPAC exercise with four naval vessels. Altogether, the 2014 RIMPAC exercise included 50 vessels, 200 aircraft and 25,000 troops from 23 countries. China was heavily criticized later for sending an additional, uninvited electronic surveillance vessel to the rim of the exercise area, supposedly to gather additional intelligence (LaGrone 2014).

Overall, most observers confirm that China has demonstrated both an increase in training quality and enhanced exercise realism. According to Cole, it is quite remarkable that the Chinese have increasingly participated in multinational naval exercises and actively taken part in multilateral anti-piracy missions. To Cole, this points to a new level of self-assurance among China's leaders:

> After many decades of isolation from navies it once considered superior, the PLAN now conducts joint exercises with them, apparently no longer concerned about being embarrassed by its backwardness or revealing its shortcomings (Cole 2010: xviii).

However, compared to the participation pattern of other naval powers of a similar standing, China's activity with regard to multinational exercises must still be considered rather timid. Blasko writes:

> Many of these events have been maritime search-and-rescue exercises and the size of the forces involved and the length of the deployments have been modest. Chinese ships also appear to make an average of five or six foreign-port calls each year. Thus, out-of-country deployments for exercises and port calls also appear to be extremely modest for a military of the PLA's size (Johnson et al. 2009: 40).

Thus, while training methods and results are undoubtedly yielding tangible results and beginning to transform the old Leninist war-fighting methods, Blasko warns against assuming China's quick alignment with Western military training standards:

> Although many capabilities have improved dramatically from previous levels, the PLA leadership understands that many obstacles remain to be overcome for the Chinese military to reach its long-term objectives and the levels demonstrated by other advanced militaries in actual combat (Blasko 2013).

One virtue that is clearly attributable to China's military planners is their ability to analyze and learn from mistakes. The 2003 submarine accident involving the "*Ming*" class hull no. 361 seems to have had a serious influence on operational practice and training within the PLAN, resulting in a major realignment of responsibilities and a stronger regularization and improvement of maintenance and training. "The deaths of the seventy personnel onboard were almost certainly due to inadequate training, with shoddy equipment maintenance apparently contributing to the tragedy," writes Cole (2010: 115). This accident led to a major leadership overhaul, including the dismissal of the PLAN commander, the PLAN political commissar, the North Fleet commander, the North Fleet political commissar as well as at least eight senior officers. For the first time, a submariner, Vice Adm. Zhang Dingfa, was selected as the succeeding PLAN commander. He even became a member of the CMC, which until then had been dominated by the ground forces (Cole 2010: 115).

Of course, self-awareness and self-critique are central ingredients for making substantial progress. As Johnson et al. report, their development has become a widespread feature in Chinese military training:

> Rigorous evaluation and critique are also now a central part of training in the PLA. At one ground-force training base, for example, a 40-page report is written after each exercise to assess the results, and only 10 percent of the report can focus on positive aspects; the remaining 90 percent must discuss issues and problems identified during the exercise (Johnson et al. 2009: 51).

To be sure, the successful operations of several Chinese flotillas off the Gulf of Aden are indicative that significant progress has already been made. According to Cole, however, it is noteworthy that

> [...] both in person and in press reports, [Chinese] naval officers grade their service and personnel relatively harshly. This may in turn impart a degree of caution on the part of Beijing in employing the military in pursuit of national security goals (Cole 2010: 142).

Indeed, the awareness of the Chinese of their limitations can be assumed to serve as a powerful restraint against military adventurism.

5.5 Summary: The PLAN's Overall Capability Profile

As shown in Sect. 5.2, the Chinese surface fleet does not seem out of proportion, especially when taking into account the size of China's claimed maritime territory and coast length. It is also not extraordinarily large in relative terms when compared with most other regional surface fleets within the Asia Pacific security complex. The submarine fleet, however, is exceptionally large, both in absolute and in relative terms, and of course high redundancy offers strategic advantages in the event of a conflict. However, contrary to popular opinion, rather than growing, the submarine fleet has in fact shrunk markedly over the past two decades. Moreover, China shares its tendency to have a preponderance of submarines with other Leninist or post-Leninist militaries such as North Korea and Russia, and the U.S. as a global naval power also places a heavy emphasis on a large submarine fleet. That said, the very high number of submarines contrasts sharply with the reportedly low overall activity rate of this fleet, despite the fact that newer classes seem to be much more quiet and capable than the older, more easily traceable ones.

In China, most of the ongoing shipbuilding programs in the surface as well as submarine fleets observable today can best be understood as replacement procurement rather than fleet enlargement, because large numbers of older units will be due for decommissioning by the time the new ones enter service, or shortly thereafter. The introduction of modern C4ISR capabilities, however, will likely result in an enhanced overall fleet capability level in the medium term.

Some degree of fleet enlargement, at least in the surface fleet, will be inevitable in the long run if China is indeed aiming to establish a carrier force of three or four aircraft carrier battle groups. China's present attempts to incorporate this new carrier capability into the fleet will therefore necessarily have to lead to a fleet enlargement. However, such ambitions are in line with similar approaches in the other BRIC navies. So far, China is the only BRIC state, as well as the only UN Security Council permanent member state, that has no carrier capability, whereas lesser powers such as Spain, Italy, and even Thailand, operate at least one. Compared with China, India's aircraft carrier capability is already far more advanced. India also receives Western as well as Russian technical support, which it is further able to access for various other shipbuilding programs.

In terms of the overall naval combat proficiency level system described in Sect. 3.2, the PLA Navy possesses the necessary equipment for conducting missions of up to combat proficiency Level VI ("Anti-submarine warfare with submarines"). In order to reach Level VII ("Naval strike and limited air control"), China's nascent aircraft carrier capability, including its air wing and carrier battle group, needs to be fully developed, which will take at least a few years to accomplish, though it will likely take even longer.

In practice, it is hard to determine the actual capability level of China's Navy in anti-submarine warfare. While it is certain that China's submarine fleet has reached combat proficiency level IV ("Anti-surface warfare with submarines"), as evidenced by its demonstrated ability to surface in close proximity to Western warships before being detected, the actual level of its "Anti-submarine warfare with surface ships" (Level V) and "Anti-submarine warfare with submarines" (Level VI) is much less clear. Detecting modern diesel submarines from surface vessels using bow-mounted sonars has become increasingly difficult, even for the U.S. Navy, and submarine hunting operations with helis typically require two dedicated helis per surface ship for radio triangulation—a feature that is currently not in place with most Chinese surface vessels. Compared to its closest strategic rivals, China is at a disadvantage when it comes to using cues from multiple sources, including satellite imagery; with regard to the U.S. and its allies in the region, it is behind in naval aviation.

Regarding proficiency Level VI "Anti-submarine warfare with submarines," there is currently no reliable and publicly available data on the actual detection performance of Chinese submarines against modern Western, Japanese or South Korean submarines. Due to the difficulty of integrating cues from very different sources and the dedicated training requirements it seems likely, however, that ASW could be one of the areas where the PLAN still has some development potential.

According to the comprehensive naval ranking criteria employed by Todd and Lindbergh, the PLAN as of 2014 must be considered a Rank 4 "Regional power-projection navy" with upward development tendency, and is almost certain to become a Rank 3 "Multi-regional power-projection navy" in the medium term (i.e. during the next five to ten years), depending on the progress of China's indigenous aircraft carrier program and the development of the respective air wings. After reaching Rank 3, further progress up the ladder would require the building of an offshore base network, a huge investment that would hinge on China's economic development and the leadership's firm, long-term dedication to reaching such a politically controversial goal. As of today, it does not seem very likely that this is what China's navy is aiming at. It does seem clear however that the regional neighbors will have to get used to the idea of another Rank 3 navy—akin to the Indian and Russian naval forces—taking root within the Asia Pacific during the next decade.

References

'AMD Building Catamarans for China's Navy'. 2008. *The Sydney Morning Herald* online, February 1, 2008. http://news.smh.com.au/business/amd-building-catamarans-for-chinas-navy-20080201-1pgi.html Accessed April 19, 2014.

Andrew, Martin and Carlo Kopp. 2012. Advances in PLA-N Carrier Aviation: Technical Report APA-TR-2012-0402. (First posted April 12, 2012, last updated 27 Jan 2014). http://www.ausairpower.net/APA-PLAN-CV.html. Accessed November 30, 2014.

Bell, Samuel. 2009. *The Impact of the Type 094 Ballistic Missile Submarine on China's Nuclear Policy* (Master thesis, Naval Postgraduate School, Monterey, California) http://www.researchgate.net/publication/235015531_The_Impact_of_the_Type_094_Ballistic_Missile_Submarine_on_China's_Nuclear_Policy. Accessed October 21, 2014.

Blasko, Dennis J. 2013. Going through the motions. *Jane's Defence Weekly* (el. ed.), September 27, 2013.

Bokhari, Farhan. 2014. China, Pakistan Set for Submarine Deal by End of Year, Say Officials. *Jane's Defence Weekly* (el. ed.), February 3, 2014. http://www.janes.com.

Bussert, James C. 2003. Foreign Navies Combat System Dis-Integration: Problem may be more widespread than officials acknowledge. *Signal* online. March 2003. http://www.afcea.org/content/?q=node/275. Accessed June 14, 2014.

Bussert, James C. 2013. China Destroyer Consolidates Innovations, Other Ship Advances. *Signal* online. http://www.afcea.org/content/?q=node/11972. Accessed August 27, 2014.

Bussert, James C. and Bruce A. Elleman. 2011. *People's Liberation Army Navy Combat Systems Technology, 1949-2010*. Annapolis: Naval Institute Press.

Chen, Yujie 陈宇杰. 2003. Eluosi haijun wuqi fazhan fenxi (俄罗斯海军武器发展分析 | An analysis of Russian naval weapons development). *Xiandai junshi* (现代军事) 2003/01: 19-21.

'China's aircraft carrier: 58,500 tonnes of coincidence'. 2012. *The Economist* online. September 26th, 2012. http://www.economist.com/blogs/clausewitz/2012/09/chinas-aircraft-carrier. Accessed August 24, 2014.

'China's Military Rise. The dragon's new teeth: A rare look inside the world's biggest military expansion'. 2012. *The Economist* (el. ed.), April 07, 2012. http://www.economist.com/node/21552193. Accessed April 18, 2014.

Cole, Bernard D. 2010. *The Great Wall at Sea. China's Navy in the Twenty-First Century*. 2nd ed. Annapolis: Naval Institute Press.

Cole, Michael J. 2011. Varyag 'To Fall Under Central Command.' *Jane's Defence Weekly* (el. ed.), August 16, 2011. http://www.janes.com.

Cole, Michael J. 2013. The DF-21D or "Carrier Killer": An Instrument of Deception? *The Diplomat,* April 22, 2013. http://thediplomat.com/2013/04/the-df-21d-or-carrier-killer-an-instrument-of-deception/. Accessed February 24, 2015.

Collins, Gabriel and Michael C. Grubb. 2008. A Comprehensive Survey of China's Dynamic Shipbuilding Industry: Commercial Development and Strategic Implications. *China Maritime Studies* No. 1, U.S. Naval War College, Newport: Naval War College Press. http://www.usnwc.edu/Research---Gaming/China-Maritime-Studies-Institute/Publications/documents/CMS1_Collins-Grubb.aspx. Accessed September 24, 2013.

Dunnigan, James. 2010. Desperate France is giving them away. *Strategy Page*. October 9, 2010. http://www.strategypage.com/dls/articles/Desperate-France-Is-Giving-Them-Away-10-9-2010.asp. Accessed December 31, 2014.

Erickson, Andrew S. 2010. Chinese sea power in action: The Counterpiracy mission in the Gulf of Aden and beyond. In *The PLA at home and abroad: Assessing the operational capabilities of China's military,* eds. Roy Kamphausen, David Lai and Andrew Scobell, 295-376. Carlisle, PA: Strategic Studies Institute, U.S. Army War College. http://www.strategicstudiesinstitute.army.mil/pubs/display.cfm?pubID=995. Accessed February 26, 2015.

Erickson, Andrew S. and Andrew R. Wilson. 2007. China's Aircraft Carrier Dilemma. In *China's Future Nuclear Submarine Force,* eds. Andrew S. Erickson et al., 229-269. Annapolis: Naval Institute Press.

Fisher, Richard D. 2007. The Impact of Foreign Technology on China's Submarine Force and Operations. In *China's Future Nuclear Submarine Force,* eds. Andrew S. Erickson et al., 135-161. Annapolis: Naval Institute Press.

Fisher, Richard D. 2013. Details emerge on Chinese 'Type 032' submarine. *Jane's Defence Weekly* (el. ed.), July 24, 2013. http://www.janes.com.

Foster, Robert. 2012. Russia to sell, co-produce Lada-class submarines to China. *Jane's Navy International* (el. ed.), December 21, 2012. http://www.janes.com.

'Frictions between Russia & China on Imitation of Russian Navy Equipments'. 2009. *Kanwa Asian Defence* No. 62, Dec. 2009: 15.

Friedman, Norman. 2006. *The Naval Institute Guide to World Naval Weapon Systems*. 5th ed. Annapolis: Naval Institute Press.

Grubb, Michael C. and Gabriel Collins. 2008. Chinese Shipbuilding: Growing Fast, But How Good Is It? *US Naval Institute Proceedings* March 2008: 44-51.

Hardy, James. 2011. China says ex-Varyag trials 'achieved objectives. *Jane's Defence Weekly* (el. ed.), September 1, 2011. http://www.janes.com.

Hardy, James et al. 2013. Images suggest China's first carrier under construction. *Jane's Defence Weekly* (el. ed.), August 9, 2013. http://www.janes.com.

Harress, Christopher. 2014. Ukraine Crisis Could Force France to Cancel Multibillion Dollar Ship Contract; Russia Prepares for Court Action. *International Business Times* online. October 11, 2014.

Heinsohn, Gunnar. 2003. *Söhne und Weltmacht: Terror im Fall und Aufstieg der Nationen* (Sons and World Power). Zürich: Füssli.

Holmes, James R. and Toshi Yoshihara. 2008. *Chinese Naval Strategy in the 21stCentury. The Turn to Mahan,* London/New York: Routledge.

Holslag, Jonathan, 2010. *Trapped Giant: China's Military Rise.* The International Institute of Strategic Studies (IISS), London: Routledge.

Howarth, Peter. 2006. *China's Rising Submarine Power. The PLA Navy's Submarine Challenge.* New York et al.: Routledge.

Hsu, Shao-hsuan. 2011. Beijing, Taipei concerned about effeminate recruits. *Taipei Times* online. March 5, 2011. http://www.taipeitimes.com/News/front/print/2011/03/05/2003497398. Accessed March 11, 2011.

'Indian Navy seeks EMALS system for second Vikrant-class aircraft carrier'. 2013. *Naval Technology.* May 30, 2013. http://www.naval-technology.com/news/newsindian-navy-seeks-emals-system-second-vikrant-class-aircraft-carrier. Accessed October 7, 2014.

Jane's Fighting Ships (el. ed.). 2014. http://www.janes.com.

Jane's World Navies (el. ed.). 2013. http://www.janes.com.

Japan Ministry of Defense. 2011. *Defense of Japan 2011.* http://www.mod.go.jp/e/publ/w_paper/2011.html. Accessed October 22, 2014.

Jean, Grace. 2013. Pacific Operational S&T Conference and Exhibition 2013: China hopes to develop next-gen carriers. *Jane's Defence Weekly* (el. ed.), March 13, 2013. http://www.janes.com.

Jiang, Rong 姜戎. 2004. *Lang tuteng* (狼图腾 | Wolf totem). Wuhan: Changjiang wenyi chubanshe.

Johnson, David E. et al. 2009. *Preparing and Training for the Full Spectrum of Military Challenges: Insight from the Experiences of China, France, the United Kingdom, India, and Israel.* Santa Monica: RAND Corp.

Johnson, Reuben F. 2013a. Russian Officials Deny Chinese Deal on Su-35s, submarines. *Jane's Defence Weekly* (el. ed.), March 26, 2013. http://www.janes.com.

Johnson, Reuben F. 2013b. Kamov role in WZ-10 design suggests China still dependent on Russian innovation. *Jane's Defence Weekly* (el. ed.), March 14, 2013. http://www.janes.com.

Johnson, Reuben F. and James Hardy. 2013. Ukraine Preparing to Offer China Use of NITKA Carrier Training Facility. *Jane's Defence Weekly* (el. ed.), November 14, 2013. http://www.janes.com.

Krishnan, Armin. 2008. *War as Business: Technological Change and Military Service Contracting.* Aldershot: Ashgate.

Kristensen, Hans M. 2007. China's Submarine Fleet Continues Low Patrol Rate. FAS Strategic Security Blog from the Federation of American Scientists, February 6, 2007. http://blogs.fas.org/security/2007/02/post_2/. Accessed September 24, 2013.

LaGrone, Sam. 2010. China Reveals Aircraft Carrier Ambitions. *Jane's Defence Weekly* (el. ed.), December 30, 2010. http://www.janes.com.

LaGrone, Sam. 2014. China Sends Uninvited Spy Ship to RIMPAC. USNI News online, July 18, 2014. http://news.usni.org/2014/07/18/china-sends-uninvited-spy-ship-rimpac. Accessed October 12, 2014.

Lanteigne, Marc. 2013. Fire over water: China's strategic engagement of Somalia and the Gulf of Aden crisis. *The Pacific Review* Vol. 26 No. 3: 289-312. http://dx.doi.org/10.1080/09512748.2012.759265.

Leitenberg, Milton. 1987. *Soviet Submarine Operations in Swedish Waters, 1980-1986*. New York et al.: Praeger.

Lennox, Duncan. 2011. Analysis: China's ASBM project: keep calm and carry on. *Jane's Defence Weekly* (el. ed.), February 16, 2011. http://www.janes.com.

Li, Mingjiang. 2010. China and Maritime Cooperation in East Asia: Recent Developments and Future Prospects. *Journal of Contemporary China* 19(64), March 2010: 291-310.

Li, Xiaobing. 2009. *A History of the Modern Chinese Army*. Lexington: The University Press of Kentucky.

Li, Nan and Christopher Weuve. 2011. Chinese Aircraft Carrier Development: The Next Phase. In *Chinese Aerospace Power. Evolving Maritime Roles,* eds. Andrew S. Erickson and Lyle J. Goldstein, 209-224. Annapolis: Naval Institute Press.

Liu, Huaqing 刘华清. 2011. *Liu Huaqing huiyilu* (刘华清回忆录 | The memoirs of Liu Huaqing). 4th ed. Beijing: Jiefangjun chubanshe.

Loewenthal, Robert G. 2007. Cold War Insight into China's New Ballistic-Missile Submarine Fleet. In *China's Future Nuclear Submarine Force,* eds. Andrew S. Erickson et al., 286-303. Annapolis: Naval Institute Press.

Medeiros, Evan S. et al. 2005. A New Direction for China's Defense Industry. RAND Project Air Force. http://www.rand.org/pubs/monographs/MG334.html. Accessed March 2, 2011.

Miller, Christine E. and Robert G. Papp. 1986. Soviet Naval Personnel and Schools. In *The Soviet Navy. Strengths and Liabilities,* eds. Bruce W. Watson and Susan M. Watson, 36-46. Boulder, Col. and London: Westview Press and Arms and Armour Press.

Minnick, Wendell. 2014. China Touts Anti-stealth Radar. *Defense News* online, October 4, 2014. http://www.defensenews.com/article/20141004/DEFREG03/310040023/China-Touts-Anti-stealth-Radar. Accessed October 25, 2014.

Moss, Trefor. 2011a. Briefing: PLAN of Action. *Jane's Defence Weekly* (el. ed.), February 4, 2011. http://www.janes.com.

Moss, Trefor. 2011b. Analysis: Interpreting China's Carrier Ambitions: *Jane's Defence Weekly* (el. ed.), September 8, 2011. http://www.janes.com.

ONI. 2009. The People's Liberation Army Navy: A Modern Navy with Chinese Characteristics. Office of Naval Intelligence (ONI), Suitland/Maryland. https://www.fas.org/irp/agency/oni/pla-navy.pdf. Accessed September 26, 2013.

Rahmat, Ridzwan. 2015. PLAN to deploy range of warships in Indian Ocean, says China's defence ministry. *Jane's Defence Weekly* (el. ed.), January 29, 2015. http://www.janes.com/article/48464/plan-to-deploy-range-of-warships-in-indian-ocean-says-china-s-defence-ministry. Accessed January 31, 2015.

'Raytheon's APG-79 AESA Radars'. 2014. *Defense Industry Daily*, September 5, 2014. https://www.defenseindustrydaily.com/new-apg79-aesa-radars-for-super-hornets-0411/. Accessed October 3, 2014.

Sakhuja, Vijay. 2011. *Asian Maritime Power in the 21st Century: Strategic Transactions – China, India and Southeast Asia*. Singapore: Pentagon Press.

Saw, David. 2008. Evolution & Revolution – Chinese Naval Developments. *Asian Defence & Diplomacy* Nov./Dec. 2008: 27-40.

Scobell, Andrew et al. (eds.). 2011. *Chinese Lessons from Other Peoples' Wars*. Carlisle: Strategic Studies Institute, U.S. Army War College.

Scott, Richard. 2011a. Conventional wisdom. *Jane's Defence Weekly* (el. ed.), April 7, 2011. http://www.janes.com.

Scott, Richard. 2011b. China Starts Sea Trials on First Aircraft Carrier. *Jane's Defence Weekly* (el. ed.), August 10, 2011. http://www.janes.com.

'Secret Sanya – China's new nuclear naval base revealed.' 2008. *Jane's Intelligence Review* (el. ed.), April 21, 2008. http://www.janes.com.

Shambaugh, David. 2004. *Modernizing China's Military. Progress, Problems, and Prospects*. Paperback ed. Berkeley et al.: University of California Press.

Shambaugh, David. 2013. *China Goes Global. The Partial Power*. Oxford: Oxford University Press.

Sheldon-Duplaix, Alexandre. 2011. Deploying the Varyag and Developing Aircraft Carriers. *China Analysis* 03/2011: 11-13. European Council on Foreign Relations (ECFR). http://www.ecfr.eu/page/-/China%20Analysis%200311.pdf. Accessed October 21: 2014.

Shirk, Susan. L. 2007. *China: Fragile Superpower*. Oxford: Oxford University Press.

'Structure of 'Wuhan aircraft carrier". 2009. *Kanwa Asian Defence*, December 2009: 12-13.

Tien, Chen-Ya. 1992. *Chinese Military Theory: Ancient and Modern*. Stevenage, Herts: Spa Books.

Till, Geoffrey. 2009. *Seapower: A Guide for the Twenty-First Century*. 2nd ed. London and New York: Routledge.

Tsouras, Peter. 1986. Soviet Naval Tradition. In *The Soviet Navy. Strengths and Liabilities*, eds. Bruce W. Watson and Susan M. Watson, 3-25. Boulder, Col. and London: Westview Press and Arms and Armour Press.

Wang, Cong 王聪. 2009. Backgrounder: Brief history of China's People's Liberation Army Navy. *Xinhua News Agency*, April 15, 2009.

Wertheim, Eric. 2007. *The Naval Institute Guide to Combat Fleets of the World*. 15th ed. Annapolis: Naval Institute Press.

Wertheim, Eric. 2013. *The Naval Institute Guide to Combat Fleets of the World*. 16th ed. Annapolis: Naval Institute Press.

Yung, Christopher D. 2011. Sinica Rules the Waves? The People's Liberation Army Navy's Power Projection and Anti-Access/Area Denial Lessons From the Falklands/Malvinas Conflict. In *Chinese Lessons From Other People's Wars*, eds. Andrew D. Scobell et al., 75-114. Carlisle: Strategic Studies Institute, U.S. Army War College.

Yung, Christopher D. and Ross Rustici. 2010. China's Out of Area Naval Operations: Case Studies, Trajectories, Obstacles and Potential Solutions. With Isaac Kardon and Joshua Wiseman. Institute for National Strategic Studies China Strategic Perspectives No. 3. Washington D.C.: National Defense University Press. http://www.ndu.edu/inss/docUploaded/ChinaStrategicPerspectives3.pdf. Accessed February 25, 2011.

Chapter 6
Strategic Implications of China's Naval Build-Up

瞅着军装，我在想，将来某一天，军装肯定会成为人类文化博物馆里的老古董。不过，这一天还很遥远。
"When I look at my uniform I tend to think that some day in the future, uniforms will certainly have become antiquities exhibited in museums of human culture. But these days are still far away."
—Admiral Liu Huaqing in his memoirs, my transl. Cited from Liu (2011: 730).

It should come as no surprise that China—the world's leading trading nation, home to one-fifth of the world's human population, a recognized nuclear power, permanent member of the UN Security Council, and the most dynamically growing country among the four BRIC states—ultimately aims to enhance its naval power profile in accordance with its overall power profile. In China, the latter is referred to as "Comprehensive National Power" (综合国力). In fact, when taking China's traditional role in the Asia Pacific region into account, it seems that it is the period of relative military weakness from ca. 1850 onwards that should be regarded as an historic anomaly, not China's current rebound.

This point should be granted to Chinese analysts such as Yan Xuetong (阎学通, *1952), although his suggestion in a 2001 article that China supposedly already enjoyed "superpower status" several times in history—in periods as remote as the Han and Tang dynasties, no less—seems to excessively stretch generally recognized definitions of the term (cf. Yan 2001: 33–34). There can however be no doubt that the Chinese civilization has indeed traditionally exerted a cultural hegemony within the East Asian sphere, which during the latter half of China's approximately 3,000 years of recorded history included Japan, Korea, and Vietnam, with Literary Chinese effectively functioning as the written lingua franca of the entire region's intellectual elite. Neither can it be disputed that Imperial China, beginning from the Han Dynasty, enjoyed regional power status at several times. This state of affairs was not an historical constant, however. Contrary to popular perceptions of an unchanging, monolithic, and externally powerful Chinese statehood throughout history, a more sober look suggests that periods of Chinese military strength,

S. Kirchberger, *Assessing China's Naval Power*, Global Power Shift,
DOI 10.1007/978-3-662-47127-2_6

expansionism, and political unity were contrasted by almost equally extended periods of weakness, fragmentation, and disintegration, especially during those periods which saw foreign invaders seize large territories, found rival states within the Chinese heartland, or even take over the whole of China.[1]

An early and limited period of Chinese maritime exploration during the Ming dynasty from 1405 to 1433, when Zheng He's "treasure fleet" journeyed as far as East Africa, did not result in a durable Chinese maritime role and was completely abandoned for reasons that are still debated by historians (cf. Dreyer 2006: 2–4). As a recent monograph has pointed out, the ubiquitous Chinese narrative of three, four, or even five millennia of unbroken national history may not be a reflection of historical reality, but rather the result of conscious efforts by political thinkers to create a glorious counter-image in the face of discontinuity, fragmentation, and insecurity at the respective time of writing (Vogelsang 2012: 14; cf. also Rozman 1999: 387–388). Widespread narratives of a natural, inevitable, or historically unchanging Chinese military great power status in the Asia Pacific region should therefore be taken with a grain of salt.

Notwithstanding these observations, it seems that the regional power status of the last and greatest, but ethnically non-Han Chinese dynasty, the Qing, represents to modern nationalist-minded Chinese the gold standard against which a Chinese national "revival" or "rejuvenation" (复兴) should be measured. At the apex of its power during the eighteenth century, China reached its greatest territorial expansion, and for the first time incorporated far-flung areas such as Tibet, Xinjiang and Taiwan fully into the Chinese state. China's period of relative military weakness and technological backwardness during the past 100 years, by contrast, is seen by modern nationalist-minded Chinese, and perhaps in fact by most of their compatriots, as a period of national "humiliation" and "bullying" by external powers, which can only come to a definite end if all the remaining contested territories are brought back under Beijing's control. Whether or not these historical narratives are plausible, the fact remains that by virtue of their near-unchallenged repetition and reiteration in school text books, mainstream news media, documentaries, movies, and books, they exert a powerful influence over the mindsets and expectations of the general Chinese populace. China's military rise is often portrayed in this context as a necessary precondition for achieving "national

[1] In fact, from the end of the Later Han dynasty in 220 AD until the end of the Chinese Empire in 1911, periods during which the Chinese heartland was *not* under unified state control amounted to at least one third of the entire time span. Vogelsang, using more rigorous standards, even arrives at ca. 750 years of disunity, or 44 % (cf. Vogelsang 2012: 23). China was invaded and completely occupied by foreign troops in the thirteenth century, when it became part of the Mongol Empire—arguably the mightiest Eurasian great power in pre-modern times; and then again in the mid-seventeenth century, when the Northeast Asian tribe of the Manchu founded the last imperial dynasty, the Qing.

rejuvenation," in order for the country to never again become a victim of "bullying" (cf. e.g. Liu 2009: 1–2).[2] In this line of thinking, which Scobell (2012: 715) calls the "underdog mentality" of China as the "aggrieved power," factual instances of Chinese aggression against neighboring countries—such as the 1950 annexation of Tibet or the 1979 "punitive" war against Vietnam—are described as either non-violent or as motivated by self-defense.[3]

In light of the high public visibility of the more extreme Chinese positions on this matter, it is important to note that populist and nationalist views represent only part of the spectrum of thinking on international relations in China today. There are liberal institutionalists in China as well, and the middle ground is occupied by the quasi-official ideology of China's "peaceful development" (和平发展), which originated with Zheng Bijian's (郑必坚, *1932) earlier concept of a "peaceful rise" (和平崛起) and which emphasizes the necessity of maintaining a positive economic growth climate through "harmonious" foreign relations.[4] In 2006, when working with the International Department of the Central Committee of the CCP, present-day Taiwan Affairs Office Minister Zhang Zhijun (张志军, *1953) published a piece entitled "China's peaceful development and the international community" for the Party's main theory magazine *Qiushi* ("Seeking Truth"). He described the peaceful aims of China's military modernization as follows:

> China's defense build-up is directed at protecting national security and unity, and to ensure that the economic build-up progresses smoothly. China does not join military alliances or arms races, does not aim at creating a sphere of influence, and does not establish military installations abroad. China is against the proliferation of nuclear weapons and weapons of mass destruction, and supports the abolition and prohibition of nuclear weapons. China has already clearly pronounced that it will not use, nor threaten to use, nuclear weapons against any non-nuclear armed country or locality, and this policy will not be changed in the future. At the end of 2005, China has conducted the 10th force size reduction since the founding of the People's Republic; during the past 20 years, China has reduced troops by two million soldiers. China's peaceful development is not only good news for China, it is good news for the world. A stable, open, and prosperous China that proceeds firmly on the road of peaceful development will be able to make an even more positive contribution towards securing

[2] On a curious side note, the Chinese preoccupation with "national rejuvenation" as part of the so-called "Chinese Dream" (中国梦) has seemingly given rise to a new "National Rejuvenation Index." Created by one Yang Yiyong, it is said to be composed of 28 different measures of progress grouped in six weighted categories. According to reports, the value of Chinese "rejuvenation" as per this index was deemed to stand at 65.3 % at the end of 2012 (Larson 2013).

[3] As Yahuda (2012: 40) points out: "The version of history propagated in China with regard to its neighbours is that China has always been benign. The fact that Koreans, Vietnamese, Japanese, Burmese and others have historical records that suggest otherwise is ignored or denied. As a result China's leaders have a poor record in recognising and appreciating the perspectives and viewpoints of neighbouring countries, including those of Southeast Asia."

[4] Although realist thinking thus far remains the dominant strand in domestic Chinese strategy debates, liberalist viewpoints gradually began to emerge in the 1990s (cf. Rozman 1999: 386). Leonard (2008: 88–93) provides a good English-language overview of the circumstances under which the "Peaceful Rise" approach originated. For a discussion of China's expanding regional multilateralism, see Zhao (2011).

peace and stability in the world and promoting common development (Zhang 2006: 62, my transl.).

However, as Hughes points out in his analysis of modern Chinese nationalist and geopolitical writing, there remains a curious incongruity between arguments like the one presented above and the numerous justifications of Chinese expansionism in the South China Sea, the East China Sea, or regarding Taiwan, which are often put forward with an apparent sense of Chinese "moral exceptionalism":

> Throughout the Chinese texts it is [. . .] possible to find the kind of moral argument once used by *geopolitik* thinkers to claim that their own country's expansion was different from the 'imperialism' and 'hegemony' of the United States and 'the West' because it was aimed at restoring justice in an unjust international order. China's use of force and expansion is thus always judged to be defensive. [. . .] Just as Japanese thinkers insisted that the Greater East Asian Economic Co-prosperity Sphere was not an empire because its constituent units were either integral parts of the most advanced state, or willing supplicants who wanted to benefit from following Japan's lead, so Taiwan and the South and East China Seas are claimed as integral parts of China. Asserting sovereignty over such territories is thus no more than a form of restorative justice (Hughes 2011: 605, italics in the original).

This leads to the question: What kind of international role is a rising China actually aiming at? Is it set on becoming a "global power" rivaling the U.S. military might, as some nationalist Chinese writings seem to suggest? Or are its objectives still much more limited and regionally oriented, in line with Shambaugh's contention in a recent monograph (2013: 310–317) that China remains a "partial power"? In addition, what are the military-strategic implications of the ongoing Chinese military modernization process from the viewpoint of other regional actors, especially when taking naval modernization into account? This chapter will discuss these questions from various regional viewpoints.

6.1 China's Emerging Naval Posture

In Sect. 3.1.1, a very rough ranking of the Asia Pacific naval forces according to their overall capability profile was presented according to the criteria outlined in Fig. 3.1. According to this system, China, as of 2014, can be conceptualized as falling within the spectrum of Rank 4 "Regional power-projection" navies, and it is currently in the process of transforming itself into a Rank 3 "Multi-regional power-projection" force. China is thus still roughly on par with Japan, South Korea and Australia in terms of its overall naval capability profile, and will eventually be on par with India and Russia if its upward development tendency continues. Given a much smaller maritime area to defend, Taiwan's navy can likewise still be seen as a Rank 4 force, albeit with a downward development tendency due to a weakened submarine capability. The U.S. Navy's 7th Fleet is clearly still the only Rank 1 "Global-reach power-projection" force, not only in Asia, but in the entire world, and it is unlikely to be challenged in that role not least because of the enormous long-term investment of financial resources that would be required to first achieve and then maintain such a status. The various smaller navies on China's southern periphery, who find themselves pitted against China's naval forces in the South China Sea territorial disputes, are much less capable overall. Malaysia is probably the most capable, with Vietnam and the Philippines remaining far behind China in overall capability.

However, while individual force strengths and capability levels are important to gain a basic understanding of the actors and their range of military options, it is important to note that military alliances formed between unequal partners can change their strategic situations markedly and balance out individual shortcomings. Moreover, since China cannot realistically hope at present to prevail on its own against the U.S. in a symmetric naval conflict, and because any military conflict between two nuclear powers invariably poses the risk of uncontrollable escalation (which is why such a war has never been witnessed in history), such a scenario would most likely be avoided at all costs both by China's leaders as well as the U.S. government.

Since the technological gap between China's navy and the more advanced naval powers of the region—which are all without exception U.S. military allies or U.S. strategic partners—can be expected to continue for some time yet, especially in the field of naval electronics and C4ISR, and given that even Russia is apparently still ahead of China in some critical naval technology fields (cf. Sects. 4.6 and 5.3), it is likely that China will continue to emphasize technological developments which enable asymmetric warfare strategies against the key vulnerabilities within the prospective opponents' line of defense. Rather than trying to compete symmetrically with the extremely costly defense procurement and out-of-area basing strategies practiced by the U.S. in keeping with its global commitment, China will likely continue to build defensive capabilities with a view to asymmetrically attacking the critical systems which enable information sharing in a network centric warfare scenario. This strategy deters attacks by threatening to raise the price for a conflict to a prohibitive level. "Pax Americana" is thus highly unlikely to be replaced by a "Pax Sinica" in the near, medium, or even longer term, and China's posture will probably remain informed by defensive realism.

On a technological level, this means that due to the trend among U.S. allies towards espousing a "joint battlespace," a situation has emerged where certain critical (and exceedingly expensive and power-hungry) sensor technologies are often physically located on only a few ships or aircraft and then effectively shared within a fleet, or among allied units. Since very few such systems can enable area defense and provide air cover to an entire flotilla or carrier battle group by distributing information through data links, these systems automatically become especially attractive targets for asymmetric strikes. The Chinese emphasis on missile development, including technologies such as the planned land-based supersonic anti-ship ballistic missile (ASBM), is likely driven by such calculations. As Yoshihara and Holmes note,

> A fairly advanced longe-range cruise missile costs as little as half a million dollars, a pittance even for many third world countries. Consider the lopsided financial burdens imposed on a naval power determined to force entry into waters contested by an adversary

> armed with shore-based anti-ship cruise-missile (ASCM) batteries. A single U.S. cruiser is worth around $1 billion, whereas the sums invested in one U.S. aircraft carrier would literally buy ten thousand missiles. As a weapon system, the guided missile is neither technologically novel nor difficult to manufacture. [...] Missile proliferation across the developing world demonstrates that even a modestly financed scientific and engineering community can produce effective, if crude, missiles (Yoshihara and Holmes 2010: 102).

The Chinese emphasis on conventional and nuclear attack submarine development is also in line with an asymmetric area-denial strategy. As Holslag writes, even though a group of "naval expansionists" in Chinese military circles is constantly arguing for a more offensive posture, "China's strategising is essentially defensive. Thus far it has been a conservative, security-seeking actor, aiming to protect its economic infrastructure along the coast and guard territorial sovereignty." However, the same author points out that the means employed have nevertheless "become more and more offensive: long-range anti-ship missiles, attack submarines, and eventually probably intermediate-range manoeuvrable ballistic missiles and new surface combatants" (Holslag 2010: 70–71).

Building or enhancing a submarine capability, however, is a strategic option that is not only open to China. With the advent of improved air-independent fuel cell propulsion systems (AIP), ever more capable and quiet diesel submarines increasingly offer effective area-denial capabilities, and thus, deterrence at a comparably modest procurement cost compared with nuclear attack submarines, once the necessary infrastructure for maintaining them has been built up. Even though the learning curve for establishing a submarine maintenance and training infrastructure can initially be rather steep, usually requires intense foreign assistance, and invariably represents a heavy financial burden for the national economy, several of the traditionally less capable navies of the 'Asia Pacific security complex'—notably Vietnam, the Philippines, and Bangladesh—have recently opted to begin the process of building such a capability. It is almost certain that concern over China's ever more assertive role in the contested South China Sea areas, which includes a basing strategy involving a major submarine base on Hainan and the ongoing fortification of occupied islands and reefs as airbases, has been a prime driver behind the submarine procurement plans of both Vietnam and the Philippines, and their recently strengthened strategic orientation towards the U.S.

6.2 Russia and China: An *Entente Cordiale* in the Making?

The current unipolar world order, which brought about an era of American supremacy in the Asia Pacific, is widely perceived in China's academic circles and strategy think tanks as a thing of the past, something to be overcome and ultimately replaced with a new, multi-polar structure. Not only is American interventionism against non-democratic regimes described as intrusive, unlawful, and hegemonial. In Asia, American influence is perceived as being directed primarily against China's rise, with containment of China the main strategic goal (Liu and Ren 2014: 264–266).

Without challenging American military supremacy directly, China has therefore made various efforts to strengthen bilateral relations with other powers, and even begun to experiment with founding rival international organizations, likely in an effort to compete with the U.S.-led system of international institutions based on shared Western values and norms. The founding of the Shanghai Cooperation Organisation (SCO) in 2001—largely on China's initiative—coincided with the beginning of the U.S. "War on Terror," which brought about a greatly enlarged U.S. military presence in Central Asia. The SCO is an outgrowth of the preceding "Shanghai Five" process, a series of annual talks that have been held between China, Russia, Tajikistan, Kazakhstan, and Kirghizstan since 1996. It is jointly led by China (which seems to be the member most interested in furthering the process) and a much more reluctant Russia (Song 2014: 90–93, 98–100). The SCO has since grown to include Uzbekistan as a full member, and has accorded Mongolia, Pakistan, India, Iran and Afghanistan observer status, while Belarus, Sri Lanka and Turkey were accepted as "dialogue partners." At the same time, the U.S. request for observer status was turned down (Fels 2013: 235). The major common concern behind the SCO's foundation was a set of shared non-traditional security concerns between China, Russia, and the Central Asian states mainly related to combating "religiously motivated terrorist groups seeking national separation" (Song 2014: 86). Other areas of prospective co-operation include raw materials, investment, cultural exchange, and economic development in general. However, many observers today conclude that the SCO still lacks substantial co-operation. So far, the incongruity of the Central Asia-related interests between the two main powers, Russia and China, has inhibited many theoretically possible synergies. As Song writes,

> [...] not much progress has been made in terms of concrete arrangements for functional cooperation. For example, the goal of trade and investment facilitation is far from being reached and a free trade area is still considered taboo. Even in the relatively developed field of security cooperation, SCO solidarity was challenged when members could not reach an agreement on joint support for Russian military action against Georgia in 2008. In an even more humiliating episode, the organization could do little when Kyrgyzstan plunged into domestic turmoil in 2010 (Song 2014: 101).

Nevertheless, this Chinese-led initiative for limited regional integration through a newly founded international organization can be perceived as an attempt at creating an alternative to Western/U.S. dominated spheres of influence in the international realm. It has also shown that China's growing economic might could be a powerful lever for exercising influence over lesser-developed countries in its direct vicinity (Hieber 2012: 167–168). Commenting on the growing Chinese influence in Central Asia, Brzezinski already observed in 1997:

> A Chinese sphere of regional influence is thus in the making. A sphere of influence, however, should not be confused with a zone of exclusive political domination, such as the Soviet Union exercised in Eastern Europe. It is socioeconomically more porous and politically less monopolistic. Nonetheless, it entails a geographic space in which its various states, when formulating their own policies, pay special deference to the interests, views, and anticipated reactions of the regionally predominant power. In brief, a Chinese sphere of influence—perhaps a sphere of deference would be a more accurate formulation—can be

> defined as one in which the very first question asked in the various capitals regarding any given issue is 'What is Beijing's view on this?' (Brzezinski 1997: 166).

It is likely that some observers would argue that the above remarks could have been made today about other regions in the Asia Pacific as well, especially some parts of Southeast Asia.

6.2.1 Russian Support for China's Military Modernization

The rivalry between China and Russia over Central Asia (as evidenced by the SCO discussions over the question of India's full membership, which Russia supports) does not necessarily inhibit substantial co-operation between these two powers in other realms. Given the current set of constraints and incentives confronting China in the international arena, China can be expected to orient itself ever more strongly towards Russia in the field of military technological cooperation. From China's perspective, the Western arms embargo is a persistent obstacle to a more substantial strategic and military cooperation between China and the Western powers. It seems that unless China's political system should undergo radical political reform—which seems unlikely in the near to mid-term—this will not change, and will more or less force China to seek alternative options.

The current Russian navy can be conceptualized as one of altogether five Rank 3 "Multi-regional power-projection" navies, alongside India, Brazil, Spain, and Italy. Neither the Russian Navy nor the Russian shipbuilding industries have fully recovered from the dire straits resulting from the demise of the Soviet Union, but ambitious plans to strengthen the fleet were announced in early 2013. Described as a Russian "naval and maritime redevelopment plan," this project reportedly encompasses naval shipbuilding infrastructure upgrades and building programs for approximately 50 modern naval surface vessels as well as 20 submarines until 2025 at a total cost of about US$42 billion (Dunai and Anderson 2013). Against the backdrop of ongoing negotiations for another major arms deal between Russia and China, these ambitious Russian naval build-up plans offer multiple synergies for China to exploit. Russia's tensions with the Western powers following its 2014 invasion of Ukraine, and its continuous support for Eastern Ukraine's pro-Russian separatists, could contribute to an increased willingness on Russia's part to enhance and significantly deepen the existing security co-operation with China in order to counterbalance the negative effects of the Western powers' economic sanctions. China's considerable international leverage as a permanent member of the UN Security Council, a nuclear power, the world's leading trade nation, and the nation with the greatest stock of foreign currency reserves makes it a highly attractive political and economic partner to Russia in the face of Western condemnation. Moreover, China offers unparalleled market opportunities to Russian arms sales as well as Russian energy and natural-resources exports. The incentives for both sides to forge a stronger strategic partnership are thus currently very strong. Western countries should be aware of this tendency, and fully realize how the need to uphold Western security concerns and political values inevitably, albeit inadvertently, contributes to an unfolding Sino-Russian entente. From a U.S. perspective, such a scenario was already identified as a threat by Brzezinski in 1997:

> Potentially, the most dangerous scenario would be a grand coalition of China, Russia, and perhaps Iran, an 'antihegemonic' coalition united not by ideology but by complementary grievances. It would be reminiscent in scale and scope of the challenge once posed by the Sino-Soviet bloc, though this time China would likely be the leader and Russia the follower. *Averting this contingency, however remote it may be, will require a display of U.S. geostrategic skill* on the western, eastern, and southern perimeters of Eurasia simultaneously (Brzezinski 1997: 55, emphasis added).

The strong and lasting legacy of Soviet and Russian naval technology evident in the Chinese fleet today, both in the surface and submarine designs of even the most modern and indigenously built ship classes, has several strategic implications for China and the rest of the world. First of all, marked progress in China's indigenous arms production capabilities notwithstanding, China finds itself in a position of ongoing dependence on Russia for the uninterrupted supply of spare parts and technical support in several critical naval technology fields. Tangible proof of this dependency can be found in the form of two permanent Russian "support enclaves" formed for the imported *Sovremenny* and *Kilo* class vessels (cf. Sect. 5.4). Those observers who expected Russian naval exports to be a transient phenomenon that would soon be succeeded by China's becoming one of the "first tier" naval arms producers were proven wrong in 2013, when it became clear that China was planning to import even more fully Russian-designed naval vessels—likely four *Lada* class diesel submarines—as well as 24 advanced *Su-35* fighter aircraft (cf. Foster 2012; Johnson 2013b).

As in any hierarchical relationship with aspects of dependency and strategic vulnerability, friction between partners can frequently occur, and the case of China and Russia is no exception (cf. Sect. 4.6). Russian officials have occasionally expressed their misgivings about the sale of high-end technology to China over the past few years, and allegations of Chinese reverse-engineering of critical weapon systems such as aircraft or fire-control radar systems have repeatedly caused alarm in Russian industrial circles (cf. Johnson 2013a). Nevertheless, these allegations have not led to the suspension or reduction of Russian arms exports to China, and the overriding strategic interests of both Russia and China are likely to outweigh such points of contention in the overall cost-benefit analysis.

6.2.2 Possible Consequences of the Ukraine Crisis

As of early 2015, the future political orientation of Ukraine, China's other major supplier of critical arms technologies, remains uncertain. The Russian invasion and annexation of the Crimea peninsula in the spring of 2014 and the ongoing conflict with pro-Russian secessionist movements in Eastern Ukraine have led to a crisis with potentially grave strategic consequences for the EU, NATO, Ukraine, and Russia. As an observation from Brzezinski made during the 1990s suggests, the geostrategic importance of Ukraine, and especially Crimea, from Russia's point of view may have been underestimated by Western observers prior to the invasion:

> Prior to 1991 [the year Ukraine declared independence from the Soviet Union], the Black Sea was the point of departure for the projection of Russian naval power into the Mediterranean. By the mid-1990s, Russia was left with a small coastal strip on the Black Sea and with an unresolved debate with Ukraine over basing rights in Crimea for the remnants of the Soviet Black Sea Fleet, while observing, with evident irritation, joint NATO-Ukrainian naval and shore-landing maneuvers and a growing Turkish role in the Black Sea region. Russia also suspected Turkey of having provided effective aid to the Chechen resistance (Brzezinski 1997: 93).

Following Ukraine's democratic (and pro-EU) Euromaidan revolution in early 2014, Russian fears of losing access to the Black Sea Fleet port of Sevastopol may thus have prompted the invasion, which quickly led to Crimea's factual annexation.

These developments have strategic consequences for China as well. Before the annexation of Crimea, China had frequently been in a position to negotiate for additional technological assistance from Ukraine, and could thereby effectively counterbalance some Russian attempts to limit strategic technology transfers. Though by no means as substantial in quantitative terms as Russian arms transfers, technical assistance provided by Ukraine has arguably been of an even more strategic nature in *qualitative* terms. To be precise: It was Ukraine that sold China the unfinished *Varyag* aircraft carrier hull and offered technical assistance for its outfitting as well as for flight training (the Chinese flight training base at Wuhan was even modeled after the Ukrainian NITKA facility on Crimea); it was the Ukrainian Kvant Design Bureau which sold China a phased-array radar system complete with technical support, which would become the basis for China's key component of an indigenous *AEGIS*-style combat system; it was Ukraine that sold China *Kilo* class submarine documentation (albeit for a different batch) when Russia refused to train Chinese maintenance personnel; and it is widely assumed that Ukraine gave China access to a prototype Russian-made *Su-33* fighter plane for the development of China's indigenous *J-15* fighter (cf. Sects. 4.6, 5.3 and 5.4; Cole 2011a). Moreover, Ukrainian gas turbines produced under license in China are currently the only such advanced naval propulsion systems available for powering China's major surface vessels, and form an integral part of all modern Chinese destroyer designs. Ukraine has also supplied China with military turbo-jet engines, air-to-air missiles, *Zubr*-class hovercraft vessels, and self-propelled howitzers (Grevatt 2011b). Several of the above-mentioned examples are critical technologies of strategic importance.

The strategic nature of the Chinese-Ukrainian arms procurement relationship had been formalized in 1995 with a military-technical co-operation agreement, which was followed by regular meetings of a military-technical co-ordination committee. In 2007, Ukrainian Antonov Aviation formed a joint venture with Shaanxi Aircraft named Saan-Anton Aircraft Engineering Centre for joint development of the transport aircraft cited above. A strategic partnership agreement between China and Ukraine was then signed on June 20, 2011 by then-state presidents Yanukovych and Hu. It envisioned deepening "collaboration in sectors such as high technology, aerospace, infrastructure, trade, energy, agriculture and investment" (Grevatt 2011b).

The Russian annexation of Crimea and the pro-Russian Eastern Ukrainian secession movement have now brought most of the geographic areas where Ukrainian naval expertise was traditionally concentrated under factual Russian control. This is not likely a coincidence. Rather, it seems that the fact that major portions of Ukraine's sizeable military-industrial complex are physically located in Southeastern Ukraine may have been a major incentive for the Russian invasion in the first place, apart from the perceived Russian need to seize the fleet harbor of Sevastopol. It occurred, namely, precisely at a moment when it seemed likely that a democratically elected Ukrainian government might opt for a closer relationship to Western powers, and even consider EU or even NATO accession. Such a development would have had severe security-related repercussions for the Russian armed forces and arms industries alike. Ukrainian-produced weapon systems currently in service with Russia include Motor-Sich aircraft and helicopter engines; missiles and space rockets, including ballistic missiles (*SS18*), air-to-air, surface-to-air, and cruise missiles; components for the *Soyuz* and the International Space Station; advanced ship propulsion systems; armored personnel carriers and tanks; and various components for Russian fighter aircraft. As Moore summarizes:

> Ukrainian scientists and engineers know many of Russia's deepest military secrets, and in fact fathered some of them. Moreover, a quick look at a map and the realization that Ukraine's tank/armor industry is in eastern Ukraine's Kharkiv, its shipbuilding industry is in the south's Mykolaiv, its space/rocket/missile industry is in Dnepropetrovsk and its Motor Sich is in Zaporizhia, both in southeastern Ukraine, and it may be clear why it is these same regions of Ukraine that are experiencing the worst (Russian-instigated) anti-Kiev violence at present (Moore 2014).

The same expert therefore concludes that the annexation of Crimea was probably driven mainly by concern in Moscow for a possible loss of access to Ukrainian components for Russian arms projects, and possible security leakages to Western powers:

> For Moscow, the loss of Ukraine to the EU (or worse, to the North Atlantic Treaty Organization as well), would mean the potential loss of all of this capacity, and the need to replace it rapidly, not to mention the loss of military secrets that could help competitors of the Russian military industrial complex (Moore 2014).

It could be added that a recent point of contention between Russia and Ukraine before the invasion concerned the Ukrainian aircraft carrier training facility NITKA located near Saky on Crimea, which Russia had leased in the past from Ukraine, because Russia lacks a comparable facility of its own. Between 2008 and 2010, Ukraine's Tymoshenko government had denied Russian pilots access to NITKA following Russia's invasion of Georgia, but after president Yanukovich came into office, Ukraine resumed the leasing agreement (Johnson 2011). Nevertheless, right before the 2014 crisis, reports dating from November 2013 indicated that Ukraine was in advanced negotiations with China for the lease of NITKA, and that Ukraine had been negotiating simultaneously with both China and India as possible alternatives to Russia due to difficulties in reaching a financial agreement (Johnson and Hardy 2013).

At the time of this writing, it looks as though the Russian annexation of Crimea could be permanent, and that Russian-supported secession movements in the southeastern parts of Ukraine might lead to a situation where even more Ukrainian territory ends up falling under factual Russian control. This also means that the remainder of Ukraine will in the long run likely either surrender to Russian political influence, or alternatively, orient itself towards the Western powers for protection against Russian transgressions (Kofman 2014). Either way, Ukrainian technical expertise could be more or less lost to China as an alternative source of advanced naval technologies, and also as a balancing option against Russian dominance in the hierarchical arms procurement relationship China finds itself in. This will likely leave China even more dependent on Russian technical support than before, and means that, for better or worse, China will not be in a position to risk endangering this critical source of support in the near to mid-term. This may thus contribute to greater political alignment between Beijing and Moscow, possibly extending to other policy arenas.

Viewed from this perspective, technological co-operation with Russia has very real consequences for China's strategic outlook and for its range of options when choosing partners for long-term military and political co-operation. The preceding observations can be expected to retain their validity as long as the current Chinese and Russian political systems remain substantially unchanged; the Western arms embargo against China is upheld; and Russia continues to be seen by Western powers as a potential military threat. They will also continue to hold until China gains the ability to supplant all critically important Russian systems with fully indigenous designs, which could however take a rather long time to accomplish. The current negotiations over a major new Sino-Russian arms deal for the procurement of *Lada* class submarines and *Su-35* fighter airplanes support this conclusion.

6.3 The China–Pakistan–India Triangle

6.3.1 China's Strategic Partnership with Pakistan

Whereas China finds itself in a dependent position towards Russia in co-operative arms production, the reverse is true with Pakistan, China's "all-weather friend" in South Asia. As already outlined in Sect. 4.6, Pakistan has effectively become China's most important naval export customer for high-end weapon systems, allowing China's shipbuilding industries to gain experience and expertise in catering to a demanding customer's needs. However, the co-operation is not limited to weapon systems transfers, but encompasses transfers of technology, infrastructure upgrades, and military training. Pakistan's shipbuilding infrastructure has profited from the Chinese modernization of the Karachi shipyard, and from the 2006 agreement to build the fourth and last of a batch of Chinese-designed *Sword* (*F-22P*) class frigates locally. Apart from the ongoing procurement of major surface vessels, there are plans for future joint development of submarines. In addition to hardware transfers, Pakistani Air Force technical personnel and pilots have received training in China following the cooperative development of the *JF-17* multirole combat aircraft (cf. Ansari 2009).

Using the analytic tool described in Sect. 3.1, Pakistan's navy, despite its much smaller fleet size, is best conceptualized as falling within the spectrum of the Rank 4 "Regional power-projection" navies, given the quality of available weapon systems, crew training levels, and extensive international mission and exercise experience. Since Pakistan's major security objective is "to balance India's overwhelming dominance," its major strategic naval concern is to counter a possible sea blockade of its harbors in the event of a conflict with India (Rajagopalan 2011: 201). Since the much larger and also more capable Indian Navy could conceivably impose an effective blockade, asymmetric deterrence options are especially important from Pakistan's point of view, especially submarine warfare and mine warfare.

Pakistan is not as dependent on China as China is on Russia, however, because Pakistan has the option of purchasing from other suppliers as well, including even the U.S. It is also mainly limited by its own financial constraints, an issue which can sometimes be worked around when there is an overriding strategic interest on the part of the supplier nation, and which therefore does not pose an absolute barrier to the purchase of relatively modern Western arms technologies. For instance, the U.S. has sold two refurbished *Perry* class frigates equipped with *Phalanx* CIWS to Pakistan, and reports indicate that Pakistan is interested in acquiring further naval and electronics technology from the U.S. because of concerns with the quality of, among other things, Chinese-built airborne radars (Hassan 2009: 35–37). However, an ever larger portion of advanced Pakistani weapon systems is in fact being sourced from China, and comparative cost advantages, as well as China's willingness to invest heavily in Pakistani state-owned production facilities such as the Karachi shipyard and the Pakistan Aeronautical Complex, and to offer very favorable loan conditions, all suggest that this cooperation will likely continue (Grevatt 2009). This also means that the increasing technological commonalities between the PLAN and Pakistan's navy will automatically enhance options for inter-fleet co-operation.

For years there have been indications that Pakistan may have systematically (and illegally) provided China with access to some advanced Western systems and critical materials, although the actual extent of such activities is not entirely clear. It is quite possible that French submarine AIP technology was leaked via Pakistan to China in order to aid the development of China's indigenous AIP systems. Pakistan enjoyed a unique bargaining position with the U.S. during the

Afghanistan campaign due to the U.S. need to station forces in Pakistan. This gave Pakistan's government many opportunities to play various contenders against each other, as well as access to a range of U.S. and European arms technologies which are certainly of interest to China (cf. Sect. 4.6.5 for a more detailed account).

Pakistan's geographic location is an important consideration for China, because the solid nature of the bilateral security partnership opens up the possibility that China will be able to gain direct access to the Indian Ocean from China's westernmost Xinjiang province via Pakistani territory. The Chinese-built Pakistani deep sea port of Gwadar, which will also be managed by a Chinese company, was already offered to China as a possible location for a future Chinese naval base. Pakistan authorities would welcome a Chinese naval presence there as it would help deter India from enforcing a naval blockade in the event of a crisis. Moreover, according to reports seen by *Jane's Defence Weekly*, "China has promised Pakistan that it will spend USD1 billion to develop a road and train network linking Gwadar to western China" (Bokhari 2013).

If reliable all-weather transport links from Gwadar harbor into China's Xinjiang Autonomous Region can indeed be established, this would then enable China to shorten its oil transport routes from the Middle East and reduce time intervals from the usual six weeks to a possible 48 hours—a significant cost advantage. This would moreover reduce the strategic risk posed to China by the so-called "Malacca Dilemma" concerning its dependence on narrow maritime trade routes. The Chinese-sponsored construction activities in Pakistan-controlled Kashmir detected by Indian military intelligence in 2010 could very well be evidence of such a plan (Bedi 2011). However, Chinese officials have so far taken pains to publicly deny that China has intentions of accepting the Pakistani offer to make Gwadar a Chinese naval base in the Indian Ocean, likely because of concerns over India's reaction (cf. Mahadzir and Hardy 2011).

China's and Pakistan's mutual interests thus provide several opportunities for meaningful synergies, not least due to Pakistan's chronic shortage of funds. There are constraints, however, due to the strategic triangle China-Pakistan-India and also Pakistan's other strategic military partner, the U.S. To Pakistan, China's support is an important lever for keeping pressure on India, and the possibility of developing the port of Gwadar into a Chinese naval base in the Indian Ocean would no doubt be a compelling option for China to pursue in order to expand the geographical reach of its navy, should the need arise in the future. Such a move, however, would greatly provoke India, and markedly enhance already existing "China Threat" perceptions in that country. This, in turn, could further fuel support for Indian arms procurement programs, and generally heighten mutual tensions and mistrust (cf. Yoshihara and Holmes 2010: 173).

6.3.2 Indian Concerns over China's Presence in South Asia

From a technological and structural standpoint, the Asia Pacific (and BRIC) navy that most closely resembles the PLAN is clearly the Indian navy. Just like its Chinese counterpart, it is characterized by a strong Soviet/Russian technological legacy. India has also started to integrate Russian technologies with various Western and indigenously developed Indian systems. However, as of today, more than 60 % of India's military hardware is still sourced from Russia. India is furthermore the only country so far that Russia has granted access to the Russian military-operated GLONASS satellite navigation system. At the 14th Indo-Russian summit in 2013, both countries agreed to pursue further co-operation in missile and naval technologies, as well as to a plan that prolongs the existing co-operation in joint development of *Brahmos* anti-ship cruise missiles in order to produce more lethal derivates (Bedi and Grevatt 2013).

Compared with China, the Indian navy thus finds itself in a much more favorable position when it comes to forming co-operative arms production partnerships. To start with, India has received far more advanced Russian arms technology than China ever has, including an unprecedented 10-year lease of a completely functional nuclear-powered attack submarine in 2012, with talks of a second lease ongoing, and the purchase of the extensively refurbished aircraft carrier *Vikramaditya*. India has also been able to import U.S. gas turbines and French state-of-the-art diesel submarines during the past few years alone, and received design support for its indigenous aircraft carrier program from Italy. The proposed lease of a second nuclear attack submarine from Russia could be related to an attempt by India to secure vital Russian technical assistance for the nuclear propulsion system of India's indigenous nuclear ballistic missile submarine, the *Arihant*, which is currently under construction (Bedi and Grevatt 2013). India has also been strengthening its military ties to the U.S. over the past decade, with significant implications for access to high technology and critical materials, e.g. for carrier development. In September 2013, both governments "formally agreed to jointly develop and produce military equipment involving the transfer of advanced US technology," which, according to U.S. officials, could include the co-development of a next-generation anti-tank guided missile as well as electromagnetic aircraft launch system (EMALS) technology for India's next-generation carrier (Bedi and Hardy 2013). Even before the two countries signed the *New Framework for the U.S.-India Defense Relationship* in June 2005, more than 25 bilateral military exercises had already taken place between them (Blazevic 2009: 65). This culminated in India's first participation in the RIMPAC exercises in 2014.

Since India has consistently dedicated much smaller funds to the development of its navy over the past 20 years than China has during the same period, its naval build-up has not progressed as rapidly by comparison though—this despite India's greater access to high technology, and with the exception of its more advanced indigenous aircraft carrier program. Moreover, many older units in India's fleet are aging rapidly, and a history of delays and cost overruns will likely inhibit their smooth replacement. Nevertheless, better access to Western as well as Russian training and support has probably given the Indian navy some leverage regarding operational readiness and actual mission experience. Due to an already existing aircraft carrier capability, the mission spectrum and naval capability level that is within reach for India's navy is greater than China's as of 2014, which is why India is included together with Brazil, Spain, Italy, and Russia in the small group of Rank 3 "Multi-regional power-projection navies" in the classification system outlined in Sect. 3.1.

All in all, both countries continue to face similar technological challenges and similar patterns of dependency upon Russian technical support and Russian arms exports. Like China and Brazil, India ultimately aims to decrease its degree of dependence on outside supply sources, and to achieve a higher level of indigenous arms development. This is evident in the currently ongoing indigenous nuclear submarine and aircraft carrier programs. India profits from the fact that its naval rise is generally not perceived as a threat by most other advanced naval powers (with the exception of Pakistan, and perhaps China) and also that it has traditional political links to several Western countries as well as Russia.

India's greatest area of strategic concern is its troubled relationship with its neighbor Pakistan, China's close strategic ally. Indian maritime security concerns were heightened following the bloody terrorist attack on Mumbai in November 2008, which was committed by a Pakistani Islamist group that entered India illegally from the sea, and which resulted in more than 160 casualties ('How Mumbai Attacks Unfolded' 2008; Masood 2009). Another major area of concern is the conflict with Pakistan over divided Kashmir. India's relationship with China is also far from being tension-free, for there are unresolved territorial disputes in the Himalayan region, and India's continuous support for the exiled Tibetan leader, the Dalai Lama, who resides in India, remains a thorny issue, not least due to the fact that the Tibet question has been consistently labeled a "core interest" (核心利益) of China, alongside Taiwan and Xinjiang, by China's leadership (Swaine 2011: 8). Ongoing Chinese military infrastructure construction activities in Tibet next to disputed border areas have likewise raised concerns in India. As Bedi (2011) reports, "China has constructed high-altitude roads, railway lines and military airfields adjoining the undefined line of actual control [. . .] that has served as the de facto Indo-Chinese border since a 1962 conflict." This has prompted India to undertake similar activities on its side of the contested border:

> In an effort to match the PLA build-up in Tibet, India has deployed two squadrons of Sukhou Su-30MKI fighter aircraft to northeastern Assam state. It is also raising two mountain divisions and renovating airfields, bridges and roads in the region (Bedi 2011).

There are thus several concrete indications of heightened tensions, even though Indian and Chinese maritime interests do not for the most part directly conflict with each other, since there are no overlapping maritime claims. Nevertheless, Indian observers have reacted with alarm to the extended Chinese maritime presence off the Gulf of Aden. As Yoshihara and Holmes observe,

> [. . .] vocal members of New Delhi's strategic community do not view the Chinese naval entry into its backyard with equanimity. Indeed, some of the more alarmist, at times almost panicky, Indian commentary suggests near-deafness to China's soothing overtures. Leaders of the Indian foreign policy and defense establishments voice deep ambivalence about China's antipiracy mission in the Guld of Aden. Many of them depict the operation as China's first step onto a slippery slope toward a permanent naval presence in the Indian Ocean (Yoshihara and Holmes 2010: 173).

In early 2015, a spokesperson of China's Ministry of National Defense made it clear that China intends to further intensify its deployment of naval vessels, including submarines, to the Indian Ocean area. A port call made by a *Type 039 "Song"* class submarine to Colombo, Sri Lanka in September 2014 was noted with concern by some regional observers (Rahmat 2015).

Notwithstanding such assessments, and the general Indian anxiety over China's technical support for Pakistan's armed forces, there seems to be a strong likelihood that a policy of mutual laissez-faire will prevail on the oceans as long as no side attempts to enlarge its maritime reach into the other's zone of interest. However, any further expansion of the Chinese naval presence into waters that are perceived to be part of the traditional Indian sphere of influence, e.g. by building naval facilities in the Maldives or Sri Lanka, would likely enhance "China Threat" perceptions in India, and would almost certainly be readily exploited by India's political, military, and industrial leaders to justify further naval build-up activities. As already mentioned in the preceding section on Pakistan, it is likely that the Chinese hesitance to accept the Pakistani invitation to develop Gwadar into a Chinese naval base is best explained by concerns over the Indian reaction. Similarly, Beijing would perceive a stronger Indian naval presence in the South China Sea area as a challenge to its regional supremacy. This was demonstrated in October 2014 when India signed an extensive co-operation agreement with Vietnam, which included Indian naval arms sales complete with a financing solution, and a Vietnamese offer of two oil-exploration blocks in disputed waters in the South China Sea within the Vietnamese EEZ, but also within an area claimed by China (FlorCruz 2014). Earlier it had been reported that the defense co-operation would also include Indian training for Vietnamese submarine crews. Unsurprisingly, the agreement immediately prompted strong Chinese protests.

Apart from India's legitimate security concerns regarding Pakistan, this author finds that current Indian threat perceptions of China's naval activity in the Indian Ocean may be somewhat overblown. This follows from the fact that India is a nuclear power which finds itself on increasingly good terms with the U.S.; that the Indian navy cannot be considered inferior to the PLAN in either technical or operational terms; and also that there are no directly conflicting maritime claims between the two countries. Moreover, India and China have been conducting joint naval exercises since 2003, are both co-operating within the SCO, which India hopes to join as a full member with Russian support, and both signed a Memorandum of Understanding on Defence Cooperation in 2006 (Blazevic 2009: 65–66). Overall, there seems to be little incentive for China, which already faces multiple complicated strategic scenarios closer to home, to add yet another power to its list of potential maritime adversaries. As Blazevic notes, "if China does indeed harbor a 'tripartite strategy' of diplomacy, forward naval bases and greater naval presence, it is far from realization and its near-term aspirations in the Indian Ocean will be limited" (Blazevic 2009: 64). According to all indications, this seems to be a reasonable assessment.

6.4 The Korean Peninsula and China

The Korea question may be, as some observers have noted, the most volatile issue among the various potentially explosive conflicts within the 'Asia Pacific security complex'. This is due not least to the fact that North Korea is in possession of nuclear weapons, and its leadership frequently engages in a dangerous game of "aggressive brinkmanship diplomacy" (Hauswedell 2012: 77–78). Geographically, the Korean Peninsula is the northernmost outpost of China's "first island chain." Its proximity to the Northern Chinese coastline and its huge land border with China make it a geostrategic factor of prime importance.

Historically, the Korea question resulted from the Korean War 1950–1953, during which China sided with the communist North, while the U.S. decided to back the anti-communist South. This war, which was never formally concluded, also indirectly brought about the de-facto independence of Taiwan from the Chinese mainland, as the U.S. military support granted to Taiwan as a consequence of the American involvement in the Korean War was able to successfully deter a long-planned Chinese invasion of the island. Both the Korea and Taiwan questions are thus connected, and the role of the U.S. as a guardian power of both South Korea and Taiwan was instrumental in bringing about and maintaining the national divisions of Korea and China.

Since about a quarter of U.S. troops in Asia are stationed permanently in South Korea, which is thus the most exposed U.S. outpost engaged in securing the post-war order dubbed "Pax Americana," any military crisis in Korea would mean the automatic involvement of the U.S. on a massive scale.

6.4.1 China's Uneasy Partnership with North Korea

As North Korea is China's only treaty ally in the Asia Pacific, the ultimate fate of North Korea's totalitarian polity is a major strategic concern for Beijing. China, for its part, would be theoretically compelled to come to North Korea's aid due to Article 2 of the 1961 *Sino-North Korean Mutual Aid and Cooperation Friendship Treaty* (cf. Hauswedell 2012: 77–78, 83). However, according to Rüdiger Frank's recent analysis, there are indications that China is seen less as a guardian power or partner by North Korea than as a prime security risk, not least due to North Korea's extreme degree of economic dependence on China, which raises the issue of its strategic vulnerability (Frank 2014: 379; cf. also Chung and Choi 2013: 258).

About 70 % of North Korea's total foreign trade is conducted with China, and oil imports from China are responsible for about 90 % of North Korea's total energy supply. The trade dependence is similarly pronounced in the realm of consumer products and food, where the share of Chinese imports is approx. 80 % and 45 % of the domestically available goods, respectively (Godehardt 2013: 1). In light of this dependence, the peculiar Chinese reluctance to publicly condemn North Korea during the 2010 crisis ultimately caused South Korea to align itself more closely to the U.S. after a period of slight estrangement and greater openness towards China. This ultimately contributed to even stronger U.S.-South Korean military co-operation (Hauswedell 2012: 82).

The potential usefulness of North Korea as a military ally to China regarding its own primary areas of concern is another matter. Based on the evaluations in *Jane's World Navies* (2013), North Korea's navy should probably be seen as a Rank 6 "Inshore coastal defence" force that is primarily able to conduct asymmetric deterrence within its own 12 nm zone. In general, the submarine fleet seems to be the most viable arm of the navy. Although reliable data is scarce on the actual state of readiness of North Korea's armed forces, including its navy, it seems reasonable to assume that the actual military assistance North Korea's navy could hope to render to China in the event of a major conflict would be limited. This is supported by the high demands continuously made on North Korea's own homeland defense, its apparent inability to conduct missions in the outer reaches of even its own EEZ, and the generally extremely dire situation of North Korea's national economy, suggesting that much of the equipment may be in a bad or non-operational state and that logistics could be compromised. However, this does not mean that asymmetric surprise attacks by North Korean vessels would not be able to inflict damage on an adversary. For instance, there is overwhelming evidence that the tragic *Cheonan* incident in 2010 was caused when a North Korean midget submarine fired a torpedo and sank a South Korean corvette during transit to an exercise area (cf. 'A guilty verdict for North Korea' 2010).

Regardless of applied international pressure or material incentives, North Korea's nuclear weapons program is not likely to be suspended. In light of recent U.S. regime-changing interventions in Libya and Iraq, North Korea's leaders likely conceive of their nuclear arsenal as their only insurance against a similar incursion (Hauswedell 2012: 79). China meanwhile likely has an overriding interest in prolonging the status quo on the Korean Peninsula. Since South Korea is a key U.S. ally and home to many U.S. military bases and listening stations, China has no interest in seeing a Korean reunification under South Korean terms, which would result in the expansion of South Korean territory up to the Chinese border and give the U.S. military greater leverage (Frank 2014: 382). As long as this remains the case, China needs North Korea to remain a neutral puffer zone. China's geostrategic concerns regarding the fate of Korea are quite legitimate, and should be given due consideration by other powers (Hauswedell 2012: 77). They also explain why China was unwilling to publicly condemn North Korea for sinking the *Cheonan* and attacking South Korean islands in 2010.

China may in fact mainly perceive North Korea as a bargaining chip for balancing U.S. influence in Northeast Asia rather than as a dependable military ally. Just the same, erratic behavior on the part of North Korea's leaders sometimes serves the strategic purpose of binding some U.S., South Korean, and Japanese military capacities. However, given the volatile state of North Korean leadership politics evident from the widespread and bloody purges of high-ranking leadership personnel in 2013, 2014 and 2015, any regime crisis in North Korea would also have the unfortunate potential of turning into a military crisis, depending on which North Korean leadership faction prevailed. Given North Korea's nuclear arsenal, China likely has no interest in seeing such a crisis materialize or escalate. Therefore, China can ultimately be expected to be co-operative in the event of a North Korean crisis. In all probability, it would work towards avoiding or containing a conflict rather than fueling it, though only under the premise of not allowing the regime itself to be overthrown completely. Given the seemingly erratic behavior that has been displayed by North Korean leaders during the past few years, which on several occasions has provoked and alienated even China, it is likely that North Korea all in all poses more problems than opportunities for China, and that "down the road, North Korea will be more of a strategic liability and uncomfortable neighbor than trusted ally" (Chung and Choi 2013: 259). According to Frank (2014: 379), from a North Korean perspective the same may be true with regard to China.

Due to the Korea question, China finds itself entangled in a strategic quadrangle with South Korea, North Korea, and the U.S. The strategic considerations connected to this problem certainly factor into other strategic considerations and issues as well, especially regarding the Taiwan question, where Taiwan, China, and the U.S. form a strategic triangle. From the point of view of naval development, North Korea is thus mainly a complicating factor, which indirectly influences China's complex strategic relationship with the U.S. and some of its allies.

6.4.2 China and South Korea

South Korea's overriding strategic interest consists in containing the North Korean military threat, and, ultimately, in contributing to eventual regime change in the North. While a reunification under South Korean terms may seem politically desirable, it is economically problematic to many South Korean observers due to the expected economic burdens (cf. the discussion in Frank 2014: 371–380). Therefore, unless North Korea first undergoes a substantial political and economic transformation that would subsequently favor a successful Korean reunification, it is likely that South Korea's strategic alliance with the U.S. will remain a central pillar of its security strategy. Despite increased economic and cultural exchanges with China, this alliance has only been reinforced due to China's unwillingness in 2010 to publicly condemn unilateral North Korean acts of aggression against South Korea (Holslag 2010: 101–102). The events of 2010 have led to an marked increase of negative South Korean views regarding China's growing military power (Glaser and Szerlip 2011: 28).

As of 2014, the South Korean Navy must be considered one of the regional navies that is at least on par with the PLAN in terms of both equipment quality and operational capability (cf. Fig. 6.1).

Among South Korea's destroyers, the three *KDX-III* (*Sejong the Great*) class vessels are equipped with the *AEGIS* combat direction system (marked in green).

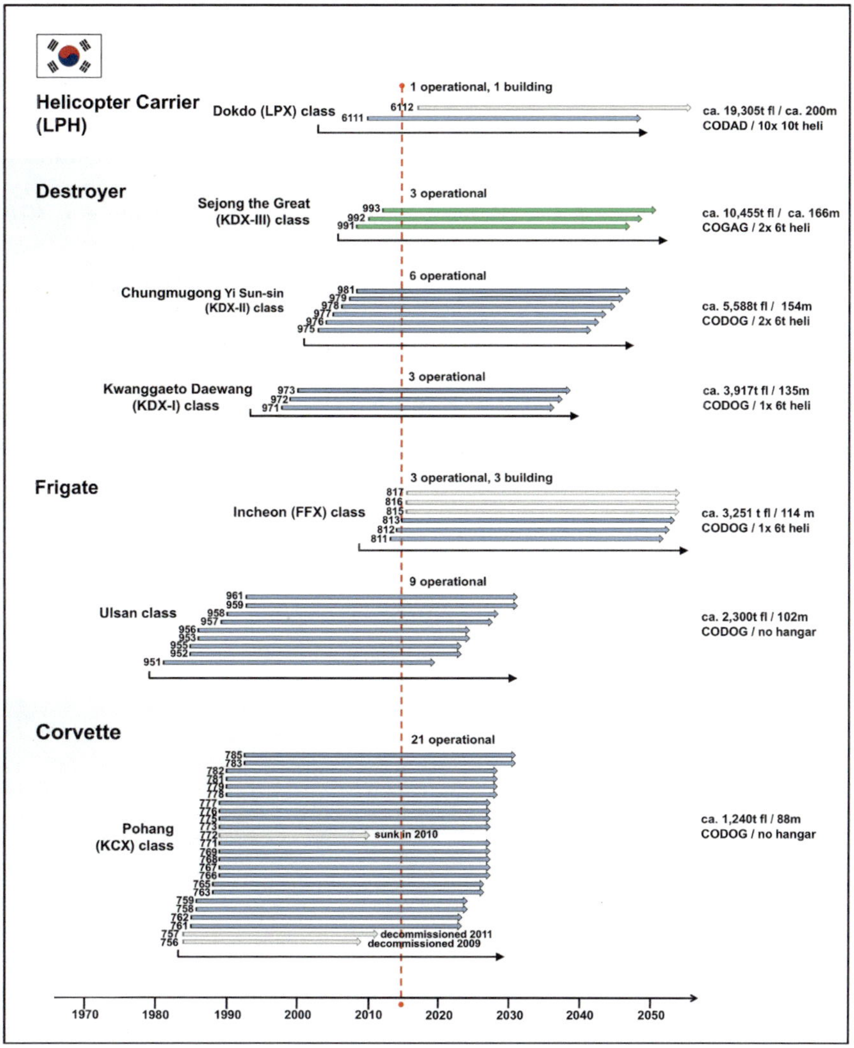

Source: Drawn using data from Wertheim (2013) and *Jane's Fighting Ships* (2014)

Fig. 6.1 Type and age structure of the South Korean surface fleet, 2015

Since the 1990s, South Korea has added a state-of-the-art submarine capability to its fleet from scratch by importing German submarine designs from HDW (cf. Fig. 6.2). The first-of-class of the *Chang Bogo* class submarines was built in Germany, while the following two were already assembled in South Korea from material packages. All follow-on vessels were constructed locally to the HDW design.

The follow-on class of fuel-cell-equipped *Type 214* submarines (marked in green) currently under construction will further enhance the operational capability level of the South Korean submarine fleet, and offer a powerful deterrent against any surface vessels. This class is being built locally with technical assistance from the design shipyard HDW.

The combination of American *AEGIS* combat system technology in the surface fleet with what is likely the most capable conventional submarine technology available gives South Korea's navy a technological advantage over most other regional navies.

Compared with both South Korea and Japan, two navies with a strong U.S. technological legacy, the Chinese Navy can thus be said to be at a comparative disadvantage as of late 2014. While the true status of China's area-defense capabilities is not entirely clear, it is unlikely fully operational yet (cf. Sect. 5.3.1). Chinese capabilities designed to one day successfully overwhelm the *AEGIS* system's defenses, such as anti-ship ballistic missiles (ASBM), are mostly still under development. Moreover, the relatively high naval combat proficiency level that is within reach for South Korea's (and Japan's) naval forces stems not only

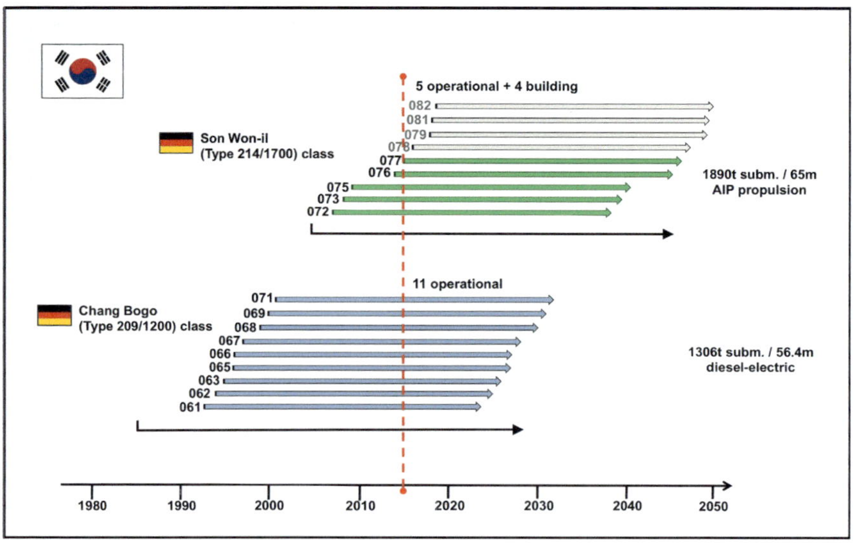

Source: Drawn using data from Wertheim (2013) and *Jane's Fighting Ships* (2014)

Fig. 6.2 Type and age structure of the South Korean submarine fleet, 2015

from superior hardware, but likewise from the long tradition of co-operation and data exchange among Asian U.S. allies, and from regular exercises with U.S. forces, which continue to maintain a very strong military presence in both South Korea and Japan. South Korea has received extensive American assistance in setting up a U.S.-developed naval command system, the LNTDS (cf. Friedman 2006: 13). Another factor worth considering is the high state of readiness of South Korea's military due to a history of tensions with North Korea ever since the Korean War. This threat has led in that country to a high degree of overall "militarization," as defined and measured by the *Global Militarization Index* (GMI, cf. Sect. 3.6). South Korea is also the world's leading shipbuilding nation, both in quantitative and qualitative terms, and therefore in a good position from the point of view of logistics, maintenance, and overall ability to sustain its fleet.

The strong U.S. technological influence in naval surface vessel design, South Korean access to *AEGIS* technology, as well as the strong German technological influence in recent submarine projects thus contribute to the very robust overall power profile of the South Korean fleet. South Korea's navy seems well disposed to effectively defend itself in case of a maritime conflict with North Korean forces, and would be a formidable addition to a U.S.-led military alliance of Asian powers against either North Korea or China if tensions were to come to a head, which is fortunately a rather unlikely scenario.

In general, however, maritime relations between South Korea and China are much less problematic than between China and Japan. This is largely the case because South Korea and China have no comparable territorial conflict to solve; because there is no unresolved war crime history between both countries to work through; and because South Korea is dependent on China's unparalleled political influence on North Korea's leadership for the 'six-party talks'. Growing economic, trade, and cultural relations between the two countries furthermore bear witness to China's considerable "soft power" vis-à-vis South Korea.

6.5 The Taiwan Question

The question of Taiwan's eventual political reunification with or official independence from the Chinese Mainland is without a doubt the number one maritime concern for China. This is evidenced by the fact that unlike the South China Sea or East China Sea islands issues, Taiwan has been consistently labeled a "core interest" (核心利益) in official Chinese documents since the 1990s, or at least as long as the term has been in use. In fact, as Swaine points out in an insightful study of the issue:

> A close examination of the historical record, along with personal conversations with knowledgeable senior U.S. officials, confirms that thus far the Chinese government has officially, and repeatedly, identified only three closely related issues as specific core interests: the defense of China's sovereignty claims regarding Taiwan, Tibet, and Xinjiang (Swaine 2011: 8).

According to Swaine, China's reference to a particular issue as a core interest "conveys a high level of resolve, and to some extent a warning of sorts to other powers." It also signals that the Chinese leadership views such issues as "purely Chinese internal affairs not subject to dispute or negotiation." (Swaine 2011: 10). To make this point as clear as possible, China went so far as to formally promulgate an "Anti-Secession Law" at the high tide of cross-strait estrangement in 2005 during the tenure of Taiwan's president Chen Shui-Bian. This law must be interpreted in conjunction with the much older U.S. Public Law 96-8—the *Taiwan Relations Act* (TRA) of January 1979—which was passed by the U.S. Congress in response to the U.S. government's decision to shift diplomatic recognition from Taiwan to China, and to simultaneously cancel the mutual defense treaty with Taiwan's government. The TRA somewhat ambiguously states that the U.S. government will "consider any effort to determine the future of Taiwan by other than peaceful means, including by boycotts or embargoes, a threat to the peace and security of the Western Pacific area and of grave concern to the United States" (Taiwan Relations Act 1979, Sect. 2, 2.4). The United States' de facto role as Taiwan's guardian power against Chinese aggression, stemming from the U.S. decision to thwart a Chinese invasion of Taiwan during the early 1950s, has thus resulted in the formation of the strategic Beijing–Taipei–Washington triangle. Nevertheless, the intentional vagueness of the TRA text regarding the exact nature of a possible U.S. military response to a Chinese aggression against Taiwan leaves much room for speculation, and all sides must therefore be prepared for a range of more or less possible turns of events by considering a variety of different military-strategic scenarios.

6.5.1 Taiwan as a Factor in China's Naval Build-Up

The geostrategic rationale behind China's pursuit of unification with Taiwan has already been discussed at some length in Sect. 2.3. Given the political background sketched above, it is to be expected that a contingency plan for Taiwan features strongly in all PLAN modernization plans for the foreseeable future, and will likely continue to do so despite the tendency towards markedly improved cross-strait relations during the past few years. This is exactly what U.S. Navy War College analyst William Murray was able to observe regarding China's fleet modernization patterns:

> My own theory is that it is still all about Taiwan. [...] I look at the types of ships they are building, and at what they are putting on them, and there is nothing that requires a great deal of endurance and not a lot of replenishment capability (quoted in Moss 2011).

The *Type 022 "Houbei"* class missile catamarans, for example, which China has been building in large numbers during recent years, would be especially useful for asymmetric deterrence operations in coastal waters near Taiwan, but much less useful in extended operations at the outskirts of China's EEZ (cf. Sect. 5.3).

In his in-depth study on the historical development of China's military-industrial complex, Cheung argues that China's military modernization remained a low priority until the early 1990s, when the new "pragmatic diplomacy" of Taiwan's president Lee Teng-hui suddenly prompted fears in Beijing that Taiwan might move towards formally declaring independence:

> [...] Taiwan's quest for independence in the mid-1990s led Beijing to embark on a far-reaching shake-up and modernization of the PLA and defense industrial complex. It committed considerable resources and devoted high-level political attention to raising technological competence and promoting innovation (Cheung 2009: 19).

From China's perspective, the absolute low point in its military readiness seems to have been reached during the 1995/96 Taiwan "missile crisis," which prompted an unexpectedly strong reaction from the U.S. (Shambaugh 2013: 276). The combined "humiliation" of China relating to the U.S. government's decision to grant Lee Teng-hui a U.S. visitor visa despite prior assurances to the contrary, and the apparent inability of China to influence Taiwan's election results in the desired way through demonstrations of force, must have been a sobering experience for China's leaders, and driven home their effective powerlessness when it came to changing circumstances in China's favor (cf. Shirk 2007: 187–190). As a result of this experience, and after having observed U.S. military prowess in the Gulf War and in Serbia, China's leaders decided to implement a program to speed up military modernization, and to accelerate the industrial infrastructure development necessary for sustaining it (Li 2009: 289–290). Thus, it can be said that the actions of Taiwan's leaders during the 1990s and thereafter may have inadvertently contributed to the more rapid pace of China's military build-up.

6.5.2 Defending an Island: Taiwan's Quest to Preserve the Status Quo

The major military threats Taiwan would face in the event of a crisis were listed by then ROC Vice Minister of Defense Andrew Yang in 2011 as follows:

- *Surveillance*: China's around-the-clock surveillance and reconnaissance using satellites, OTHR, early-warning and maritime patrol aircraft, UAVs, etc. leave little opportunity to hide force deployment and other military dynamics
- *Blockade*: China's enhanced naval capabilities allow for a partial or even total sea blockade of Taiwan and disruption of its SLOCs
- *Precision strike*: Improved accuracy of PLA missiles (ballistic and cruise missiles) endanger ships, aircraft, and strategic installations on the island
- *Occupation of outlying islands*: China's amphibious capabilities could allow for invasions of Taiwan's outlying island territories and even Taiwan itself.

In the event of a conflict, the ROC Navy—in conjunction with the ROC Air Force—would therefore be charged with thwarting a naval blockade, inhibiting amphibious landing operations, inhibiting as well as conducting mine warfare operations, and conducting asymmetric attacks against Chinese vessels. The requirement would be to survive a first strike and mount a counterattack ("resolute defense"), counter a blockade and keep open the SLOCs ("anti-blockade"), prevent the enemy from crossing the Taiwan Strait in co-operation with the other military branches ("joint interdiction"), and help fight an amphibious invasion force ("ground defense"). Preferably, asymmetric deterrence capabilities would already be demonstrably built up as a "credible deterrence" in order to discourage an aggressor from mounting an attack in the first place (Yang 2011).

As for Taiwan itself, the ROC Navy has relatively capable and modern naval forces. At the time of this writing, its approximate position in the system of naval ranks described in Sect. 3.1 would still be that of a Rank 4 "Regional power-projection" navy, despite a slowly degrading submarine capability.

However, as Fig. 6.3 illustrates, the fleet age structure overview reveals some further areas of concern. Some of Taiwan's major naval assets are either quickly nearing the end of their maximum service lives, or have already long passed that

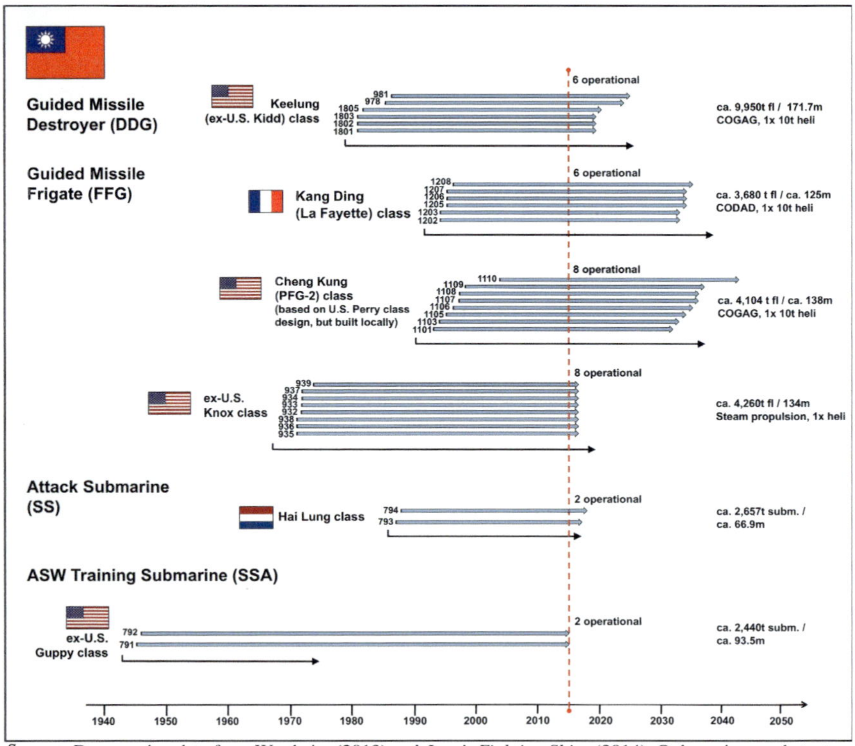

Source: Drawn using data from Wertheim (2013) and *Jane's Fighting Ships* (2014). Only major combatants shown

Fig. 6.3 Major surface and subsurface combatants of the Republic of China Navy, 2015

stage, but are nevertheless kept in service due to difficulties in replacing them. The largest and most capable operational vessels are probably the secondhand *Keelung* (ex-U.S. *Kidd*) class destroyers, which the U.S. sold to Taiwan as an "interim solution." According to Wertheim (2013: 709–710) they were delivered with a *Mk 14* weapons direction system, NTDS, and are equipped with data links 4a, 11, and 14. Their armament includes *Standard Missiles* (*SM-2 Block IIIA*) and *Harpoon missiles* (*RGM 34-84 L Block II*). However, as Rahmat and Hardy (2014) pointed out in a recent analysis, these "1970s-era warships, displacing almost 10,000 tonnes at full load, are said to be unsuitable for defending the Taiwan Strait in light of new weapons and swifter naval platforms being acquired by China."

Among the more modern operational surface vessels are six French-built *Kang Ding* (or *La Fayette*) class frigates and eight indigenously built, but U.S.-designed *Cheng Kung* (or *PFG-2*) class frigates, which were based on the outmoded U.S. *Oliver Hazard Perry* class design. The *Kang Ding* class vessels are equipped with the French *TAVITAC 2000* combat direction system, *Sea Chaparral* SAM, and French radars plus the indigenously developed *Hsiung Feng II* SSM (Wertheim 2013: 711).

To make up for the impending loss of its aging surface vessels, Taiwan's naval planners have opted to build a fleet of up to 12 small (500 t, 60 m) twin-hull wave-piercing missile corvettes armed with Taiwan's indigenous *Hsiung Feng II* and *III* missiles and a U.S. *Phalanx* CIWS, the *Tuo Chiang* class (not shown in Fig. 6.3). The first-of-class has been commissioned in late 2014, and up to 12 vessels are planned altogether. A larger fleet of small, swift, heavily armed vessels built at comparatively low cost indeed seems like a reasonable and cost-efficient solution in light of Taiwan's strategic goals, especially compared to possible further transfers of used and very large U.S. vessels (Rahmat and Hardy 2014).

Nonetheless, U.S. military technology transfers have been, and will continue to be, substantial. As Friedman already reported in 2006, Taiwan received extensive American assistance in the area of data links, a necessary precondition for effective command and control, which also facilitates integration with the U.S. fleet:

> The U.S. Navy is managing development of an integrated, all-service command and control network for Taiwan under the Po Sheng program. Lockheed Martin is system integrator. Work began in the fall of 2004, with early operational capability expected in November 2005. The core is apparently integration of the service headquarters with a national command center, which would maintain a common operational picture. Later the system is to be extended to deployed units, and it is to include a combat management capability. Po Sheng includes data links, and it may include Taiwanese purchase of Link 16 (Friedman 2006: 7).

So far, this co-operation has not led to Taiwan gaining access to *AEGIS* technology, however. In contrast to e.g. South Korea and Japan, which have likewise received massive U.S. arms transfers in the past, Taiwan has been sold less advanced technologies in packages carefully calibrated to not unduly upset Beijing.

The greatest area of concern from the naval hardware perspective is without a doubt Taiwan's degrading submarine capability. Taiwan is in danger of losing the submarine arm by the end of this decade at the latest, if its two most modern submarines, which will have passed the 30-year mark of their service life, cannot be replaced. However, according to recent reports, the situation could already be even graver. In a recent report, Phipps cites the Taiwanese defense journalist Eric Shih, who observed that the two *Hai Lung* class submarines are already "no longer serviceable and at least one of the vessels' fuselages is bent" (quoted in Phipps 2013).

Taiwan's strategic goals of offering a credible deterrence and countering a sea blockade will become much more difficult to achieve once its submarine capability degrades. This would also negatively affect ASW training for surface vessels and aircraft, and result in a reduced overall naval combat proficiency level of the entire force. In the face of constantly increasing Chinese combat proficiency, Taiwan's Navy and Air Force would thus be challenged to keep up a credible deterrence on their own. A promise made by U.S. President George W. Bush already in 2001 of supplying diesel submarines to Taiwan now seems impossible to honor, given that U.S. shipyards abandoned this technology decades ago, while the handful of other conventionally-powered submarine manufacturers worldwide has collectively declined to trade with Taiwan due to Chinese political pressure. As a result of this de-facto embargo, it is not surprising that in the fall of 2014, a plan was announced costing US$4.9 billion to begin development of an indigenous Taiwanese submarine. Depending on the approval of this budget, development of four 1,500 t submarines is expected to begin in 2016, with the first boat ready to enter service by 2025. Phipps and Hardy, however, argue that this ambitious plan faces what could in the end prove to be "insurmountable hurdles," even if the budget is approved. They further point out that "[g]iven the complexity of the platforms and the complete absence of indigenous submarine building experience, the working figure of USD4.9 billion is likely to grow exponentially" (Phipps and Hardy 2014). Nevertheless, the impending capability gap seems to be forcing Taiwan to take action now in the hope of overcoming the technological obstacles step by step, and to press forward even if success is uncertain.

On the other hand, at least according to the assessment of William Murray, submarines might not be such a decisive factor in the end for Taiwan's ability to offer a credible deterrence against invasion. As Murray remarks: "there are other, purely defensive and non-escalatory, alternatives that could more quickly offer Taiwan equal or better deterrence and at lesser cost." Such alternatives would include "small missile craft and cruise missile technology that offer much better deterrence and would be part of a larger 'porcupine strategy' that would make Taiwan 'patently useless to attack'" (quoted in Phipps and Hardy 2014). Notwithstanding this viewpoint, since ASW training for surface vessels is also dependent upon a submarine capability, the idea of willingly giving up the submarine arm would nevertheless seem to be less than ideal from Taiwan's point of view.

As long as China continues to be preoccupied with resolving bugs in its indigenously developed area defense capability, and Taiwan's submarine force remains somewhat operational, Taiwan's naval capability level will be sufficient for its strategic goals given the much smaller maritime area it needs to defend. Moreover, due to the geographical proximity of Mainland China, Chinese military preparations for an attack on Taiwan would likely be noticed by U.S. and Taiwanese surveillance and reconnaissance, making a surprise attack rather difficult. Likewise, should a first strike be survived, Mainland China's Southeastern coastal areas are vulnerable to Taiwanese counterattacks. As of 2014, however, it seems rather likely that U.S. military support would be needed for Taiwan to successfully deter a full-fledged and sustained Chinese invasion attempt.

In theory, there is an extremely wide range of possible scenarios in which a Taiwan crisis could play out (cf. Viner 1997: 184–185). Academics occasionally discuss highly implausible scenarios, such as a theoretical Chinese attack on Taiwan with nuclear, biological, or chemical weapons to quickly subdue resistance. Such a move, however, would not only physically exterminate the desired object, but would certainly also bring about a major war against another nuclear power and thus annihilate most, if not all, of the developmental success China has achieved over the past few decades. Bringing up such extreme possibilities for discussion moreover seems rather moot not least in light of China's strict No First Use (NFU) policy, already declared in 1964, and China's overall relatively rational international behavior. Nevertheless, the fact remains that while Taiwan unilaterally renounced the use of force against the Mainland after martial law was lifted in 1991, China has so far not responded in kind, and continues to deploy additional missiles aimed at Taiwan's coastal areas.

Given the resulting strategic uncertainty, Taiwan's own ability to offer a credible deterrence should not be deemed inconsequential, despite a de facto U.S. commitment to act on behalf of Taiwan's defense (Shirk 2007: 185). Since China is trying to implement an asymmetric anti-access/area denial strategy against a possible U.S. intervention on behalf of Taiwan in order to raise the cost of an engagement, the existence or lack of a realistic self-defense capability on Taiwan itself would likely have a strong impact on the willingness of U.S. leaders to come to Taiwan's aid. However, it may be argued that just as China's leaders cannot credibly relinquish the goal of reunification in the face of nationalist sentiment without losing domestic credibility (even though this nationalist sentiment was most likely created by official propaganda in the first place), the U.S. also cannot afford to remain passive in the event of a Chinese military attempt to conquer Taiwan by force. As Brzezinski wrote already in 1997, shortly after the "missile crisis":

> [...] American acquiescence in an attempt at the forcible reintegration of Taiwan, sought by the use of military power, would be so devastating to America's position in the Far East that America simply could not afford to remain militarily passive if Taiwan were unable to protect itself. In other words, America would have to intervene not for the sake of a separate Taiwan but for the sake of America's geopolitical interests in the Asia-Pacific area (Brzezinski 1997: 188–189).

This assessment still seems plausible. Given the rapid pace of rapprochement between both sides of the Taiwan Strait since the beginning of the Ma administration in 2008, and the high mutual economic benefit derived by Taiwan and China from Cross Strait investments and economic and cultural exchanges, a dramatic turn of events has fortunately become increasingly less likely over the past few years (cf. Zhang 2011). Both sides have reportedly established secret military communication channels since 2008 in order to avoid unintentional escalations of small incidents, and joint maritime search-and-rescue exercises have been performed by local maritime police forces from both sides of the Taiwan Strait (cf. Cabestan 2011: 9). In the absence of a formal peace treaty, however, domestic political turmoil on either side of the Strait could still lead to a setback, and the dangers posed by rapprochement are all too evident to a large portion of Taiwan's populace. As recently as Spring 2014, the Ma administration had to face widespread protests against its pro-Mainland policy, and the ratification of a new Cross-Strait Service Trade Agreement had to be halted after student protesters occupied Taiwan's parliament, the Legislative Yuan, for several weeks after the government denied a public clause-by-clause review of the document (Chung 2014).

Taiwan continues to face the same dilemma that other small countries located next to a hegemonic power have experienced at one point or another in history. The choice between appeasement and deterrence is not obvious, and the success of either survival strategy is uncertain. Pursuing both approaches at the same time, however, does seem to have worked, at least in some documented cases. The historical case study of Finland's struggle to maintain independence from the Soviet Union after 1917 offers a few instructive points in this regard. Originally a Grand Duchy of the Russian Empire, Finland had declared its independence after the October Revolution and struggled hard to resist forceful attempts at reintegration for several decades. After he guided Finland through a near hopeless situation in World War II, namely when the Soviet Union mounted an overwhelming surprise attack in the so-called Winter War of 1939/1940, Finland's great strategist C.G. Mannerheim concluded in his memoirs that Finland might have avoided this invasion attempt altogether, had its lack of military preparedness, which suggested easy victory to the Soviet leaders, not encouraged it (Mannerheim 1953: 516). Moreover, after the attack began, Europe's major powers deemed defending Finland a lost cause and thus declined to offer military support. By the end of World War II, when Finland's forces had managed to prevail and thus secure Finland's political independence, democratic polity, and capitalist economy,[5] Mannerheim cautioned that a small nation's vital field of national defense must never be jeopardized by the bickering of rival political parties. Leaders in power, he writes, should not forget that an "absolute prerequisite for an independent foreign policy is the existence of a strong defence force," especially in the case of small countries,

[5] This unlikely outcome should be measured against the fact that Finland is the only country directly bordering on Russia (with a land border of more than 1,300 km) that never became communist, part of the Warsaw Pact, or part of the Soviet Union. When comparing this with the fate of Poland, Latvia, Lithuania, or Estonia, or with the recent course of events in Ukraine, which brought back memories of Russia's earlier aggression against Georgia in 2008, the significance of this military achievement becomes apparent. Taiwan's military situation is much less dire compared with pre-War Finland in terms of military capability and alliance status, because it falls under de facto U.S. guardianship. Finland, on the other hand, had to prevail alone during the entire Winter War and was initially severely outgunned before accepting German military support during the "Continuation War" (1941–1944).

and must above all realize the dangers of relying on "promises and illusions" (Mannerheim 1953: 515). These remarks echo a sentiment recorded centuries earlier in a military treatise written by the Chinese strategist Liu Ji (劉基, 1311–1375), the main military advisor to the founder of the Ming dynasty:

> Sages are very careful not to forget about danger when secure, not to forget about chaos in times of order. Even when there is peace in the land, it will not do to abandon the military altogether. If you lack sufficient foresight, you will be defenseless (Liu Ji, *Baizhan qilüe* 百戰奇略, transl. by Cleary 2005: 107).

Following what Pugh calls the 'small nation's deterrence strategy' therefore probably remains essential for maintaining the status quo in the Taiwan question. Here, Taiwan's aim would be "to threaten to exact such a price for victory that, even if achieved, it is not worth its cost to the aggressor" (Pugh 1986: 21). According to Yang (2011), this is precisely the aim of Taiwan's military.

Cautiously managing the complex triangle of U.S.-China-Taiwan relations, and being realistic about present and future threats and opportunities arising from continued rapprochement with the Mainland, are preconditions for the success of such a 'small nation's strategy'. Besides military deterrence and political rapprochement, Taiwan has the option to use its considerable soft power to undermine perceptions of hostility among the Mainland Chinese populace. By enhancing mutual understanding among Chinese and Taiwanese elites and furthering cross-strait cultural, economic, and touristic exchanges as a confidence-building measure, with more Mainland Chinese visitors gaining first-hand experience of Taiwan's social and political reality, Taiwan could gradually counteract the effects of official PRC propaganda, and even indirectly contribute to positive political change on the Mainland (Yang 2011). The transformative effect of exposure to an alternate lifestyle on the mindset of people living in divided nations was clear in the case of a divided Germany. Years of exposure to Western media, and especially the visibility of West Berlin, markedly shaped the expectations of East German citizens during the momentous year of 1989 (Frank 2014: 353–355). In fact, Taiwan's successful post-war economic modernization and political transformation today seem to be perceived as attractive models by reform-minded quarters within the Chinese leadership, and related developments in Taiwan have been intensely studied by numerous Mainland Chinese experts (cf. Gilley 2008: 239–240). In addition, Mainland Chinese students enrolled at Taiwanese universities have acknowledged in various interviews the thoroughly transformative effect that their exposure to Taiwan's liberal society has had on their way of thinking (cf. Sui 2012). Thus, assuming that Taiwan is able to prolong the status quo of de-facto independence long enough, there is at least a chance that its successful transformation experience could begin to serve as a source of inspiration for future reformers of the Mainland Chinese polity.

6.6 Japan: Pacifist Tradition vs. New Assertiveness

The U.S. alliance with Japan is often described as being of prime importance for upholding the U.S.-led hegemonic order which is currently in place in East Asia. Notwithstanding Japan's huge economic importance, it has been dubbed in military terms a "geopolitical extension of American power":

> Since the end of the Chinese civil war in 1949, America's policy in the Far East has been based on Japan. At first only the site for the occupying American military, Japan has since become the basis for America's political-military presence in the Asia-Pacific region and America's centrally important global ally, yet also a security protectorate (Brzezinski 1997: 173).

Still hindered by its pacifist post-war constitution from fully participating in missions under the umbrella of U.S.-led operations, Japan's naval forces are not even technically a navy, but a so-called "Maritime Self-Defence Force" (and thus abbreviated JMSDF). Due to stipulations in Japan's security treaty with the U.S., the JMSDF moreover cannot as yet be fully counted on to participate in a potential U.S.-led military alliance against China in the event of a conflict:

> The U.S.-Japan Security Treaty stipulates American obligations to defend Japan, but it does not provide (even if only formally) for the use of the Japanese military in the defense of America. The treaty in effect codifies a protective relationship (Brzezinski 1997: 175).

However, recent indications of an impending policy shift in this regard suggest that this sticking point could perhaps become a mere formality that is easily remedied in the event of a crisis. Heightened tensions during the past few years between China and Japan over the mutually claimed Senkaku/Diaoyu Islands have helped to coalesce those political forces in Japan which have argued for years in favor of a so-called "normalization" of Japan's military affairs (cf. Nabers 2007: 4). The latent Senkaku sovereignty issue has become more pressing since an incident in September 2010, when a Chinese fishing trawler was detained for intentionally ramming a Japanese Coast Guard vessel near the contested islands (Swaine 2013: 1). Following this episode, multiple provocations by both sides exacerbated tensions over these small and uninhabited islands—even before the Japanese government's controversial decision in September 2012 to buy three of the five main islands from the Japanese Kurihara family set off a storm of Chinese nationalist outrage. A list of related incidents compiled by Swaine shows that between March and November 2012 alone, Chinese ships transgressed into the contested waters 47 times, while the Japanese Air Self-Defense Force had to react 160 times to Chinese aircraft in the vicinity (Swaine 2013: 13–17).

Japan has a strong incentive to balance China's growing naval influence largely because of this unresolved maritime territorial conflict. U.S. military, political, and economic influence on Japan is overwhelming, not least because of the large number of U.S. forces continuously stationed on Japanese territory. Unsurprisingly, the security relationship extends to the realm of defense technological co-operation. Like South Korea and Australia, Japan has received *AEGIS* technology for its large surface combatants, making Japanese naval forces highly interoperable with U.S., South Korean, Australian, and NATO forces. Given its size, vessel type and age structure, capability, and mission experience, the JMSDF must be considered as of 2014 a powerful Rank 4 "Regional power-projection" navy.

As indicated in Fig. 6.4, Japan has an impressive fleet of relatively large and capable surface combatants, including several *AEGIS*-equipped destroyers (marked in green). Japan, like South Korea, also operates several large helicopter carriers

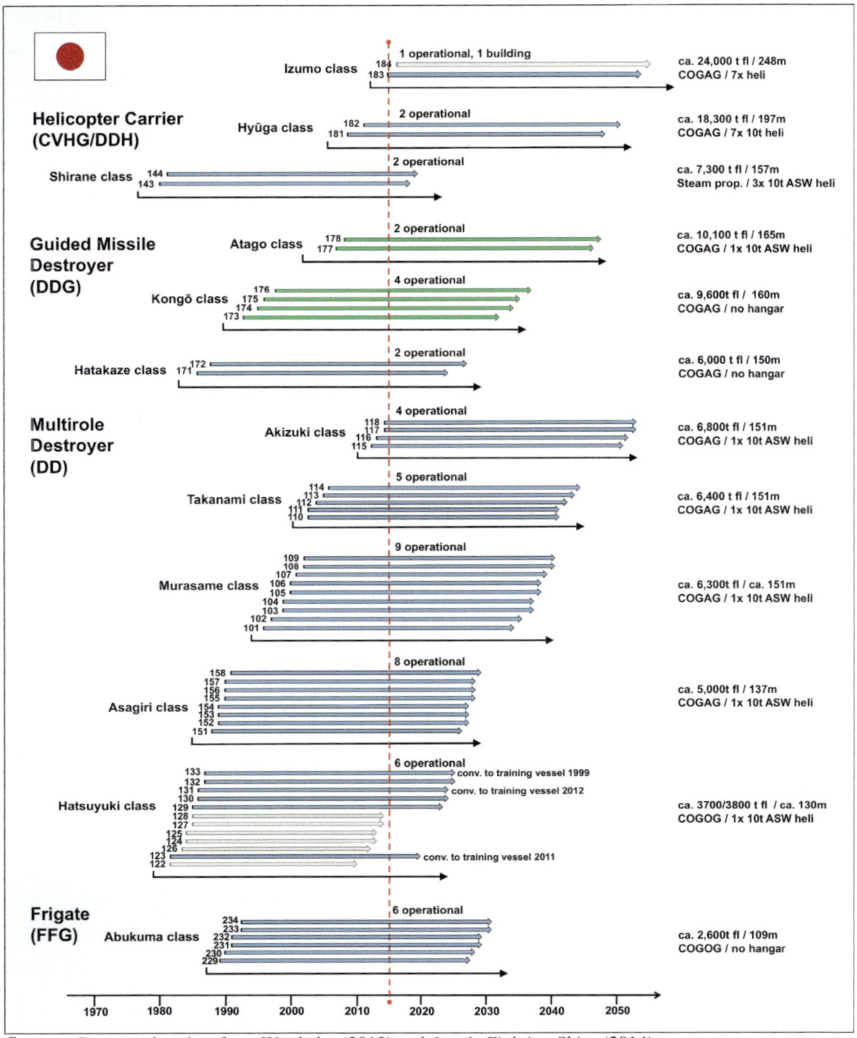

Source : Drawn using data from Wertheim (2013) and *Jane's Fighting Ships* (2014)

Fig. 6.4 Type and age structure of the JMSDF principal naval surface vessels, 2015

(officially dubbed "helicopter-carrying destroyers"), which could theoretically be converted to accommodate STOVL aircraft as well—though this does not seem to be part of the official plan (Hardy 2013). In conjunction with Japan's large and well-trained coast guard, the naval forces are in all probability sufficiently equipped to safeguard Japan's maritime interests.

The Japanese submarine force structure has already been depicted in Fig. 3.9 in Sect. 3.3. Compared with Taiwan, which has difficulty replacing its aging boats, Japan is in the reverse position insofar as it routinely retires its submarines much earlier than is customary in order to create stable build cycles for the shipbuilding industry. Since the maximum number of submarines in service was fixed at 20 boats (up from 16 until 2010), and, moreover, since arms cannot currently be exported due to Japan's constitutional restrictions, Japan routinely decommissions and scraps fully functioning submarines. This curious industrial policy has one potential advantage for Japan's naval planners, however: Without increasing building activities, the force can be quickly enlarged by simply keeping older boats in service longer. Alternatively, if Japan were to decide to allow arms exports, used Japanese submarines could become an attractive commodity on the export market. As of 2014, it appears that Australia is in advanced talks with Japan for the purchase of up to 12 new large *Soryo* class submarines, indicating both a possible shift in Japan's traditional stance on arms transfers and a move towards stronger Japanese regional integration (cf. Richards 2014; Taylor 2014).

Due to Japan's alliance with the U.S., Japan has been perceived as an extension of U.S. military power in China for quite some time. As a content analysis of Chinese journal articles from the leading military, foreign policy, and economic think tanks published during 1986–2006 indicated, Chinese perceptions of Japan as a threat have dramatically increased since 1996, the year of the Taiwan missile crisis (cf. Sasaki 2010: 574–580). Likewise have Japanese threat perceptions regarding China grown during the past few years. Japan has routinely criticized China for its military intransparency. U.S. and Japanese accounts of China's maritime behavior in the Asia Pacific region tend to contain long lists of supposedly aggressive or questionable incidents involving Chinese naval vessels. An example of such a list can be seen in Fig. 6.5.

Cole (2010) similarly mentions that, since 2001, China has interfered at least five times with U.S. airborne or seaborne intelligence-gathering efforts, once causing a (Chinese) casualty. He also remarks that Chinese vessels have frequently operated in waters claimed by Japan (Cole 2010: xviii). Takahashi notes how "Chinese flotillas have [...] repeatedly sailed through the Miyako Strait near Okinawa and voyaged to waters near Okinotorishima in the western Pacific, thus intruding into Japan's exclusive economic zone" (Takahashi 2011). On several occasions, Chinese submarines have sailed close to the U.S. aircraft carrier *Kitty Hawk*. Miller reports on one such incident:

> A U.S. Navy S-3 Viking patrol aircraft detected the submarine when it was some 450 nm from the carrier and heading in the direction of the carrier at periscope depth. The submarine came within 20nm of *Kitty Hawk* before turning away and it seems clear that its commander meant to be seen so as to make it clear that there was a new challenge to U.S. naval supremacy in the area (Miller 2002: 343, italics in the original).

Date	Incident
10/1994	A *Han* class SSN tracks the U.S. carrier *Kitty Hawk* for several days in the Yellow Sea.
1995	China fortifies Mischief Reef in the South China Sea.
1995/1996	"Taiwan Missile Crisis".
11/2004	Submerged *Han* class SSN intrudes into Japanese territorial waters, "breaching international law" (Japan Ministry of Defense 2011: 75, 82).
09/2005	Five vessels incl. one *Sovremenny* class destroyer sail near Kashi gas field (Tianwaitian in Chinese) in the East China Sea, with some circling the gas field.
10/2006	A *Song* class SSK surfaces near U.S. carrier *Kitty Hawk* in international waters near Okinawa; "the foreign submarine's approach to a U.S. aircraft carrier is a militarily noteworthy incident" (Japan Ministry of Defense 2011: 82).
11/2007	Chinese sub broaches near U.S. carrier *Kitty Hawk*.
1/2007	Successful anti-satellite destruction shot conducted.
11/2007	China refuses port entry to U.S. naval vessels including CVL *Kitty Hawk* on the very day of their planned arrival.
10/2008	Four vessels, incl. a *Sovremenny* class destroyer, pass through the Tsugaru Strait and sail South to the Pacific Ocean to circle Japan (first identified passage by Chinese surface combatants through the Tsugaru Strait).
11/2008	Four vessels, incl. a *Type 051C* class destroyer, pass between Okinawa Island and Miyako Island and head to the Pacific Ocean.
12/2008	Two maritime research ships conduct operations "not permitted under international law such as hovering and cruising within the territorial waters of Japan" near the Senkaku/Diaoyu Islands (Japan Ministry of Defense 2011: 82).
03/2009	Chinese ships including a naval vessel, a maritime research ship, a patrol ship and trawlers approach a U.S. Navy acoustic research ship operating in the South China Sea to obstruct its operations. Other incidents also in the same month.
6/2009	Five vessels incl. a *Type 051C* class destroyer pass between Okinawa Island and Miyako Island and head to the waters northeast of Okinotori Island before ostensibly engaging in drills.
3/2010	Six vessels incl. a *Type 051C* class destroyer pass between Okinawa and Miyako Islands and head to the Pacific Ocean. These vessels then reportedly advanced to the South China Sea. In the same month, exercises were reportedly conducted by six vessels to the South China Sea over 3 weeks.
04/2010	Ten naval vessels, incl. *Kilo* class submarines and *Sovremenny* class destroyers, pass between Okinawa and Miyako Islands and head to the waters West of Okinotori Island, before ostensibly engaging in exercises.

Fig. 6.5 List of Chinese aggressive maritime behavior incidents, 1994–2014

Date	Incident
07/2010	Two vessels including a *Type 051C* class destroyer pass between the main island of Okinawa and Miyako Island and advance to the Pacific Ocean. The same month, a "large-scale, multi-branch joint live-ammunition exercise was conducted and included surface vessels and naval air forces" (Japan Ministry of Defense 2011: 82).
9/2010	Major dispute between China and Japan over Chinese fishing boat that crashed into two Japanese Coast Guard patrol ships near the Senkaku/Diaoyu Islands, causing a diplomatic row after the Chinese Captain was first arrested and then released.
10/2010	China asks a JMSDF training squadron for a postponement of planned port call in Qingdao 1 day before planned arrival.
3–4/2011	Multiple incidents of Chinese helicopters that appeared to belong to the State Oceanic Administration, flying close to Japanese destroyers engaged in vigilance monitoring in the East China Sea.
03/2011	A Philippine research vessel is ordered to withdraw by Chinese government ship in the South China Sea.
05/2011	China constructs new buildings such as station poles on islands in the South China Sea.
06/2011	Eleven vessels including *Sovremenny* class destroyers and a *Jiangkai II* class frigate pass between Okinawa and Miyako Islands and advance to the Pacific Ocean. Besides target practice, they also carried out unmanned aircraft and shipborne helicopter flight training as well as replenishments at sea.
6/2011	A PLAAF aircraft crosses the Taiwan Strait center line on June 29, 2011 when shadowing a high-altitude U.S. *U-2S* reconnaissance aircraft and is chased away by two Taiwanese *F-16s*.
1/2013	A Chinese naval vessel aims its Fire Control Radar at a Japanese destroyer. Later, Chinese officials publicly deny the fact (Japan Ministry of Defense 2013: 32).
5–7/2014	*Haiyang Shiyou 981* standoff: Drilling performed by Chinese oil rig *Haiyang Shiyou-981* in waters 17 nm South of the Spratly Islands, within Vietnam's 200 nm EEZ and continental shelf, and on Vietnam's side of the median line of overlapping EEZs, gives rise to violent public protests in Vietnam.
11/2014	Satellite images show that China has built an artificial island in the South China Sea by fortifying Fiery Cross Reef. It is the fourth artificial island built by China after fortifying Johnson South Reef, Cuateron Reef and Gaven Reef.

Source: Compiled from (Cole 2010: xviii), Japan Ministry of Defense (2011: 75, 82–83) and Japan Ministry of Defense (2013)

Fig. 6.5 (continued)

China's recent decision to establish an air defense identification zone (ADIZ) over its coastal areas has likewise drawn heavy international criticism. The announcement in 2013 of this ADIZ, which includes the airspace over the contested Senkaku/Diaoyutai Islands, was portrayed as a worrisome act by most Western and Japanese mainstream media, and the U.S. made a point of having several transport airplanes cross this zone unannounced, without apparently suffering any consequences (Keck 2013a, b). All in all, such incidents—in the words of a Japanese MoD publication—"incite concern over China's decision-making and behavior concerning its military" (Japan Ministry of Defense 2011: 75).

Regarding the conflict with Japan in the East China Sea, it should be noted however that Japan's position seems fairly secure. The islands are currently under Japanese jurisdiction. Japan also has a very capable and well-trained navy which is likely able to counter any kind of Chinese aggression, both now and in the future. Moreover, Japan is a U.S. military ally and therefore under the protection of a much more powerful naval force. Thus, if Japan continues to secure its strategic position regarding the Senkakus, there is little danger that China will be able to snatch these territories from Japan by force.

In this light, it seems appropriate to ask whether any of the above listed transgressions by Chinese vessels do not in fact constitute normal practice in most navies. For instance, the U.S. Navy commonly sails through island straits en route to deeper waters, even if this means intruding upon another country's EEZ or even its territorial waters. Such behavior, furthermore, is also in accord with international law. A country's EEZ is not, in any case, usually regarded as territorial waters, and Japan's interpretation of the issue of other navies passing through its island straits is much more narrow than international custom proscribes (cf. Dutton 2007: 165).

Regarding the much-discussed intrusion in 2004 of a *Han* class nuclear submarine into Japanese territorial waters, which caused alarm in U.S. political quarters, nuclear analyst Kristensen remarks that the Pentagon at the time "did not mention that the intrusion was one of only three patrols conducted by the entire Chinese submarine force in 2004, and that no patrols at all were conducted in 2005" (Kristensen 2007). As for the establishment of the Chinese ADIZ, it is noteworthy that several countries within the region, among them U.S. allies Japan and South Korea, have long established comparable flight control zones over their claimed territories.

Similar observations about a possible negative bias in reports against China's maritime behavior have been made quite frequently by PLA functionaries and Chinese political leaders (cf. Liu and Ren 2014: 264). This therefore gives rise to the question: Is the world judging Chinese maritime behavior unfairly? And relatedly: What kind of increased maritime activity on China's part would actually be welcomed by other powers, both in the region and abroad? Observers are often at odds in regard to the latter question, which has resulted in mixed signals being sent to China. On the one hand, there is the demand for "more responsibility" on China's part for the region's maritime security; on the other hand, other voices simultaneously deplore virtually any instances whatever of increased Chinese maritime activity (cf. Li 2010: 292). An example of the incomprehension this can cause in Beijing is found in an *Economist* article from 2012, which cites Major General Yao Yunzhu (姚云竹, *1954) as follows: "We are criticised if we do more and criticised if we do less. The West should decide what it wants" (quoted in 'China's Military Rise' 2012). Her observation, indeed, does not seem to be entirely unjustified. Furthermore, it speaks to the bewilderment many Western observers continue to exhibit towards China's naval development. Another statement supposedly made by PLA Chief of General Staff, General Chen Bingde (陈炳德, *1941) when he meet with the German Minister of Defense in 2010 is even more blunt: "If one opposes the American government, he is a terrorist; if one opposes the People's Republic of China, he wins a Nobel Peace Prize" (quoted in Liu and Ren 2014: 264). Whether or not such evaluations are justified, they reveal a deep-seated discontent among China's military elite with disingenuous criticisms of China that should be taken note of.

6.7 The South China Sea Issue and China

While some of the criticisms of Chinese naval activities described in the preceding section may seem unwarranted, the fact remains that other types of maritime behavior regularly displayed by Chinese vessels are much more problematic and thus give rise to legitimate concerns regarding China's possibly hidden intentions. This is especially true for China's activities in the South China Sea. Here, China is involved in a drawn-out conflict with several regional neighbors over islands, rocks, reefs, and related claims of adjacent maritime zones. Regarding the land in the South China Sea, China's stance is clear: It claims all the land features of the South China Sea in their entirety as Chinese sovereign territory. They were first depicted as such on Chinese maps dating back to the 1930s, encircled by a so-called "nine-dashed" or "U-shaped" line. Since these maps originated in the Republic of China before the founding of the People's Republic, the Republic of China on Taiwan today makes essentially identical historical land claims in the South China Sea as the PRC. These claims continue to conflict with claims made by other powers regarding some or all the islands and reefs, especially those that lie within 200 nm of other countries' coastlines. In particular, Vietnam, China, and Taiwan claim the entire Paracel Islands, which are currently occupied by China; China, Taiwan and the Philippines claim Scarborough Reef; and China, Taiwan, Vietnam, the Philippines, Malaysia, and Brunei all make claims for all or at least several of the Spratly Islands, the majority of which are currently occupied by Vietnam. However, various rocks, reefs, and islands in the South China Sea have been occupied by all of these countries at one time or another (U.S. Department of State 2014: 13).[6]

In addition, China made it clear in several *notes verbales* submitted to the UN in 2009 that it considers these islands to be entitled to their own EEZ, territorial sea, and continental shelf, thus making an additional maritime claim (Swaine and Fravel 2011: 3). This is highly problematic from a legal viewpoint, for as the provisions

[6] A new book by Hayton (2014) gives a very detailed overview of the historical development and current status of all these various claims, with multiple maps showing the exact nature of the claims made by all sides. The historically changing positioning of China's "nine-dashed line" and the resulting legal implications are discussed at length by the above-cited U.S. report (U.S. Department of State 2014: 3–7).

made in UNCLOS (Art. 121) state, territorial sea and EEZ can only be claimed for islands able to sustain human life. This would exclude artificial structures or mere rocks, and runs contrary to China's apparent position. The provisions further explain that an EEZ can maximally extend as far as the median line between two countries' coastlines in the case of overlapping 200 nm zones, is subject to mutual negotiations, and cannot be unilaterally declared (Art. 15). Maritime zones can moreover only be claimed if sovereignty over the land territory from which they project is uncontested. This is clearly not the case in this instance:

> Because sovereignty over South China Sea islands is disputed, the maritime zones associated with these islands would also be disputed. In addition, even if China possessed sovereignty of the islands, any maritime zones generated by those islands in accordance with Article 121 [of UNCLOS] would be subject to maritime boundary delimitation with neighboring States (U.S. Department of State 2014: 23).

A review of all Chinese maps and documents put forth on the matter by China's government seems to suggest that China has failed to clarify the exact nature of its maritime claims "in a manner consistent with international law" (U.S. Department of State 2014: 23). Moreover, actual Chinese behavior indicates that China makes extensive maritime claims emanating from the contested islands which in some instances conflict with legal maritime claims that emanate directly from other countries' uncontested coastlines. Notwithstanding the contested nature of some of China's maritime and land claims in the South China Sea, it has repeatedly used force, and sometimes caused high numbers of human casualties, to uphold what it sees as its sovereign territorial right. Exploiting the fact that Vietnam found itself in an entrenched conflict with the U.S., China forcefully occupied the Paracel Islands in 1974. In 1988, a naval conflict between China and Vietnam in the Spratly area claimed more than 70 Vietnamese lives (Tœnnesson 2009: 353). In light of this history, China's more recent stance regarding other states' legitimate rights in the South China Sea is sometimes hard to follow. As Yahuda observes:

> Nearly all the rocks, reefs and cays which make up the Spratlys cannot sustain human life and would not qualify for jurisdictional claims to EEZ beyond the sovereign territorial water of 12 nm. Nevertheless the Chinese have tried to stop the exploration for oil in areas claimed by Vietnam and the Philippines that are within 200 miles of their coasts and far from any inhabitable islands claimed by China. [...] Moreover while arrogating itself the right to conduct oil exploration and drilling within its own EEZ in the S[outh]C[hina]S[ea], China has demanded that other claimants must first seek Chinese permission before doing so within their own EEZs (Yahuda 2013: 451).

Ever since it first occupied the Paracels, China has followed a continuous strategy of fortifying reefs and islands in the South China Sea, building artificial structures including military installations, forcefully asserting Chinese interests, and insisting on bilateral rather than multilateral negotiations for solving issues—even though China signed and ratified UNCLOS. Fifteen years ago, Chen Hurng-yu already noted that

> [...] the PRC is in control of the entire Paracels and seven Spratly reefs – Zhubi, Gaven [...], Johnson, Dongmen [...], Fiery Cross, Cuarteron, and Mischief reefs. On each reef Beijing has built structures for military use as well as radar installations to monitor the vessels and aircraft navigating in the area (Chen 2000: 126).[7]

Although Chinese maritime activities in the South China Sea have received more intensive press coverage in recent years consistent with growing fears of China's naval build-up, it should be noted that the basic Chinese strategy described above has clearly been in place for some time. In fact, ever since the forceful seizure of the Paracel Islands from Vietnam in 1974, China's behavior has remained fairly predictable (Chen 2000: 126). Nevertheless, as in the conflict with Japan over the Senkakus, violent incidents seem to have occurred with increasing frequency over the past few years. And, in the eyes of many regional and international observers, reports of heavily armed Chinese naval vessels intruding upon the Philippine or Vietnamese EEZs to harass unarmed fishing vessels, or to enforce Chinese oil drilling within Vietnam's EEZ and continental shelf, indicate nothing short of bullying. The irony here is unmistakable given the fact that Chinese leaders often deplore China's own history of being "bullied" by superior powers (cf. Lampton 2014: 160).

Examples of questionable Chinese practices have been regularly reported in recent years. Amurao (2012), for instance, writes that on July 11, 2012, the *Jianghu V* class frigate *Dongguan*

> [...] ran aground on Half Moon Shoal, a contested maritime territory a mere 60 nautical miles from the Philippine island of Palawan. The incident was not initially publicized by either side but its later disclosure has raised new questions about China's naval intentions and capabilities in the area less [...] than a month after both sides backed away from a two-month stand-off over another contested shoal in the same island chain (Amurao 2012).

Another related incident occurred on February 25, 2011, when three unarmed Philippine fishing boats were chased off Jackson Atoll by the frigate *Dongguan*, which even fired live ammunition in the vessels' direction. According to Amurao (2012), armed Chinese fishery boats regularly fish within the Philippine EEZ, and are moreover often escorted by military or paramilitary vessels. Philippine fishery vessels, by contrast, are harrassed by armed Chinese ships even within their own EEZ.

European military sources deployed to the region have informed me over the past two years of at least one unpublicized incident in which civilian Malaysian ships were chased off by Chinese naval vessels when trying to install marine equipment within Malaysian claimed maritime territory. Furthermore, they indicated that Brunei, traditionally the least active among the claimants in the South China Sea issue, has recently decided to strengthen its military ties with the U.S. at the expense of its longstanding security partnership with Britain. These

[7] A list of military and civilian infrastructures built by China on each of the Paracels and the other reefs, including air strips, hangars, helicopter platforms, radar stations and oceanographic research installations until 2000 is provided by Chen (2000: 126–128).

developments are all the more noteworthy because Malaysia is considered to be the most China-friendly among the countries involved in the Spratly conflict, and Brunei previously seemed more or less unconcerned with China's activities in the region.[8]

China's apparent strategy of gradual encroachment upon the conflicted island territories by continuously erecting increasingly more permanent structures leads many regional observers to question the country's intent. In October 2014, it became clear that China has proceeded to build more artificial islands, and enhance and fortify existing structures in the Spratly area (Ghosh 2014). This same year, a Chinese oil rig placed in contested maritime space led to widespread anti-Chinese riots in Vietnam (Godement 2014: 4). Such incidents call into question optimistic assessments about a supposedly "changed" and "restrained" Chinese South China Sea strategy, such as the one Li Mingjiang saw evolving in 2010 (Li 2010: 300–303). In contrast to his opinion, I tend to concur with Wertheim's assessment of the matter, as quoted by Moss:

> 'The [Chinese] strategy of developing a navy is perfectly legitimate,' said Wertheim. 'But their behaviour in the South China Sea and in the disputed waters that China claims as sovereign territory is the worrisome part' (quoted in Moss 2011).

China seems to deliberately exploit the fact that none of the navies engaged in the Spratly and Paracel territorial conflicts is presently able to match China's naval forces in either qualitative or quantitative terms, and that the U.S. has so far taken a neutral position in the conflict. China, consequently, does indeed seem to be bullying its weaker neighbors, and following a strategy of gradual encroachment. Moreover, Godement has reported from discussions with interlocutors from various Asian countries in Tokyo in June 2014 that China—judging from past incidents—apparently

> calibrates its challenges according to the perceived strength or alliance status of the other party. As one participant said: 'With Japan in the East China Sea they send unarmed fishermen; with the Philippines they send patrol boats; with Vietnam they send the PLA Navy' (Godement 2014: 2).

This, the same analysis concludes, points to a strategy of calculated risk-taking by Beijing.

China's naval modernization has thus already given it increasing leverage over the smaller naval forces in the South China Sea region (Yahuda 2012: 42). As discussed in Sect. 5.2, all other South China Sea stakeholders—except for Malaysia, which is on par with China in this regard—have a relatively unfavorable patrol vessel ratio given the size of their maritime claims. In the face of increasing Chinese oil exploration activities in Vietnam's EEZ, and given China's apparent strategy of deploying nuclear submarines (and possibly an aircraft carrier) at Hainan very close to Vietnam's coastline, it is easy to explain Vietnam's current attempts at forging

---

[8] In October 2013, China and Malaysia announced a "comprehensive strategic partnership" that includes a planned defence industrial cooperation and joint military exercises (cf. Grevatt 2013a, b).

stronger bilateral security partnerships with other powers (Russia, India, and the U.S.). These include significant arms procurement plans comprising, among other things, six Russian *Kilo* class submarines. Other regional navies have adopted similar arms acquisition and alliance strengthening strategies. While the Philippines have publicly announced a procurement strategy called "Sail Plan 2020," encompassing a submarine, landing platform dock ships, maritime patrol aircraft, a strategic sea-lift vessel, two offshore patrol vessels (OPV) and several helicopters, other regional naval forces have already either followed through with or begun developing a submarine acquisition program, including Malaysia, Singapore, Indonesia, and Thailand (Grevatt 2011a).

The gap in numerical strength and overall capability between the smaller regional forces and China is likely to widen, however, since China is currently in the process of strongly enhancing the size of its maritime surveillance fleet (Cole 2011b). The organizational streamlining of several maritime services under the State Oceanic Administration's 'China Marine Surveillance' into the newly established China Coast Guard in March 2013 is likely to result in an even more concentrated management of maritime policing. In addition, *Jane's Defence Weekly* reported in 2014 that China was in the process of procuring two exceptionally large (displacement ca. 10,000 t at full load) new cutters for this service (Hardy and Weening 2014).

China's nascent Chinese aircraft carrier capability could tip the naval power balance in the South China Sea even further in China's favor, assuming these ships are going to be patrolling this region and can be protected against asymmetric threats. The incentive will thus continue to remain strong for other Southeast Asian navies to respond by building or upgrading asymmetric capabilities as a strategic deterrent, and by trying to strengthen alliances with superior powers in the region, i.e. India and the U.S. The danger of an arms race tied to perceptions of a more provocative China is thus not to be underestimated.

Given the consistency of China's behavior in the South China Sea over many years and its fairly constant strategy of fortifying an increasing numbers of small reefs and islands, Chen is probably correct when he calls the Chinese South China Sea strategy "outwardly soft but inwardly hard." China's approach is unlikely to be changed due to pressures arising from Chinese popular expectations, despite its fundamental conflict with official Chinese notions of a "peaceful development." This inherent tension in the Chinese strategy is apparently not lost on some Chinese observers, as Hayton reports:

> Chinese officials privately recognise the legal absurdity of maintaining a claim to places like the James Shoal, which lie under water and within the EEZ of another country. Chinese diplomats are also reported to have given assurances to Indonesia that Beijing has no claim on the waters around the Natunas, even though they are partly enclosed within the 'U-shaped line'. But those same officials say they cannot formally adjust the 'U-shaped line' for political reasons – the domestic criticism would be too great – so they must continue to maintain their claim. Some of this is the calculated result of deliberate political propaganda (Hayton 2014: 264–265).

In light of the fact that Mainland Chinese politicians consistently assert that they are hostage to public opinion in this matter (even though public opinion was deliberately shaped by state propaganda over years and decades), Hayton points out how a possible future Taiwanese re-evaluation of historical claims could perhaps gradually reverse public opinion even on the Chinese Mainland. After all, these claims originated in the 1930s with the KMT-led Republic of China, and were thus inherited by both China and Taiwan. As Hayton writes:

> There are already a number of 'dissident' academics rethinking aspects of twentieth-century history. [. . .] An open and thorough examination of the haphazard process through which the [U-shaped] line came to be drawn might convince opinion-formers to re-examine some of the nationalist myths they have long declared to be gospel truth. [. . .] The key to a peaceful future could lie in an honest and critical examination of the past (Hayton 2014: 265).

However, despite advocating a peaceful approach to solving the conflicts, official Taiwanese statements distributed since 2012 to scholars around the world in the form of letters, which the present author also received, seem so far to uphold the historical claims in the East and South China Seas. It could therefore take some time before a historical re-evaluation process on Taiwan actually materializes. In the meantime, it seems that the smaller naval powers in the South China Sea area are the most likely to be negatively affected by China's naval rise, mainly because of the sheer size of China's increasingly capable patrol fleet. The effective lack of a regional security architecture in the region, which could help contain conflicts, is another factor. The ongoing construction and fortification activities suggest that China is set on quietly enlarging its presence in the region, rather than negotiating bilateral or multilateral solutions. Reversing this trend could prove very difficult.

6.8 The U.S. and the Future of "Pax Americana" in the Asia Pacific

As a result of specific historical conditions brought about by World War II and its aftermath, the U.S. has become "simultaneously the first and the only truly global power" (Brzezinski 1997: 10). After the demise of the Soviet Union in 1992, a power vacuum was created that led to the rise of the unipolar world order which still prevails today. U.S. aircraft carriers and other major military assets have effectively become symbols of the U.S. global military leadership role, which further entails upholding external security and providing nuclear deterrence for its treaty allies. Some countries have all but "outsourced" their external security concerns to the U.S. Following its involvement in the Korean War and China's civil war after World War II, the U.S. became a factual "resident power" and has established a presence in virtually every major world region. The U.S. shapes and influences regional governance nearly everywhere to at least some degree (cf. Legro 2011: 177–183).

Judging from various official pronouncements and publications, the U.S. can generally be said to have seven enduring strategic interests: protecting U.S. allies and partners from state adversaries; promoting U.S. influence in key regions;

dissuading military competition and arms races; protecting American citizens from terrorist attacks; restricting the flow of illegal trade and the proliferation of dangerous materials; ensuring the flow of commerce and key resources; and, responding to humanitarian emergencies and regional conflicts (Davis et al. 2012: 7). As a global military power and one of the world's leading trade nations, the U.S. has a fundamental interest in "[k]eeping the global commons—the air, sea, and space domains—available for use by all yet owned by none" (Davis et al. 2012: 35). The U.S. is also the only power that can globally enforce this ideal to at least some degree. It can thus provide a service of definite value, not only to its allies, but to all interested parties, including those nations heavily dependent on the maritime SLOCs for their foreign trade.

Due to its firmly entrenched network of military bases along the "first island chain," and the fact that more than 57,000 U.S. military personnel were officially deployed to 23 Asia-Pacific countries—including the Philippines, Singapore, and Australia in 2013, with the vast majority stationed in Japan—the U.S. military is a potential factor in all the region's various latent conflicts (Park 2014). The U.S. bases on Hawaii, Guam, and Diego Garcia present a formidable forward presence which is supported and secured by major U.S. strategic naval assets in the region, including six aircraft carriers and 60 % of the U.S. nuclear-powered submarine force (cf. Sect. 2.3.2). In recent years, increased anxiety among China's strategic rivals has enabled the U.S. to forge even more security partnerships than before, including one with its former enemy, Vietnam (defense agreement in 2011 and civil nuclear power transfer agreement in 2013). The heightened "China Threat" perceptions have also helped the U.S. military to improve traditional defense ties with South Korea, Japan, Australia, Singapore, and the Philippines.

When comparing military postures across countries, it has to be kept in mind that the U.S. force deployment pattern must be seen as highly atypical from a historical-comparative perspective:

> [...] for most militaries, employment of their forces is not generally assumed to entail deployment abroad. Most militaries, including China's, focus primarily on conflicts that could occur on their immediate periphery. [...] The United States, in contrast, has not conducted an MCO [major combat operation] in its own hemisphere, much less on its immediate periphery, for more than 100 years (Johnson et al. 2009: 57).

The degree of homeland vulnerability to foreign attacks in the event of a confrontation is thus markedly smaller for the U.S. than for any of its potential adversaries around the world. This alone gives the U.S. a definite strategic advantage in terms of sustainable logistics under wartime conditions.

The naval power shift in China poses no direct threat to U.S. homeland security, only to the American forward bases in the region—a fact that distinguishes the "global power" of the U.S. from that of any regional power. The rise of the Chinese navy nevertheless affects U.S. strategic planning, because, given greater anti-access/area denial capabilities developed by the PLA, movement within the "first island chain" will become riskier. This will not only influence risk evaluations for U.S. missions in those waters, but it also might lead to changes over time in deployment patterns, and even require changes in standard vessel design philosophies, e.g. by enhancing survivability and systems redundancy, as Yoshihara and Holmes predict on a rather sober note (2010: 99–100). However, as Crisp insightfully points out in an essay discussing another analysis by Holmes and Yoshihara (2008: 122), the notion that China's eventual ability to erect a "contested zone" within the First Island Chain could actually threaten U.S. military supremacy in Asia to a marked degree seems not only overblown, but inherently flawed:

> [A]re the authors really implying that *uncontested* power projection in the East China and Yellow Seas by the US Navy, extending to China's territorial waters and coastline, is essential to the credibility of its strategic pre-eminence in the region overall or even beyond? And are they implying that the US has more rights to defend its ships and the sea lines of communication passing through the Chinese side of the East China Sea [. . .], or even the Taiwan Strait, than China itself? If the answers [. . .] were to be yes then one would have to conclude that the prospects for Sino-American accommodation in East Asia are very dim. Indeed, it would confirm China's worst fears that a US-led coalition wants to contain China by denying it the ability to effectively bargain on any of its sovereign claims, and be in a position to keep China subordinate to their interests through control of China's commercial lifeline (Crisp 2010: 206, emphasis in the original).

It is nevertheless likely that China will aim to make U.S. naval operations in its maritime territory ever more risky, and thus more costly, by offering asymmetric deterrence, while simultaneously avoiding a symmetric engagement:

> It is, over time, cheaper and easier for some hostile actors to acquire long-range missiles and antiship, antiarmor, antisatellite, and other disruptive weapons and technologies than it is for the United States to defend its forward-based assets and forces against them. Indeed, the increasing vulnerability of forward military presence to antiaccess threats is often cited as the most important harbinger of coming instability in the commons (Davis et al. 2012: 34).

These points aside, the effect of China's naval build-up is not entirely problematic from the U.S. point of view. For one, the increasing "China Threat" perceptions in the whole region provide U.S. military planners (and procurement agencies in allied countries) with ample arguments in support of enhancing military capabilities, and can thus help mobilize funding for large and expensive arms procurement programs. They also provide an incentive for closer military and political co-operation among all those countries that seek to counterbalance growing Chinese influence in the region (cf. Glaser and Szerlip 2011). China's military build-up thus indirectly strengthens the U.S.-led alliance politically as well as technologically. Criticisms of U.S. "hegemonial" behavior in the international arena, which were increasingly prevalent in many developing countries as well as in Europe following the U.S. military interventions in Iraq and Afghanistan, could even slowly begin to give way to a more positive perception of the U.S. as a benevolent "guardian power" of the Great Commons, providing a welcome boost to U.S. soft power in the face of new challenges arising from Islamist terrorism around the world. Conversely, among the increasingly vocal group of Chinese military officers who routinely publish books and articles on China's military development and strategic challenges, there appears to be a uniform threat perception that paints the U.S. as a "hegemonial" and fundamentally anti-Chinese power, and as ultimately being the "real" force behind tensions in the South China Sea. These views tend to see a U.S. conspiracy at work that aims to "cause trouble around China's periphery and draw these periphery countries into a coalition against China" (Liu and Ren 2014: 270). Seen along these lines, the Taiwan issue could strike Chinese strategists as a U.S. plot to bind Chinese military capacities. Indeed, as Wertheim put it in an interview with Moss, "from a US perspective the Taiwan issue is beneficial because it distracts China's navy from looking beyond that"—a view that is consistent with the above-cited U.S. threat perceptions prevalent among Chinese military leaders (quoted in Moss 2011).

In fact, Chinese perceptions of the ambitions and goals of U.S. hegemony are quite in accord with some U.S. evaluations. As Brzezinski, for instance, writes in 1997:

> [...] for the United States, Eurasian geostrategy involves the purposeful management of geostrategically dynamic states and the careful handling of geopolitically catalytic states, in keeping with the twin interests of America in the short-term preservation of its unique global power and in the long-run transformation of it into increasingly institutionalized global cooperation. To put it in a terminology that hearkens back to the more brutal age of ancient empires, *the three grand imperatives of imperial geostrategy are to prevent collusion and maintain security dependence among the vassals, to keep tributaries pliant and protected, and to keep the barbarians from coming together* (Brzezinski 1997: 40, emphasis added).

Given the U.S. interest in preserving the current world order, the U.S. leadership cannot afford to lose ground in Asia. Therefore, notwithstanding strains on its national budget, the U.S. "Pivot to Asia" strategy voiced by President Obama in late 2011 is likely to be followed up with substantial measures. Chinese hopes of a gradual withdrawal of U.S. forces from its regional sphere of influence, moreover, are unlikely to materialize. Indeed, the possibility of a closer entente between China and Russia in the near term may lead to an even stronger U.S. military presence in Asia, if not in numbers, then in actual capability.

It is important to note, however, that in view of the close economic and financial entanglement between China and the U.S. and the manifold institutional and economic ties China has forged with the entire world, the "containment" of China as envisaged by some U.S. strategic thinkers is not feasible, or, from a U.S. standpoint, even actually desirable. In fact, as Shambaugh puts it, it may be "a complete nonstarter." He writes:

> Those who advocate it are detached from reality. China literally could *not* be contained even if it were decided that this was a wise course of action—precisely because of China's *existing integration* in the global system. The genie cannot be put back into the bottle. [...] Not a single nation on earth would go along with a containment strategy should the United States (or others) seek to pursue it (Shambaugh 2013: 315, emphasis in the original).

Similarly, Singapore's elder statesman Lee Kuan-yew has pointed out in 2009 that the common interests between both powers incline them towards meaningful co-operation, while conflict is in the interest of neither side:

> Unlike US-Soviet relations during the Cold War, there is no bitter, irreconcilable ideological conflict between the US and a China that has enthusiastically embraced the market. Both countries want stability in their relations and an international system that increases trade and investments. Sino-US relations are both cooperative and competitive. Competition between them is inevitable, but conflict is not. For China to grow its economy, it needs stability at home and peace abroad (Lee 2009).

These points seem very reasonable. In fact, rather than misguided attempts at some form of "containment," enhanced co-operation and communication, including on the military level, would seem to be the most rational and promising approach for the U.S. when dealing with a changing Chinese regional posture. The greatest danger may actually lie in the threat perceptions on both sides becoming increasingly detached from reality. After all, China is still very much caught up with solving its hereditary problems of national identity, with confronting its difficult past, and with defining its future role in the world amidst a situation of rapid and often problematic domestic change.

6.9 The European Union's Concerns Regarding Asia's Naval Power Shift

Likely due to lack of geographic proximity, the European powers do not seem directly concerned with the various issues and conflicts within the Asia-Pacific security complex. Unlike the U.S., which is active in the Asia Pacific through a multitude of permanent bases and a complex system of bilateral security partnerships, the EU does not have a comparable presence either politically or militarily. Accordingly, the idea of the "EU's irrelevance as a security actor in maritime East Asia" appears to be "currently widely shared in European foreign policy circles" (Duchâtel and Sheldon-Duplaix 2011: 31). Moreover, the EU clearly "has difficulty acting like a unitary state" (Legro 2011: 177).

After the end of the Cold War, Europe's expectation of obtaining a "peace dividend" arose, among other reasons, because of a perceived absence of external threats to its security, and a voluntary acceptance of U.S. pre-eminence in military affairs. NATO membership, which promises military assistance from the whole alliance in case of attack to any of its members, seemed to offer most European NATO members the chance to reduce their military modernization efforts and divert funds and political priorities towards economic construction. This led to a situation where most European powers' defense burdens fell well below the 2 % of GDP mark that NATO recommended to its members. Relative economic prosperity and the absence of direct security challenges arguably gave rise to a certain degree of self-centeredness in Europe, which only grew stronger as a result of the 2008 financial crisis, and which still inhibits a more substantial level of EU participation in world affairs today. This situation is reflected in the multitude of EU strategic partnerships with a host of Asian nations, and the absence of a consensus between EU member states on various critical issues:

> Formally, the EU has an 'enhanced partnership' with ASEAN, and multiple 'strategic partners' in Asia: China, India, Indonesia, South Korea, and Japan. [...] Meanwhile, the lofty aim of 'partnership in reform' with China [. . .] has not been attained and macroeconomic consultations between the world's second and third currency zones have been scarce. There are simmering battles over dumping and the overall disconnect on values is being bridged only at Europe's own expense (Godement 2014: 6).

The EU as an institution seems primarily interested in unobstructed access to Asian markets, in maintaining a positive trade and investment climate with all of its primary Asian trade partners, and in uninterrupted trade flows along the major SLOCs. The value of EU trade with Asia in 2013 amounted to €870 billion (Godement 2014: 7). Among the Asian EU trade partners, China is by far the most important. Its share of total EU trade in 2011 was only slightly smaller than the share of the EU's largest trading partner, the U.S. During 2000–2011, EU-China trade has more than quadrupled; from the point of view of China, the EU is already the largest trading partner (Maull 2013: 41–43).

Since European countries are also among the prime suppliers of advanced arms technologies to some of the Asia Pacific countries, these economic relationships gain an added security perspective that runs counter to the notion of Europe's "strategic irrelevance." Nations that have been supplied with naval surface vessels or submarines from Europe in recent years include Pakistan, India, South Korea, Australia, Malaysia, Brunei, Indonesia, Singapore, and Taiwan. Therefore, individual supplier nations have a more than cursory interest in the ongoing Asian naval build-up, and the EU is often challenged when trying to find a consensus for dealing with the arms embargo against China, which is regularly being called into question by various interested parties (Duchâtel and Sheldon-Duplaix 2011: 31; cf. also Sect. 4.6). Lastly, since a possible entente between Russia and China would directly affect Russia's Western neighbors, and thus also NATO, such a development cannot be deemed inconsequential from a European perspective.

Dealing with the complexities of EU-China relations has not been easy for Brussels. In 2003, a "Strategic Partnership" between China and the EU was forged. However, the arms embargo is consistently cited by Chinese leaders as a stumbling block for furthering this partnership. According to reports, in an internal strategy paper distributed to EU leaders in preparation of the EU summit in Brussels in December 2010, the EU's then High Representative for Foreign Affairs and Security Policy, Catherine Ashton, argued in favor of lifting the embargo, calling it a "major impediment for developing stronger EU-China co-operation on foreign policy and security matters" (cf. Rettman 2010). However, in order to allow European governments to make an informed choice regarding the future of the embargo, it would be necessary for the EU to fully analyze *all* the relevant implications of such a move, including possible scenarios for various unintended consequences. So far, it seems that the EU has found it difficult to establish institutions of information sharing and policy co-ordination in fields such as high-tech development and export controls of sensitive or dual-use technologies, which would be affected by a change of policy. As Stumbaum warns, overall there is still an "absence of a strategic assessment of China's rise" from the point of view of Europe (Stumbaum 2009: 7). Especially in comparison with the U.S. approach, which gathers and synthesizes expertise from military, industry, politics, and academia by forging avenues for the exchange of ideas and information on China's military rise, in Europe,

> [...] there is little exchange and expertise on China's military transformation among European officials and advisers in Brussels and other European capitals. Increased efforts are being made to follow debates and developments in China, e.g. by seeking personnel for the EU's Situation Centre, but intelligence is still primarily exchanged bilaterally between selected Member States (Stumbaum 2009: 11).

This unfortunately leads to information loss in some critical areas, and to internally contradictory behavior of member states that can be (and often is) exploited by the Chinese counterpart:

> There is no European overview on the amount and exact nature of dual-use technology that has been delivered to China already. Moreover, the lack of a collective European approach leads to undercutting and incoherence. It happens that non-critical goods such as paint for military applications are banned from being exported as they appear on national munitions lists, and the export of other such minor items is unnecessarily delayed, while whole diesel engines for use in submarines or space technology that can be misused to attack satellites in space can be exported without difficulty (Stumbaum 2009: 12).

Europe is distinguished by its absence of consensus on many other China-related issues of contention, and a regrettable lack of solidarity between EU member states in the event that China takes punitive measures against individual states that decide to exercise their sovereign rights, e.g. by allowing visits by persons deemed dissidents or separatists by China's authorities, such as the Dalai Lama or politicians from Taiwan. By leaving the door open for China to turn various European countries against each other even over matters of principle, strong EU positions regarding European strategic interests in Asia cannot effectively be put forth. Consequently, the EU is primarily perceived as an economic, not a political, actor in Beijing. The ongoing Euro crisis has however tainted the international perception of Europe, and further eroded European ambitions of strengthening the EU's global political role (Maull 2013: 45–46).

The euphoria resulting from the end of the Cold War prompted the hope in Europe that a more civilized way of solving conflicts would eventually become the international norm. Indeed, many Europeans expected that armed conflict would soon become a thing of the past—a sentiment echoed in the remark cited at the outset of this chapter by China's retired Admiral Liu Huaqing. In large parts of Europe, this ideal has already become a reality, for despite conflicting territorial claims much more substantial than those in the South China and East China Seas, the ongoing process of European integration has effectively ended five centuries of nearly continuous warfare between Europe's traditional strategic rivals. In honor of this momentous achievement, the EU was awarded the Nobel Peace Prize as an institution in 2012. In light of this historical feat, it is worth considering whether the model of Europe's post-war integration based on economic co-operation ultimately offers lessons to the conflict-ridden regions of East Asia. For instance, could it serve as a regional governance model for reducing tensions and replacing hostility with a voluntary adherence to mutually agreed upon rules? Could the European powers even act as mentors in this regard, transferring experience collected in Europe to other world regions, including Asia? According to a joint communiqué signed by then EU High Representative Ashton and U.S. Secretary of State Clinton, European leaders

> [...] condemn the use of intimidation, coercion, or force on territorial issues, uphold freedom of navigation and promote resolution 'according to international law', proclaim 'common security interests' with Japan, and encourage its 'proactive contribution to peace' (Godement 2014: 6).

From this liberal institutionalist viewpoint, the idea of promoting European values of conflict resolution more proactively in Asia seems only logical. And, increasing levels of trade integration of many Asian countries with each other, even between China and Taiwan, seems to make such a development at least theoretically possible. Presumably, U.S. influence could be used to facilitate the beginnings of such a process among various U.S. allies first. Commenting on the possibilities for a true rapprochement between former war enemies South Korea and Japan, Brzezinski commented:

> The United States could play the critical role in promoting that reconciliation. Many specific steps that were taken to advance first the German-French reconciliation and later that between Germany and Poland (for example, ranging from joint university programs eventually to combined military formations) could be adapted to this case (Brzezinski 1997: 191).

It is interesting to note that in 2009, during the short-lived tenure of Japan's then Prime Minister Yukio Hatoyama, ideas of establishing an "East Asian Community" modeled upon the European integration process were floated by Japan's government which explicitly mentioned the example of Franco-German post-war reconciliation as the keystone of the EU's success. In a speech at Singapore, Hatoyama called upon the Asian countries to learn from this example. After explicitly acknowledging the damage done by Japan to its neighbors during the war, he stated:

> The central idea of my 'East Asian community' initiative is based upon reconciliation and cooperation in Europe. In my initiative, I propose that countries sharing a common vision promote cooperation in various fields. This would be based on the principle of 'open regional cooperation.' Through this, our region would develop a multi-layered network of functional communities. I attach the greatest importance to the promotion of concrete cooperation in a broad range of areas such as trade, investment, finance and education. [...] As we cooperate, we will set rules for ourselves, work together, share our wisdom, and respect the rules we have made. Therefore, we will be able to not only achieve practical gains, but also build mutual trust (Hatoyama 2009).

However, skeptics would no doubt point out that the European post-war period was a unique window of opportunity for such a historical shift in thinking to occur, and that similarly compelling conditions do not exist in Asia today. Moreover, at present the various Asian countries' political leaders are not even theoretically committed to such an ideal of regional integration. Beijing consequently perceived

Hatoyama's above cited overture as primarily directed against China's rise (cf. Zhao 2011: 59).

A more direct European involvement in the Asia Pacific security complex and its affairs is likewise not necessarily welcome from Beijing's point of view. According to a French assessment,

> Beijing rejects the idea of European involvement in the maritime territorial disputes that undermine regional security, and resists European attempts to include discussions on Asian security matters on the agenda of EU-China meetings. Chinese academics generally consider Europe not to be a legitimate security actor in East Asia (Duchâtel and Sheldon-Duplaix 2011: 31).

Since most EU member states are simultaneously NATO members, and therefore formally allied with the U.S., Europe is often perceived as an extension of U.S. hegemony by Asian critics of the unipolar world order, rather than a potential "honest broker", and especially so in China. Just the same, a more substantial European involvement in Asia's conflict areas would likely be welcomed by many other players in the region, even if this were limited to merely taking a clearer stance on the various territorial issues (cf. Godement 2014: 7). However, the EU's "soft power"—as well as the attractiveness of the European model of integration—has arguably been diminished as a consequence of the ongoing European debt crisis (Maull 2013: 47).

For the time being, the chances seem remote that a similar integration process can be promoted in Asia (e.g. among China, Japan, South Korea, and perhaps also Taiwan) to the one that has taken place in Europe, which has grown to encompass 28 European countries. Meanwhile, Europeans will be increasingly challenged to clearly define the EU's own security interests in the Asia Pacific region in the twenty-first century, and to become more actively involved in efforts there to peacefully resolve territorial conflicts. For this to occur, Europe must overcome the institutional and cultural obstacles that impede efficient information sharing, aim at better policy co-ordination between EU members states, and actively strengthen EU exchanges with all its major Asian "strategic partners."

References

'A guilty verdict for North Korea: Their number is up'. 2010. *The Economist,* May 22, 2010: 57.

Amurao, George. 2012. A Shoal Too Far in South China Sea. *Asia Times* Online, July 21, 2012. http://www.atimes.com/atimes/Southeast_Asia/NG21Ae01.html. Accessed Sep. 14, 2012.

Ansari, Usman. 2009. China, Pakistan Pledge to Further Boost Defense Ties. *Defense News* (el. ed.), February 2, 2009.

Bedi, Rahul. 2011. India 'Concerned' About Chinese Projects in Pakistan-controlled Kashmir. *Jane's Defence Weekly* (el. ed.), September 8, 2011. http://www.janes.com.

Bedi, Rahul and Jon Grevatt. 2013. India, Russia in Talks Over Second SSN Lease. *Jane's Defence Weekly* (el. ed.), October 22, 2013. http://www.janes.com.

Bedi, Rahul and James Hardy. 2013. US, Indian leaders agree to joint materiel development, production. *Jane's Defence Weekly* (el. ed.), September 9, 2013. http://www.janes.com.

Blazevic, Jason J. 2009. Defensive Realism in the Indian Ocean: Oil, Sea Lanes and the Security Dilemma. *China Security* Vol 5, No. 3: 59-71.

Bokhari, Farhan. 2013. Analysis: Pakistan Modernisation Feeds China's Export Growth. *Jane's Defence Weekly* (el. ed.), April 12, 2013. http://www.janes.com.

Brzezinski, Zbigniew. 1997. *The Grand Chessboard: American Primacy and Its Geostrategic Imperatives*. New York: Basic Books.

Cabestan, Jean-Pierre. 2011. Taiwan's Deteriorating Security Environment and its Impact on Cross-Strait Relations. SWP Discussion paper for Track Two Dialogue on EU-China-Relations and the Taiwan Question, Wujiang, 23-24 June 2011. http://www.swp-berlin.org/fileadmin/contents/products/projekt_papiere/Taiwan2ndTrack_Cabestan_2011.pdf. Accessed December 12, 2014.

Chen, Hurng-yu. 2000. The PRC's South China Sea Policy and Strategies of Occupation in the Paracel and Spratly Islands. *Issues & Studies* 36, No. 4 (July/Aug 2000): 95-131.

Cheung, Tai Ming. 2009. *Fortifying China. The Struggle to Build a Modern Defense Economy*. Ithaca and London: Cornell University Press.

'China's Military Rise. The dragon's new teeth: A rare look inside the world's biggest military expansion'. 2012. *The Economist* (el. ed.), April 07, 2012. http://www.economist.com/node/21552193. Accessed April 18, 2014.

Chung, Lawrence. 2014. Beijing 'freezes' cooperation talks with Taiwan over stalled trade pact. *South China Morning Post* online. June 9, 2014. http://www.scmp.com/news/china/article/1528429/beijing-freezes-cooperation-talks-taiwan-over-stalled-trade-pact. Accessed December 10, 2014.

Chung, Jae Ho and Myung-hae Choi. 2013. Uncertain allies or uncomfortable neighbors? Making sense of China-North Korea Relations, 1949-2010. *The Pacific Review* 2013, Vol. 26 No. 3: 243-264, http://dx.doi.org/10.1080/09512748.2012.759262.

Cleary, Thomas. 2005. *Mastering the Art of War: Zhuge Liang and Liu Ji*. Boston and London: Shambala.

Cole, Bernard D. 2010. *The Great Wall at Sea. China's Navy in the Twenty-First Century*. 2nd ed. Annapolis: Naval Institute Press.

Cole, Michael J. 2011a. Varyag 'To Fall Under Central Command.' *Jane's Defence Weekly* (el. ed.), August 16, 2011. http://www.janes.com.

Cole, Michael J. 2011b. China beefing up maritime patrol fleet. *Jane's Defence Weekly* (el. ed.), May 4, 2011. http://www.janes.com.

Crisp, Michael. 2010. The Great Chinese Sea Power Debate: A Review Essay. *Journal of Contemporary China* 19(63), January 2010: 201-212.

Davis, Lynn E. et al. 2012. U.S. Overseas Military Presence: What Are the Strategic Choices? RAND Project Airforce, RAND monograph series, Santa Monica, CA: RAND Corp. http://www.rand.org/pubs/monographs/MG1211.html. Accessed 22 Nov 2014.

Dreyer, Edward L. 2006. *Zheng He: China and the Oceans in the Early Ming, 1405-1433*. Harlow: Longman.

Duchâtel, Mathieu and Alexandre Sheldon-Duplaix. 2011. The European Union and the Modernisation of the People's Liberation Army Navy: The Limits of Europe's Strategic Irrelevance. *China Perspectives* No. 2011/4: 31-41.

Dunai, Peter and Guy Anderson. 2013. Russia Outlines USD42 Billion Naval and Maritime Redevelopment Plan. *Jane's Navy International* (el. ed.), January 17, 2013. http://www.janes.com.

Dutton, Peter A. 2007. International Law and the November 2004 'Han Incident'. In *China's Future Nuclear Submarine Force*, eds. Andrew S. Erickson et al., 162-181. Annapolis: Naval Institute Press.

Fels, Enrico. 2013. The Role of the SCO in Afghanistan. In *Partners for Stability: Involving Neighbors in Afghanistan's Reconstruction – Transatlantic Approaches*, eds. Henning Riecke and Kevin Francke, 235-250. Baden-Baden: Nomos.

FlorCruz, Michelle. 2014. Vietnam and India Sign Oil, Naval Agreement Amid South China Sea Disputes, Angering Beijing. *International Business Times* online. October 29, 2014. http://www.ibtimes.com/vietnam-india-sign-oil-naval-agreement-amid-south-china-sea-disputes-angering-beijing-1715677. Accessed November 16, 2014.

Foster, Robert. 2012. Russia to sell, co-produce Lada-class submarines to China. *Jane's Navy International* (el. ed.), December 21, 2012. http://www.janes.com.

Frank, Rüdiger. 2014. *Nordkorea: Innenansichten eines totalen Staates* (North Korea: inside views of a total state). München: Deutsche Verlags-Anstalt.

Friedman, Norman. 2006. *The Naval Institute Guide to World Naval Weapon Systems.* 5th ed. Annapolis: Naval Institute Press.

Ghosh, P. K. 2014. Artificial Islands in the South China Sea. *The Diplomat,* September 23, 2014. http://thediplomat.com/2014/09/artificial-islands-in-the-south-china-sea/. Accessed October 25, 2014.

Gilley, Bruce. 2008. Taiwan's Democratic Transition: A Model for China? In *Political Change in China: Comparisons With Taiwan. Eds.* Bruce Gilley and Larry Diamond, 215-242. Boulder, Colorado: Lynne Rienner.

Glaser, Bonnie and David Szerlip. 2011. US Allies Respond: Japanese, Australian, and South Korean Reactions to China's Military Modernisation. China Perspectives No. 2011/4: 17-29.

Godehardt, Nadine. 2013. Pekings zweigleisige Politik gegenüber Pjöngjang (Beijing's dual-track policy towards Pyongyang). *SWP-Aktuell* No. 32, June 2013. http://www.swp-berlin.org/fileadmin/contents/products/aktuell/2013A32_gdh.pdf. Accessed November 16, 2014.

Godement, François. 2014. China on Asia's Mind. European Council on Foreign Relations (ECFR) Policy Brief. September 2014. http://www.ecfr.eu/publications/summary/china_on_asias_mind318. Accessed October 22, 2014.

Grevatt, Jon. 2009. Proposed Sino-Pakistani Venture Focuses on Investment Opportunities. *Jane's Defence Weekly* (el. ed.), October 20, 2009. http://www.janes.com.

Grevatt, Jon. 2011a. Philippines Eyes Submarine Acquisition by 2020. *Jane's Defence Weekly* (el. ed.), May 17, 2011. http://www.janes.com.

Grevatt, Jon. 2011b. China and Ukraine Sign Defence Pact. *Jane's Defence Weekly* (el. ed.), June 22, 2011. http://www.janes.com.

Grevatt, Jon. 2013a. China and Malaysia agree to comprehensive strategic partnership. *Jane's Defence Weekly* (el. ed.), Oct 7, 2013. http://www.janes.com.

Grevatt, Jon. 2013b. China, Malaysia to strengthen defence industry collaboration. *Jane's Defence Weekly* (el. ed.), Oct 30, 2013. http://www.janes.com.

Hardy, James. 2013. Japan Unveils Largest-ever Helicopter Carrier. *Jane's Defence Weekly* (el. ed.), August 6, 2013. http://www.janes.com.

Hardy, James and Alexander Weening. 2014. China building 10,000-tonne coastguard cutters. *Jane's Defence Weekly* (el. ed.), October 15, 2014. http://www.janes.com.

Hassan, Zafar. 2009. Military Cooperation Between Pakistan and China. *Kanwa Asian Defence* April 2009: 31-37.

Hatoyama, Yukio. 2009. Japan's New Commitment to Asia - Toward the Realization of an East Asian Community. Address by H.E. Dr. Yukio Hatoyama, Prime Minister of Japan, November 15, 2009, Singapore. http://japan.kantei.go.jp/hatoyama/statement/200911/15singapore_e.html. Accessed February 22, 2015.

Hauswedell, P. Christian. 2012. Die koreanische Halbsinsel in der strategischen Kalkulation Chinas und der USA (The Korean Peninsula in the Strategic Calculus of China and the USA). *Asien* No. 122 (January 2012): 75-93.

Hayton, Bill. 2014. *The South China Sea: The Struggle for Power in Asia.* New Haven and London: Yale University Press.

Hieber, Saskia. 2012. *Sicherheit und Frieden in Asien: Konfliktpotentiale und regionale Kooperation,* (Security and Peace in Asia). Tutzinger Schriften zur politischen Bildung. Schwalbach: Wochenschau Verlag.

Holmes, James R. and Toshi Yoshihara. 2008. *Chinese Naval Strategy in the 21stCentury. The Turn to Mahan*, London/New York: Routledge.

Holslag, Jonathan, 2010. *Trapped Giant: China's Military Rise*. The International Institute of Strategic Studies (IISS), London: Routledge.

'How Mumbai attacks unfolded'. 2008. *BBC News* online. November 30, 2008. http://news.bbc.co.uk/2/hi/south_asia/7757500.stm. Accessed November 16, 2014.

Hughes, Christopher. 2011. Reclassifying Chinese Nationalism: The *Geopolitik* Turn. *Journal of Contemporary China* (2011), 20(71), September: 601-620.

Jane's Fighting Ships (el. ed.). 2014. http://www.janes.com.

Jane's World Navies (el. ed.). 2013. http://www.janes.com.

Japan Ministry of Defense. 2011. *Defense of Japan 2011*. http://www.mod.go.jp/e/publ/w_paper/2011.html. Accessed October 22, 2014.

Japan Ministry of Defense. 2013. *Defense of Japan 2013*. http://www.mod.go.jp/e/publ/w_paper/e-book/2013/index2.html. Accessed October 22, 2014.

Johnson, David E. et al. 2009. *Preparing and Training for the Full Spectrum of Military Challenges: Insight from the Experiences of China, France, the United Kingdom, India, and Israel*. Santa Monica: RAND Corp.

Johnson, Reuben F. 2011. Ukraine, Russia at Odds Over Carrier Training Centre. *Jane's Defence Weekly* (el. ed.), November 3, 2011. http://www.janes.com.

Johnson, Reuben F. 2013a. China's Domination of Global Arms Trade 'Inevitable', Say Russian, Ukrainian Industry Insiders. *Jane's Defence Weekly* (el. ed.), March 22, 2013. http://www.janes.com.

Johnson, Reuben F. 2013b. Russian Officials Deny Chinese Deal on Su-35s, submarines. *Jane's Defence Weekly* (el. ed.), March 26, 2013. http://www.janes.com.

Johnson, Reuben F. and James Hardy. 2013. Ukraine Preparing to Offer China Use of NITKA Carrier Training Facility. *Jane's Defence Weekly* (el. ed.), November 14, 2013. http://www.janes.com.

Keck, Zachary. 2013a. China Imposes Restrictions on Air Space Over Senkaku Islands. *The Diplomat,* November 23, 2013. http://thediplomat.com/2013/11/china-imposes-restrictions-on-air-space-over-senkaku-islands/. Accessed October 25, 2014.

Keck, Zachary. 2013b. US Bombers Challenge China's Air Defense Identification Zone. *The Diplomat,* November 27, 2013. http://thediplomat.com/2013/11/us-bombers-challenge-chinas-air-defense-identification-zone/. Accessed October 25, 2014.

Kofman, Michael. 2014. An Uneasy Ménage à Trois: Reliance on Russian and Ukrainian Weapons Puts China in a Tight Spot. *Foreign Affairs* online, Dec 4, 2014. http://www.foreignaffairs.com/articles/142417/michael-kofman/an-uneasy-menage-a-trois. Accessed January 2, 2015.

Kristensen, Hans M. 2007. China's Submarine Fleet Continues Low Patrol Rate. FAS Strategic Security Blog from the Federation of American Scientists, February 6, 2007. http://blogs.fas.org/security/2007/02/post_2/. Accessed September 24, 2013.

Lampton, David M. 2014. *Following the Leader: Ruling China, from Deng Xiaoping to Xi Jinping*. Berkeley et al.: University of California Press.

Larson, Christina. 2013. Chinese Scholar Calculates That China's Rise Is 65.3% Complete. *BloombergBusinessweek Global Economics* online, November 25, 2013. http://www.businessweek.com/articles/2013-11-25/chinese-scholar-calculates-china-s-rise-is-65-dot-3-percent-complete. Accessed December 12, 2014.

Lee, Kuan-yew. 2009. Speech at the US-ASEAN Business Council's 25th Anniversary Gala Dinner, 27 October 2009, 3:48 pm at Washington, DC. http://www.nas.gov.sg/archivesonline/speeches/view-html?filename=20091105001.htm. Accessed February 20, 2015.

Legro, Jeffrey W. 2011. The Omnipower: The United States and Regional Orders. In *Regional Powers and Regional Orders*. Eds. Nadine Godehard and Dirk Nabers, 175-192. London and New York: Routledge.

Leonard, Mark. 2008. *What Does China Think?* London: Fourth Estate.

Li, Mingjiang. 2010. China and Maritime Cooperation in East Asia: Recent Developments and Future Prospects. *Journal of Contemporary China* 19(64), March 2010: 291-310.

Li, Xiaobing. 2009. *A History of the Modern Chinese Army*. Lexington: The University Press of Kentucky.

Liu, Huaqing 刘华清. 2011. *Liu Huaqing huiyilu* (刘华清回忆录 | The memoirs of Liu Huaqing). 4th ed. Beijing: Jiefangjun chubanshe.

Liu, Mingfu 刘明福. 2009. Zhongguo meng: Zhongguo de mubiao, daolu ji zixinli (中国梦: 中国的目标, 道路及自信力 | The Chinese Dream). Rev. ed. Beijing: Youyi chuban.

Liu, Yawei and Justine Zheng Ren. 2014. An Emerging Consensus on the US Threat: the United States according to PLA officers. *Journal of Contemporary China* Vol. 23, No. 86: 255-274.

Mahadzir, Dzirhan and James Hardy. 2011. Chinese Defence Minister Dismisses Pakistan's Port Plans. *Jane's Defence Weekly* (el. ed.), June 8, 2011. http://www.janes.com.

Mannerheim, Carl Gustav Emil. 1953. *The Memoirs of Marshal Mannerheim*. Transl. Count Eric Lewenhaupt. London: Cassell and Co.

Masood, Salman. 2009. Pakistan Backtracks on Link to Mumbai Attacks. *New York Times* online, February 12, 2009. http://www.nytimes.com/2009/02/13/world/asia/13pstan.html?partner=permalink&exprod=permalink&_r=0. Accessed November 16, 2014.

Maull, Hans. 2013. Die Europäische Union und die aufsteigenden Schwellenländer: Das Beispiel Volksrepublik China (The European Union and the rising powers: The Example of the PRC). In *EU-Außenpolitik in Zeiten der Finanz- und Schuldenkrise* (EU foreign policy in the age of the financial and debt crisis), eds. Ronja Kempin and Marco Overhaus, 40-49. SWP-Studien 2013/S 09, April 2013. Berlin: SWP. http://www.swp-berlin.org/fileadmin/contents/products/studien/2013_S09_kmp_ovs.pdf#page=40. Accessed February 19, 2015.

Miller, David. 2002. *The Illustrated Directory of Submarines of the World*. London: Salamander Books.

Moore, Gregory J. 2014. Ukraine: A Military-industrial Complex to Die For. *Asia Times* online, May 27, 2014. http://www.atimes.com/atimes/Central_Asia/CEN-01-270514.html. Accessed October 31, 2014.

Moss, Trefor. 2011. Briefing: PLAN of Action. *Jane's Defence Weekly* (el. ed.), February 4, 2011. http://www.janes.com.

Nabers, Dirk. 2007. Japan diskutiert Abkehr von der 'Friedensverfassung. GIGA Focus Asien No. 5/ 2007. Hamburg: GIGA. http://www.giga-hamburg.de/de/system/files/publications/gf_asien_0705.pdf. Accessed November 26, 2014.

Park, Hyun. 2014. US stationing more aircraft carriers and troops in Asia-Pacific region. *The Hankyoreh* online, January 17, 2014. http://www.hani.co.kr/arti/english_edition/e_international/620232.html. Accessed January 22, 2014.

Phipps, Gavin. 2013. Taiwan 'developing long-range cruise missile'. *Jane's Defence Weekly* (el. ed.), January 30, 2013. http://www.janes.com.

Phipps, Gavin and James Hardy. 2014. Taiwan says local submarine programme back on drawing board. *Jane's Defence Weekly* (el. ed.), October 2, 2014. http://www.janes.com.

Pugh, Philip. 1986. *The Cost of Seapower: The Influence of Money on Naval Affairs from 1815 to the Present Day*. London: Conway Maritime Press.

Rahmat, Ridzwan. 2015. PLAN to deploy range of warships in Indian Ocean, says China's defence ministry. *Jane's Defence Weekly* (el. ed.), January 29, 2015. http://www.janes.com/article/48464/plan-to-deploy-range-of-warships-in-indian-ocean-says-china-s-defence-ministry. Accessed January 31, 2015.

Rahmat, Ridzwan and James Hardy. 2014. Taiwan starts trials of 'carrier killer' corvette. *Jane's Defence Weekly* (el. ed.), October 29, 2014. http://www.janes.com.

Rajagopalan, Rajesh. 2011. Pakistan: Regional Power, Global Problem? In *Regional Powers and Regional Orders*. Eds. Nadine Godehard and Dirk Nabers, 193-208. London and New York: Routledge.

Rettman, Andrew. 2010. Ashton Pragmatic on China in EU Foreign Policy Blueprint. *EU Observer,* December 17, 2010. https://euobserver.com/china/31538. Accessed February 10, 2015.

Richards, Clint. 2014. The Australia-Japan Submarine Deal Gets Complicated. *The Diplomat,* Nov 13, 2014. http://thediplomat.com/2014/11/the-australia-japan-submarine-deal-gets-complicated/. Accessed November 30, 2014.

Rozman, Gilbert. 1999. China's quest for great power identity. *Orbis* Summer 1999: 383-402.

Sasaki, Tomonori. 2010. China eyes the Japanese military: China's threat perception of Japan since the 1980s. *The China Quarterly* 203, September 2010: 560-580.

Scobell, Andrew. 2012. Learning To Rise Peacefully? China and the Security Dilemma. *Journal of Contemporary China* (2012), 21(76): 713-721.

Shambaugh, David. 2013. *China Goes Global. The Partial Power.* Oxford: Oxford University Press.

Shirk, Susan. L. 2007. *China: Fragile Superpower.* Oxford: Oxford University Press.

Song, Weiqing. 2014. Interests, Power and China's Difficult Game in the Shanghai Cooperation Organization (SCO). *Journal of Contemporary China* Vol 23, No. 85: 85-101.

Stumbaum, May-Britt U. 2009. Risky Business? The EU, China and Dual-Use Technology. Occasional Paper No. 80. The European Union Institute for Security Studies (EUISS). http://www.iss.europa.eu/uploads/media/op80.pdf. Accessed October 9, 2013.

Sui, Cindy. 2012. Is Taiwanese democracy changing Chinese students? *BBC News Asia,* September 19, 2012. http://www.bbc.com/news/world-asia-19365133. Accessed December 10, 2014.

Swaine, Michael D. 2011. China's Assertive Behavior—Part One: On Core Interests. *China Leadership Monitor* No. 34, 2011. http://media.hoover.org/sites/default/files/documents/CLM34MS.pdf. Accessed September 17, 2013.

Swaine, Michael D. 2013. Chinese Views Regarding the Senkaku/Diaoyu Islands Dispute. *China Leadership Monitor* No. 41 (6 June 2013): 1-27. http://media.hoover.org/sites/default/files/documents/CLM41MS.pdf. Accessed January 20, 2014.

Swaine, Michael D. & Taylor M. Fravel. 2011. China's Assertive Behavior—Part Two: The Maritime Periphery. *China Leadership Monitor* No. 35, 2011. http://media.hoover.org/sites/default/files/documents/CLM35MS.pdf. Accessed September 17, 2013.

Takahashi, Kosuke. 2011. Japanese White Paper denounces 'assertive' China. *Jane's Defence Weekly* (el. ed.), August 5, 2011. http://www.janes.com.

Taylor, Rob. 2014. Australia Nears Deal to Buy Up to 10 Japanese Submarines. *The Wall Street Journal,* September 8, 2014. http://online.wsj.com/articles/australia-nears-deal-to-buy-up-to-10-japanese-submarines-1410166854. Accessed November 30, 2014.

Tønnesson, Stein. 2009 (2001). An International History of the Dispute in the South China Sea. In *China and Southeast Asia, Vol. VI: The People's Republic of China and Southeast Asia,* ed. Geoff Wade, 337-362. London and New York: Routledge.

Taiwan Relations Act. 1979. Public Law 96-8, 96th congress. Full text available at http://photos.state.gov/libraries/ait-taiwan/171414/ait-pages/tra_e.pdf. Accessed February 16, 2015.

UNCLOS. 1982. *United Nations Convention on the Law of the Sea,* http://www.un.org/depts/los/convention_agreements/texts/unclos/unclos_e.pdf. Accessed December 12, 2014.

U.S. Department of State. 2014. China: Maritime Claims in the South China Sea. *Limits in the Seas* No. 143, Dec. 5, 2014. Office of Ocean and Polar Affairs. http://www.state.gov/e/oes/ocns/opa/c16065.htm. Accessed December 10, 2014.

Viner, Kim. 1997. Potential Military Solutions for the 'Taiwan Question'. *Asian Affairs* Vol. 24, No. 3 (Fall 1997): 180-189.

Vogelsang, Kai. 2012. *Geschichte Chinas* (History of China). Stuttgart: Reclam.

Wertheim, Eric. 2013. *The Naval Institute Guide to Combat Fleets of the World.* 16th ed. Annapolis: Naval Institute Press.

Yahuda, Michael. 2012. China's Recent Relations with Maritime Neighbours. *The International Spectator: Italian Journal of International Affairs* Vol 47, No. 2 (June 2012): 30-44.

Yahuda, Michael. 2013. China's New Assertiveness in the South China Sea. *Journal of Contemporary China* Vol. 22, No. 81: 446-459.

Yan, Xuetong 阎学通. 2001. The Rise of China in Chinese Eyes. *Journal of Contemporary China* (2001), 10(26): 33-39.

Yang, Andrew N. D. 2011. Mainland China's Military Development and Taiwan's Countermeasures. Brookings: Taiwan-U.S. Quarterly Analysis No. 6. http://www.brookings.edu/research/papers/2011/08/chinas-military-development-yang. Accessed December 1, 2014.

Yoshihara, Toshi and James R. Holmes. 2010. *Red Star Over the Pacific. China's Rise and the Challenge to U.S. Maritime Strategy*. Annapolis: Naval Institute Press.

Zhang, Baohui. 2011. Taiwan's new grand strategy. *Journal of Contemporary China* 20:69: 269-285.

Zhang, Zhijun 张志军. 2006. Zhongguo de heping fazhan yu guoji shehui" (中国的和平发展与国际社会 | China's peaceful development and the international community). *Qiushi* (求是) 6/2006: 61-62.

Zhao, Suisheng. 2011. China's Approaches toward regional cooperation in East Asia: motivations and calculations. *Journal of Contemporary China* 20 : 68: 53-67. http://dx.doi.org/10.1080/10670564.2011.520846.

Chapter 7
Concluding Remarks and Summary of Results

The sleep of reason produces monsters.
—Francisco de Goya, ca. 1799

China's naval build-up, this study has found, has indeed been remarkable compared with earlier historical periods. It is almost certainly bound to pick up speed and to thus present the Asia Pacific region with new strategic challenges. In order to avoid misconceptions on all sides, many more rational analyses of empirical data are needed, ideally from a wider range of regional backgrounds and viewpoints than have been represented so far in the English-language scholarly discourse. In light of sometimes blatantly contradictory interpretations of the exact same evidence, the issue of methodology is of prime importance. One major aim of this study, therefore, was to present a set of analytical tools that would allow for a transparent evaluation of China's ongoing naval modernization according to explicit criteria. This study argues that structural/dependency and historical/systemic approaches should be combined with the predominant realist and constructivist approaches in order to capture more characteristics of China's naval rise, including strategic vulnerability and dependency issues arising from imbalances in the international arms trade, and in order to map China's naval modernization against that of its regional neighbors and other rising powers in the world.

In Chap. 2, Chinese domestic politics was examined for factors that facilitate, necessitate, inhibit, or contribute to the aim of building a more powerful navy. Among other points, it showed how the officially endorsed historical narrative of a Chinese "national rejuvenation" is deliberately used to shape expectations and strengthen national cohesion in the face of centrifugal forces affecting the Chinese populace in a time of rapid change and uncertainty. Chapter 2 also discussed how China's history and maritime geography have influenced previous and current Chinese threat perceptions of other naval powers, and contributed to strategic thinking on naval defense and naval expansionism.

China's high level of trade integration with the rest of the world in conjunction with the unique set of constraints imposed on China's leaders through domestic politics is a strong incentive to keep pursuing "peaceful development" as China's grand strategy. Strengthening those forces in China who argue for a rational and balanced foreign and security policy should thus be a major goal of regional

S. Kirchberger, *Assessing China's Naval Power*, Global Power Shift,
DOI 10.1007/978-3-662-47127-2_7

neighbors and other interested powers alike. The other powers within the "Asia Pacific security complex" will need to strike a careful balance between accepting and endorsing China's legitimate developmental goals while furthering their own security interests without resorting to an overall demonization of China's rise. China's leaders would likewise do well to take the heightened threat perceptions within the region seriously, and search for ways of building trust rather than perpetually painting China as the victim of an anti-Chinese conspiracy. China's maritime behavior, this study has found, is indeed at times ambiguous, especially when it comes to contested claims in the South China Sea and Taiwan, and China's leaders therefore do have some leeway for improving their regional perception as a "responsible stakeholder".

Chapter 3 showed how the common analytical difficulty of distinguishing between routine naval replacement procurement, which does not usually result in rising military capability, and fleet enlargement and/or qualitative capability increases resulting from new technologies, can be tackled by taking procurement cycles and project timelines more systematically into account. Capability increases can be said to ensue if new, previously non-existent levels of combat proficiency can be reached through mastery of a newly introduced threshold technology, and if the necessary training and maintenance infrastructures are also built up to enable their use. A simple system for distinguishing between various levels of naval combat proficiency, which points out the technology thresholds demarcating boundaries between various levels, was cited and described in Chap. 3. According to that system of altogether nine naval combat proficiency levels, China can be said as of late 2014 to have in all likelihood mastered the technologies associated with Level VI out of nine, and to currently be in the process of working towards mastering Level VII. Reaching this goal will require a robust and sustained financial, technological, and training effort for many years to come.

Chapter 3 also made use of a ranking system of overall naval capability based on a comprehensive evaluation of a fixed set of qualitative criteria, which can be utilized to classify whole navies into a system of ten capability ranks in total. According to this system, China's navy presents as of late 2014 a total capability profile corresponding with Rank 4—"Regional power-projection." Once the still nascent aircraft carrier capability has become fully available to China's navy, China's position in this system will then automatically rise to that of a Rank 3 "Multi-regional power-projection" navy. This can be expected to occur sometime before 2020 based on present indications, though it could also take longer in the case of unforeseen technical or operational problems. A significant rise in capability beyond that level would call for an extremely sustained and focused commitment to out-of-area missions, and much higher funding levels than seem to be currently available in view of other pressing government spending priorities, among them internal security.

Chapter 3 also addressed and discussed various economic factors influencing China's naval capability build-up plans. Threat perceptions resulting from China's often harshly criticized military spending have been found to be somewhat overblown in light of the fact that inflation effects and huge military manpower costs take up a large portion of the budget, while overall state expenditure growth rates suggest that the share of military spending may have even shrunk in recent years. If either the U.S. or the EU is taken as a yardstick, which may be warranted given the size of China's territory, population, and national economy, China's military spending does not seem to be disproportionate. China's overall degree of "militarization" as measured by the GMI also does not seem extraordinary in regional and international comparison. Moreover, China's continued ability to invest heavily in military modernization hinges upon the medium and long term availability of sizeable and growing funds, and thus on successful and uninterrupted economic growth. This remains however largely dependent on foreign trade and is thus vulnerable to disruption by factors beyond China's control. Chapter 3 discussed example cost breakdowns showing the long-term cost impact associated with previous military procurement decisions, and effects arising from huge personnel costs. The anticipated cost impact of building a network of offshore military bases rivaling the U.S. global posture—dubbed a "string of pearls"—would significantly add to this already heavy defense burden. It would also likely raise funding requirements to a prohibitive level in light of the fact that China has to address several other costly and critical modernization projects, including combating of air pollution, mitigating the problematic effects of China's increasingly skewed demographics, providing social security to an aging population, and maintaining "internal security," which in recent years has outpaced external security in terms of budgeting. These requirements naturally limit the funds that are available for military modernization, and will certainly continue to do so in the future.

Chapter 4 discussed some universal structural aspects of present-day co-operative arms production. Usually, this seems to take place within the global framework of a highly uneven system of "core" or "first tier" arms producers, who then distribute and transfer critical technologies to their close military allies and other strategic partners, while selling less advanced technologies to various peripheral customers belonging to lower rungs on the ladder of arms production capability. As a result of the Western arms embargo, China is currently excluded from trading and co-operating with most first-tier producers of naval weapon systems, and struggles to make up for this by means of various coping mechanisms. Nevertheless, China's naval shipbuilding capability has shown discernible and impressive progress against a backdrop of rather severe externally imposed constraints. Ever since the 1950s, China has had to overcome obstacles such as withheld Soviet assistance and, later, the long-term imposition of a comprehensive U.S./EU arms embargo. In 2014, China remains much more limited than most other rising naval powers in its ability to find international partners for co-operative arms production. Ukraine and Russia have consequently become the chief suppliers of advanced naval technologies to China, especially in several critical technology fields—notably gas turbine propulsion and naval electronics.

The step-by-step analysis conducted in Chap. 5 regarding China's fleet structure and related infrastructures and institutions demonstrated the impact of these constraints on China's fleet development. It seems that China has only recently begun building larger classes of ships to the same design, an indication of a more mature and effective build strategy. By the same token, many indigenous Chinese naval weapon systems still appear to be derived from reverse-engineered foreign (Russian, Ukrainian, French, or Italian) systems, legally or illegally obtained. This would indicate that while China can copy and adapt existing naval technology, it has not yet reached the ability to innovate at the technological frontier. A strong technological legacy of Russian electronics and weapon systems is still noticeable even aboard the latest Chinese ship designs. Undoubtedly, the resulting dependency and a continued need for Russian technical support in China's naval infrastructure could be deemed an area of strategic vulnerability. Indeed, the increasing estrangement between Russia and the Western powers following Russia's 2014 invasion of Ukraine, and the likely loss of Ukrainian support as an alternative to Russian expertise after Russia's assumption of most of Ukraine's military-industrial complex in early 2014, give rise to a set of incentives that could result in an even closer Sino-Russian entente. This development will merit closer analysis.

Chapter 6 attempted to highlight some strategic consequences of China's naval build-up from various regional perspectives. All in all, China seems to be primarily building its naval capability for an asymmetric area-denial strategy, with effective interdiction of Taiwan's independence as the larger goal. At present, there is no conclusive evidence that China aims to achieve global military reach, and actual indications of a planned global role of the Chinese Navy, either in terms of technology or basing, remain scarce. China instead seems to have its sights on becoming a regional power, in line with the tenets of defensive realism, rather than rivaling U.S. global hegemony—notwithstanding hawkish rhetoric from some Chinese military quarters, and judging from China's actual military procurement decisions and observable maritime behavior in recent years. Indications sometimes interpreted to the contrary include China's aircraft carrier program and the so-called "string of pearls" theory concerning future Chinese overseas bases. The final evaluation of these still hypothetical capabilities must remain tentative as of this writing. It should further be noted that before the composition of the first carrier's air wing is actually known, no reliable assessment can be made regarding its likely operational doctrine. Given the immense challenges of carrier protection, however, the author of this study deems it unlikely that China's leaders would put such a highly symbolic (and extremely costly) asset at risk in a high-threat environment, not least due to the near-catastrophic psychological effect its loss could have on the Chinese populace. Rather than using it in the context of a hypothetical Taiwan contingency, or a hypothetical conflict with U.S. ally Japan over the Senkaku/Diaoyutai Islands, where its military usefulness would be highly questionable, the carrier would more likely be used as a platform for maritime surveillance and air support in the vast South China Sea areas. There, it would be much less vulnerable to asymmetric attack from China's weaker maritime rivals than in any of China's other maritime conflict areas. It should be noted that the concept of deploying elite troops against the opponent's weakest forces to ensure overwhelming victory has been a central tenet of classic Chinese strategic thinking ever since *Sunzi's Art of War*. Thus, the psychological effect of deploying China's carrier—its prime symbol of "national rejuvenation" and offensive military power—to the South China Sea is possibly a measure to intimidate and constrain weaker naval powers in the region, and could thereby eventually help China assert its factual supremacy in those waters.

An unavoidable consequence of China's military build-up and of the harsh restrictions imposed upon Taiwan in weapons acquisition due to Chinese diplomatic pressure is that the military balance in the Taiwan Strait is likely to shift more strongly in China's favor. Some aspects of the asymmetric area denial capabilities being developed by China—such as the ASBM—could render possible U.S. military intervention into a conflict much more risky than in past decades. Taiwan is therefore under pressure to continue presenting itself as an unattractive target through military deterrence, while working for political rapprochement in order to avoid provoking hostilities. As the sole remaining foreign supplier of major weapon systems to Taiwan, the U.S. will be challenged to carefully weigh the pros and cons of individual arms sales to the island, and, in the face of enhanced Chinese deterrence and surveillance capabilities, will likely have to adapt its naval strategy in the Asia Pacific region somewhat. This could affect vessel deployment and basing patterns as well as ship design. However, enhanced threat perceptions of a stronger Chinese Navy seem to already have resulted in a counterbalancing movement among all the Asian U.S. allies. South Korea, Japan, Australia, the Philippines, and lately, even Vietnam and India are likely to align themselves more strongly with the U.S. in order to gain protection from the perceived threat presented by both North Korea and China. Thus, rather than criticizing the U.S. presence in Asia, traditional ideas of a "Pax Americana" are currently being reinforced in many Asian countries, traditional U.S. treaty allies and newly found strategic partners alike. In terms of sharing naval technology and enhancing naval presence, this has already resulted in decisions by the U.S. to share *AEGIS* technology with South Korea, Japan, and Australia, and a commitment to move 60 % of its naval forces to the region in the context of a "Pivot to Asia" doctrine. This strengthening of alliances, however, is likely to further exacerbate Chinese threat perceptions of an "encirclement" of China's strategic areas, and will no doubt be utilized by Chinese military planners when bargaining for additional funds. An arms race in the Asia Pacific can therefore not be ruled out, and there are already some indications of such a race beginning to take shape. A stronger Chinese naval presence in contested waters within the South China Sea area will almost certainly provide a greater incentive for enhancing naval combat capabilities across the region, especially among the relatively weak navies such as Vietnam and the Philippines, and result in the build-up of more submarine fleets in the Asia Pacific region, leading to an overall enhanced degree of militarization.

Overall, as Chap. 6 discussed, strategic developments precipitated by China's naval build-up would have the following implications: a likely (and in fact already observable) strengthening of U.S. alliances with its traditional allies, but also with new partners such as India and Vietnam, as well as reinforced ties with the Philippines; a stronger likelihood that Japan will become more active in military affairs and possibly even become an arms exporter; the likelihood that Taiwan will be compelled to strike a careful balance between conciliatory gestures and credible deterrence in the face of a growing military threat; and, last but not least, the likelihood that Russia and China will exploit possible strategic synergies. There is thus currently a risk that a new anti-Western coalition will be forged between China and Russia, and possibly including North Korea, against a U.S.-led alliance that continues to encompass an ever larger number of regional actors reaching from South Korea, Japan, Singapore, and the Philippines southwards to Australia and westwards to India. Since China, Russia, and the U.S. are all nuclear powers, incentives for restraint remain exceedingly high. Therefore, a new type of Cold War scenario seems more likely than an actual great power conflict.

From the perspective of a European observer, such as the present author, the above-described security dynamics are a cause of concern. As European history has repeatedly shown, small crises can escalate under the right circumstances, and mutually reinforcing threat perceptions can easily lead to tensions if nationalist sentiments, domestic political instability, arms races, and contested territories combine in a number of countries within the same security complex. Unfortunately, this seems to be the case in Asia today, and in view of the weak regional security architecture, international institutions will be challenged to contain conflicts in the event of a sudden escalation. Military—even nuclear—deterrence is therefore likely to remain one of the chief mechanisms for containing actual outbreaks of hostilities in the Asia Pacific, at least between the major powers. In light of the multiple interrelated issues facing the region today, more in-depth analyses concerning all these diverse developments, including regional threat perceptions, arms procurement and cooperative production patterns, and political and military alliance building, are therefore greatly needed.